Texts in
Philosophy
Volume 2

Probability

and

Inference

Texts in Philosophy Series Editors
Vincent F. Hendricks   vincent@ruc.dk
John Symons   jsymons@utep.edu

Probability

and

Inference

Essays in Honour of
Henry E. Kyburg, Jr.

edited by

William Harper
and
Gregory Wheeler

ISBN 978-1-904987-18-5
Published by College Publications
Scientific Directors: Dov Gabbay, Vincent F. Hendricks and John Symons
Managing Director: Jane Spurr
Department of Computer Science
King's College London
Strand, London WC2R 2LS, UK

http://www.collegepublications.co.uk

Original cover design by Richard Fraser
Cover produced by orchid creative www.orchidcreative.co.uk
Printed by Lightning Source, Milton Keynes, UK

CONTENTS

vi

Contributors

Gert de Cooman
Systems Group
Department of Electrical Energy, Systems, and Automation
Universiteit Gent.
gert.decooman@ubent.be

Clark Glymour
Department of Philosophy, Carnegie Mellon University, and
The Institute for Human and Machine Cognition.
cg09@andrew.cmu.edu

William Harper
Department of Philosophy, University of Western Ontario.
wlharp@uwo.ca

Henry E. Kyburg, Jr.
Departments of Philosophy and Computer Science, University of Rochester,
and The Institute for Human and Machine Cognition.
kyburg@cs.rochester.edu

Isaac Levi
Department of Philosophy, Columbia University.
levi@columbia.edu

Ronald P. Loui
Department of Computer Science and Engineering
Washington University at St. Louis.
loui@cs.wustl.edu

Enrique Miranda
Department of Statistics and Operations Research
Universidad Rey Juan Carlos.
enrique.miranda@urjc.es

John Pollock
Department of Philosophy, University of Arizona.
pollock@arizona.edu

Teddy Seidenfeld
Departments of Philosophy and Statistics
Carnegie Mellon University.
teddy@stat.cmu.edu

Choh Man Teng
The Institute for Human and Machine Cognition.
cmteng@imhc.us

Mariam Thalos
Department of Philosophy, University of Utah.
mariam.thalos@philosophy.utah.edu

Gregory Wheeler
Centre for Artificial Intelligence
Department of Computer Science, Universidade Nova de Lisboa.
grw@fct.unl.pt

Jon Williamson
Philosophy, University of Kent.
j.williamson@kent.ac.uk

Introduction

WILLIAM HARPER AND GREGORY WHEELER

Recent work in philosophy, artificial intelligence, and the decision sciences has brought a renewed focus to the role and interpretation of probability in theories of uncertain reasoning. Henry Kyburg has long resisted the now dominate Bayesian approach to the role of probability in scientific inference and practical decision. The sharp contrasts between the Bayesian approach and Kyburg's program offer a uniquely powerful framework within which to study several issues at the heart of scientific inference, decision, and reasoning under uncertainty.

Some of Kyburg's innovative ideas, such as his concern with interval valued probabilities, the insistence that probabilities be based upon statistical knowledge, suspicion about Bayesian updating by conditionalization, have since been absorbed into theories of probability and uncertain inference, while other features of his distinctive view remain controversial.

The papers in this volume at demonstrate the scope of Kyburg's views on probability and inference, the clashes between his theory and orthodox Bayesian approaches, and offer some measure of the impact of his ideas on probability and inference.

A few recent developments appear in this volume, including Gert de Cooman and Enrique Miranda's elegant new results within the theory of imprecise probabilities, which offers new insight on de Finetti's exchangeability theorem; John Pollock's new result, the Y-function, within the theory of nomic probability; Ron Loui's sketch of a new theory of agent negotiation; Choh Man Teng's remarks on consistency maintenance within evidential probability and a proposal for how to perform updating; Bill Harper's application of the Kyburgian theory of acceptance to model the structure of Newton's reasoning in *Principia*; Clark Glymour's objection on complexity grounds to the psychological claim that actual causal reasoning is Bayesian reasoning; Miriam Thalos's remarks on the source of Kyburg's pragmatism; Jon Williamson's advocation of Objective Bayesianism, with replies to both Kyburg and de Cooman and Miranda; Issac Levi, and also Teddy Seidenfeld, each launching objections to evidential probability, followed by a new reply by Henry Kyburg. And Gregory Wheeler revisits the lottery paradox, and its original motivation, which serves double duty as an introduction to the volume.

This volume is based upon papers by Isaac Levi, Ron Loui, Teddy Seidenfeld, and Miriam Thalos, which were presented in October 9-10, 2004 at a Symposium honoring Henry at the University of Rochester. Each of those papers is reprinted here, and most of the remaining contributors to this volume were also in attendance that autumn weekend, where the idea for this volume was hatched. Thanks to each of the contributors, and to the members of symposium organizing committee, Randy Curren, Gabriel Uzquiano, Len Schubert, and Mitsunori Ogihara. We would also like to thank Ken Ford for his support.

William Harper would like to thank Henry E. Kyburg for his intellectual inspiration, generous support and timely advice over all these years.

Gregory Wheeler would also like to thank Henry E. Kyburg for his sage advise, and for shaping his thinking about applied logic, probability, and scientific inference. Thanks also to the Department of Computer Science at the New University of Lisbon for providing an excellent environment for research, Luis Moniz Pereira for supporting this project, and the Portuguese Science and Technology foundation (FCT) and the Leverhulme Trust for financial support. And Phiniki, where to begin?

Finally, we are grateful to Dov Gabbay at King's College for supporting this project, and for Jane Spur's endless patience, hard work, and good cheer for seeing it to completion.

A Review of the Lottery Paradox

GREGORY WHEELER

Henry Kyburg's *lottery paradox* (1961, p. 197)[1] arises from considering a fair 1000 ticket lottery that has exactly one winning ticket. If this much is known about the execution of the lottery it is therefore rational to accept that one ticket will win. Suppose that an event is very likely if the probability of its occurring is greater than 0.99. On these grounds it is presumed rational to accept the proposition that ticket 1 of the lottery will not win. Since the lottery is fair, it is rational to accept that ticket 2 won't win either—indeed, it is rational to accept for any individual ticket i of the lottery that ticket i will not win. However, accepting that ticket 1 won't win, accepting that ticket 2 won't win, ..., and accepting that ticket 1000 won't win entails that it is rational to accept that no ticket will win, which entails that it is rational to accept the contradictory proposition that one ticket wins and no ticket wins.

The lottery paradox was designed to demonstrate that three attractive principles governing *rational acceptance* lead to contradiction, namely that

1. It is rational to accept a proposition that is very likely true,

2. It is not rational to accept a proposition that you are aware is inconsistent, and

3. If it is rational to accept a proposition A and it is rational to accept another proposition A', then it is rational to accept $A \wedge A'$,

are jointly inconsistent.

The paradox remains of continuing interest because it raises several issues at the foundations of knowledge representation and uncertain reasoning: the

---

[1] Although the first published statement of the lottery paradox appears in Kyburg's 1961 *Probability and the Logic of Rational Belief*, the first formulation of the paradox appears in "Probability and Randomness," a paper delivered at the 1959 meeting of the Association for Symbolic Logic and the 1960 International Congress for the History and Philosophy of Science, but published in the journal *Theoria* in 1963. It is reprinted in Kyburg (1983).

relationships between fallibility, corrigible belief and logical consequence; the roles that consistency, statistical evidence and probability play in belief fixation; the precise normative force that logical and probabilistic consistency have on rational belief.

1   Rational Acceptance

Nevertheless, many commentators think the lottery paradox is a blunder in reasoning rather than a puzzle about the structure of collections of rationally accepted statements. Some suggest that the challenge is to explain away the initial appeal of rationally accepting a statement—the assumption being that rational acceptance is either untenable or incompatible with other entrenched epistemic principles. Others think that the notion of rational acceptance is under-specified in the setup for the lottery paradox— motivating some to argue that there are substantive differences between the lottery paradox and other epistemic paradoxes, notably David Makinson's *paradox of the preface* (1965).[2]

Although there is a consensus in this group that one should deny that it is rational to accept that a ticket of the lottery loses, there is less agreement over why this should be so. Some think that there are features of "lottery propositions"[3] that clash with norms for accepting, believing or claiming to know a proposition.[4] Another view is that "lottery contexts"[5] are self-undermining,[6] or require an agent to reject all alternatives before accepting that a particular ticket will lose.[7] Still another theme is the claim that statistical evidence is an insufficient basis for rational belief,[8] either because of missing explanatory information,[9] missing causal information,[10] or because logical consistency is believed to be fundamental to the notion of rationality but cannot be guaranteed on the basis of statistical evidence alone.[11] Fi-

---

[2]For instance, John Pollock (1986, 1993). See Neufeld and Goodman (1998) for a critical discussion of Pollock's approach, and Kvanvig (1998) for an overview of the epistemic paradoxes. Also, an historical note: Makinson formulated the paradox of the preface without any prior knowledge of Kyburg's lottery paradox (Makinson, July 2005 conversation).

[3]Harman (1973), Cohen (1988), Vogel (1990), Baumann (2004), J. P. Hawthorne (2004)

[4]Harman (1967), Stalnaker (1987), DeRose (1996), Williamson (1996, 2000), Christensen (2004).

[5]Levi (1967), Cohen (1988, 1998), DeRose (1996), Lewis (1996).

[6]Bonjour (1985), Pollock (1986), Ryan (1996), Evnine (1999), Douven (2002).

[7]Goldman (1986).

[8]Harman (1967), Cohen (1988), Nelkin (2000).

[9]Harman (1968)

[10]See Shoeman's (1987) lucid discussion of a version of this proposal in Thompson (1984).

[11]Levi (1967), Lehrer (1980), Stalnaker (1987).

nally, the notion of rational acceptance itself, when it is conceived in terms of high, finite, *subjective* probability is often shown to be problematic for accepted sentences whose probability is any value less than unity.[12]

So although there is a common assumption that the lottery paradox results from a confusion over rational acceptance, there is little agreement on precisely what that confusion is. Part of the reason for this impasse is that most of the contemporary literature has developed without engaging Kyburg's original motivations for a theory of rational acceptance. This isn't necessarily a bad development, as we shall consider in section 2. But one consequence has been a consensus view that simultaneously under- and over-constrains the lottery paradox. The paradox has been under-constrained by its popularity in traditional epistemology, where common-sense intuitions about knowledge ascriptions and ordinary language arguments hold sway, giving commentators an unusually wide berth. But the paradox has been over-constrained too, by well-intended but misguided formal attempts to draw rein. For most of the formal constraints proposed to resolve the paradox are explicitly rejected by Kyburg's view of rational acceptance.

The lottery paradox was neither motivated by ordinary language epistemology, nor was it the result of misunderstanding the structural properties of measure functions. The theory of rational acceptance was devised as part of a comprehensive theory of uncertain inference, and Kyburg's lottery ticket thought experiment was intended not to be a paradox about rational acceptance, but rather to highlight the trouble with unrestricted aggregation of well-supported but less than certain statements. It was a capstone to an argument *against* viewing evidential relations in measure theoretic terms.

The problem that aggregation poses to an account of uncertain inference is a surprisingly deep one—or so I will argue in this review. Despite the volume of literature on the lottery paradox, or perhaps because of it, the full force of this problem is sometimes missed.[13] I shall offer an account for why I think this distortion occurs, and propose remedies in the form of two different types of constraints.

In this section I offer an historical remedy by reviewing what the theory of rational acceptance was designed to do, for it appears that in ignoring the details of how Kyburg arrives at his view of rational acceptance many also miss the deeper point behind the paradox—or worse, discover an unusual property of rational acceptance and conclude that it must be the result

---

[12]Hintikka and Hilpinen (1966), Kaplan (1981), Halpern (2003), Douven and Williamson (2006).

[13]See Neufeld and Goodman's (1998) and Hunter (1996) for insightful discussions on this point.

of a superficial error. In section 2 I consider how to approach the lottery paradox formally without necessarily embracing the constraints imposed by evidential probability. There are many good reasons one might have for doing this. But the main conclusion of this section is that even if you do not accept the constraints of evidential probability, you are still faced with choices about how to handle particular problems that evidential probability was designed to solve. The third section applies the remedies on offer in the first two sections to a recent proposal to resolve the lottery paradox by blocking statistical evidence from serving as sufficient grounds for reasonable belief. The argument for this proposal rests upon an informal and untenable view of statistical inference. The effect, I believe, is to strap statistical inference with properties it does not have, and to punt on the hard problem of specifying the relationship between statistical distributions and reasonable belief. My claim is that this proposal is representative of much of the recent literature on the lottery paradox in the sense that the points I cover in this review are applicable to several proposals mentioned in the introduction.

1.1 Practical Certainty and Full Belief

On the Kyburgian view, uncertainty is a fundamental feature of statements with empirical content. The aim is not to provide a "logic of cognition" that specifies complete and precise rules for belief change, but instead to formulate the logical structure of the set of conclusions that may be obtained by non-demonstrative inference from a body of evidence.

There are important differences between rationally accepting p because the evidence makes it very probable that p, and rationally accepting p because an agent assigns a high probability to the event (a degree of belief that) p. On the first reading, a statement is assigned a (possibly) imprecise probability based upon all evidence available about p. All parameters for making this inference are presumed to be objective; the result is a unique, if imprecise, probability assignment. This outline conforms to the Kyburgian conception of rational acceptance.

On the second view each agent represents the statements he believes about a problem domain as events of a probability space over which a probability distribution is defined. Changes to probability assignments are effected by changing this distribution, subject to the mathematical constraints of the probability measure. Changes are then propagated through the entire algebra representing the event space. These probability assignments are traditionally construed as precise point values; the assignment is determined entirely by an individual agent, so probability assignments for the same problem domain may vary among agents. We will return to the differences between these two conceptions of probability in the next subsection.

The basic idea behind Kyburgian rational acceptance is to represent the maximum chance of error that is considered tolerable by a real value ϵ close to zero,[14] and to judge rationally acceptable those statements whose chance of error is less than ϵ. To disambiguate between these two readings of high probability, a statement whose chance of error is less than ϵ is considered to be a statement of *practical certainty*. A threshold for acceptance may then be defined in terms of ϵ, namely $1 - \epsilon$.

Evidential probability construes non-demonstrative inference as a relation from a set Γ of accepted evidence statements, called *evidential certainties*, to a practical certainty. The set of practical certainties is the *knowledge* that is based upon that evidence. The evidence itself may be uncertain, but it is supposed that the risk of error is less than the knowledge derived from it. Membership to the evidence set is determined by a threshold point δ for acceptance that is less than or equal to the threshold point ϵ for accepting practical certainties. Thus, the notion of rational acceptance is fundamental to constructing the distinction between evidence and knowledge on this account.

One consequence of this stratified view of evidence and knowledge is that the rule of adjunction is unsuitable for aggregating practical certainties. Another consequence is that there is a sharp distinction drawn between evidential relations and classical closure operations.

> It is not ... that the premises of the lottery paradox (ticket i will not win; the lottery is fair) do not *entail* its contradictory conclusion, but that the grounds we have for accepting each premise do not yield grounds for accepting the conjunction of the premises, and thus do not force us to accept the proposition that no ticket will win. (Kyburg 1997, p. 124)

These two points—the denial that the rule of adjunction is suitable for aggregating rationally accepted statements, and that evidential relations are not isomorphic to logical closure operations—are important. Although there are commentators who likewise reject one or both of these conditions,[15] Kyburg's view is a consequence of his theory of uncertain inference.

The lottery paradox is Kyburg's shorthand warning of the perils that the aggregation properties of boolean algebras present to models of eviden-

---

[14]Note that "ϵ" is to be construed as a fixed number, rather than a variable that approaches a limit. This is in sharp contrast to an approach advanced by Ernest Adams (Adams 1966, 1975) and Judea Pearl (1988, 1990) designed to draw a connection between high probability and logic by taking probabilities *arbitrarily* close to one as corresponding to knowledge. For a discussion of the comparison between these two approaches with respect to preferential semantics (System P), see (Kyburg, Teng and Wheeler, 2006).

[15]For instance, (Dretzky 1971), (Nozick 1981), (McGinn 1984), (Sorensen 1988), (Kitcher 1992), (Korb 1992), (Foley 1993), (Hunter 1996), (Brown 1999), (Hawthorne and Bovens 1999), (Wheeler 2005, 2006).

tial relations. For him it is a bumper sticker, not a puzzle. Nevertheless, some have viewed the clash between these logical properties and the conditions for rational acceptance encoded in the lottery as a *reductio* argument against the very concept of rational acceptance. So let's turn to review the motivations for the theory of rational acceptance.

1.2   Uncertain Inference

The first exchanges between Kyburg and his critics over the lottery paradox pitted the notion of rational acceptance against the constraints imposed by *probabilism*.[16] Probabilism maintains that the mathematical structure of the probability calculus reveals the structure of the problem one intends to model with probabilities—or, specifically, it maintains that the probability calculus reveals the proper structural constraints on our degree of beliefs about the problem we are considering.

From the point of view of a probabilist, the lottery paradox illustrates the bankruptcy of rational acceptance. Rudolf Carnap (1968) thought that acceptance of a statement should be viewed as a rough approximation of assigning a high probability to it, and Richard Jeffrey (1956, 1970) pressed this point by charging that developing a theory to accommodate rational acceptance amounts to building a rigorous theory around "loose talk".

The Carnap-Jeffrey view of rational acceptance as an approximation of one's assignment of high, precise probability is a feature of orthodox Bayesianism, and it is a view tacitly accepted by many commentators on the lottery paradox. It should be made clear, however, that the Kyburgian theory of rational acceptance results from a thorough rejection of probabilism.

Evidential probability is designed to represent the structure of probability assignments made on the basis of known statistical frequencies. The very aim of the project clashes with orthodox Bayesianism, which does not provide an account of this relationship. On the contrary, orthodox Bayesianism is conceived to get around explicating this relationship by dispensing with the supposition that probability is based upon known statistical distributions. The problem with probabilism is that this move is embraced as a matter of principle.

Strong partisanship over interpretations of probability remains commonplace even if behind the times. There is little gained by doing battle by selectively picking through the pile of positive results and counter-examples that have accumulated from the clashes between the various accounts on offer. Rather the aim now is to construct generalized frameworks for un-

---

[16] In particular, see Richard Jeffrey's (1956; 1965; 1970); Rudolf Carnap's (1962, 1968); Isaac Levi's (1967); Henry Kyburg's (1961; 1970a; 1970b). See also Carl Hempel's (1960).

certainty, such as the theory of *imprecise probability* (Walley 1991, 2000; de Cooman and Miranda, this volume). A goal of this research is to systematize the disparate counter-examples against, and motivations for, various theories of uncertain reasoning, so that different interpretations of probability fall out as special instances.

However, there are features of evidential reasoning that remain difficult to capture in this class of theories. Two properties distinguish evidential probability from orthodox Bayesian theories[17], and it remains an open question whether the full benefits evidential probability exploits from having these two properties can be reproduced within the theory of imprecise probabilities. The first property is that evidential probability allows genuine learning to occur about variables that are assigned vacuous priors $[0,1]$, whereas the second feature is that the underlying logic of evidential probability is genuinely *nonmonotonic*. But more important than the fact that evidential probability has these two properties is the reason why they appear. Each is a consequence of looking at the structure of arguments that take as premises statistical distributions and have as their conclusions probability assignments based upon the structure of that statistical knowledge. Let's review some features of this program.

Statistical statements within evidential probability are all expressions of direct inference of the form

$$\ulcorner \%\overline{x}(T(\overline{x}), R(\overline{x}), [l, u]) \urcorner,$$

which expresses that given a sequence of variables $\overline{x}$ that satisfies R, the proportion/relative frequency that $\overline{x}$ also satisfies T is between l and u. It is imagined that statements of this form represent known statistical distributions, and are collected in the set of evidential certainties.[18]

So the evidential probability based upon a set Γ of evidential certainties is $[l, u]$ that this ball a is white, because a is known to be drawn from urn U and it is known only that between l and u of all balls in U are white.

---

[17]But not necessarily from objective Bayesian accounts. See Jon Williamson's contribution for a defense of objective Bayesianism.

[18]Syntactically, '%' is a 4-place binding operator that connects open formulas of a sorted first-order language, which occur in the first two positions, and real numbers, which occur in the last two positions. The first argument position, marked by $T(\overline{x})$, denotes a target formula, whereas $R(\overline{x})$ denotes a reference formula. The object language for evidential probability is sorted, so $\ulcorner \%\overline{x}(T(\overline{x}), R(\overline{x}), [l, u]) \urcorner$ is well-formed only if $\ulcorner T(\overline{x}) \urcorner$ is from a specified set of target formulas and $\ulcorner R(\overline{x}) \urcorner$ is from a specified set of reference formulas. This regulation of the language blocks the construction of wffs where target formulas appear in the position of reference formulas, and blocks the construction of wffs from arbitrary predicates like Nelson Goodman's 'grue' color predicate. Statements of this form can be given interpretations that cover several standard forms of statistical inference. For further discussion, see (Kyburg and Teng 2001).

When there is precise knowledge that n of the N balls in U are white, the evidential probability is $[\frac{n}{N}, \frac{n}{N}]$. When there is no knowledge about the composition of U, the evidential probability is $[0, 1]$.

In this example U is the *reference class* from which a is *randomly drawn*. The evidential claim about a is this: since a is a random member of the population of balls in U, and there is known bounds $[l, u]$ on the proportion of white balls in U, then the probability of a being white is between l and u.

In practice an individual a may belong to several reference classes with known statistics, and so we must select which class we should use to base our inference.[19] Ideally we would like to have precise information about a, which translates to a small reference class and a narrow bound on the frequency interval. But in practice there is a trade-off between precision and specificity, since typically a small reference class will broaden the frequency bounds, and narrow bounds on the frequency interval typically forces a larger, less specific reference class.[20]

Inverse inference within evidential probability is treated as a special case of direct inference by exploiting a property called *rational representativeness*. Suppose we observe a sample of n balls, drawn with replacement, from an urn of black and white balls of unknown number, and we wish to draw an inference about the proportion of white balls in U from those n observations.

A sample of n draws is *rationally representative* at 0.1 if and only if the difference between the observed proportion of white balls in the sample n and the actual proportion of white balls in the urn is no greater than 0.1. If 100 draws are made, 62 of which are white, the sample is rationally representative at 0.1 just in case the proportion of white balls in the urn is between 0.52 and 0.72. If we know that the urn is composed of exactly as many white balls as black, then the sample n would *not* be rationally representative at 0.1, for we would have then expected the proportion of white balls to be between 0.4 and 0.6.

There are properties of rationally representativeness that may be exploited to make informative inferences. By using a normal distribution approximation for the binomial frequencies, we may infer that between 95% and 100% of all samples of n-fold sequences ($n = 100$) of independent outcomes from a binomial process are rationally representative at 0.1. Assuming that our sample of 100 balls is a random member of the class of all 100-fold trials, we may then infer non-trivial conclusions about the popula-

---

[19]See Choh Man Teng's contribution for a discussion of the principles for resolving conflicts among competing reference classes.

[20]See John Pollock's contribution for an alternative account of direct inference.

tion. Specifically, we may infer that the evidential probability (based on Γ) that the sample of 62 white balls of 100 is rationally representative at 0.1 is $[0.95, 1]$. In other words, the evidential probability is no less than 0.95 that the proportion of white balls in the urn is between 0.52 and 0.72. If our threshold for rational acceptance is 0.95 or less, then we would simply say that it is rational to accept that the proportion of white balls is between 0.52 and 0.72.

But what grounds do we have for making the substantive assumption that our sample of 100 balls *is* a random member of the collection? The policy of evidential probability is to treat randomness as a default assumption (Kyburg and Teng 1999, Wheeler 2004) rather than an explicit condition that is satisfied before an inference is carried through. This appears to match practice: in making a statistical inference, randomness conditions are assumed to hold on the basis of passing tests designed to detect bias. Asserting that a sample satisfies epistemic randomness translates to the absence of evidence that the sample is biased. Readers may be uncomfortable with assuming that epistemic randomness holds without direct evidence. But notice that if we could observe the rational representativeness property directly then we would not need to perform an inverse inference, for having direct evidence that a sample is representative is in effect is to know the statistical distribution of the population already.

The underlying point is that we don't need to accept directly that the sample is unbiased, but instead assume that the sample is random unless, and until, we obtain specific information that uncovers a bias. Thus, within evidential probability, new evidence is treated as irrelevant unless relevancy is demonstrated. This is the feature of the theory that allows it to escape Seidenfeld and Wasserman's dilation results (Seidenfeld *et. al.* 1993, Seidenfeld, this volume): the new evidence presented by the coin flip is classified as irrelevant under evidential probability, and therefore would be ignored.

By exploiting the random representativeness property and treating epistemic randomization as a default assumption (when combined with a testing regime for bias), evidential probability yields a genuine inductive logic, one that features a nonmonotonic consequence relation. An inference from Γ to an evidential probability statement may be undone by new, non-contrary information: we may generate new statistics relevant to the parameter of interest, or we may augment the theory with information that suggests bias in a sample, thus undercutting particular inferences based on *that* sample. The account also allows for learning from new evidence in cases where there is no prior evidence for that parameter. When there is no statistical distribution for a parameter of interest, the probability assigned is $[0, 1]$.

Nevertheless, we can add information to Γ and update this probability.

Both of these behaviors are in marked contrast to orthodox probabilistic accounts, which are inferentially monotone because measure functions are monotone functions, and which are unable to update vacuous prior because updating is performed by conditionalization. Some other novel features of this approach to probability include a theory of measurement (Kyburg 1984) that, among other things, provides an account of theory selection based partly upon epistemic criteria (Kyburg 1990), an imprecise evidential decision theory based upon "full belief" (Kyburg 1988, 1990),[21] and a "non-factive" theory of knowledge (Kyburg and Teng 2002).

The irony of Jeffrey's "loose talk" remark is that he's right: Kyburg's program was initially conceived to remove the illusion of rigor created by precise but arbitrarily assigned prior values, and to provide a theory of uncertain inference, i.e., a theory for the assignment of probability based upon evidence. It should be clear, however, that Kyburg's critique of probabilistic methods for representing evidential relations is not restricted to orthodox Bayesian views. Despite the advances made on the construction of a unified account of uncertainty, there remains an insightful intersection of key features of evidential probability and open problems in the theory of imprecise probabilities—e.g., non-trivial learning from ignorance, dilation, and restrictions on conditionalization.

2   Formal Solutions

Suppose that we divorce the lottery paradox from its original motivations—as many commentators are inclined to do. One question we might ask is whether there are formal constraints on the candidate solutions described informally in the literature. If it is impossible to formally represent a particular strategy, then this impossibility result would seem to discredit that strategy.

However, evaluating the force of impossibility results is subtle, for it raises both methodological and substantive issues about formal knowledge representation. Since the lottery paradox is an applied logic problem rather than a pure logic problem, solutions must be judged by the fit between the properties of the problem domain (rational acceptance) and the properties of the formal framework under consideration (evidential probability, classical epistemic probability, lower probability/belief functions, lower previsions, first-order probability logics, default logics, modal/hybrid logics, and so on).

There are three factors that govern the evaluation of a proposed solution

---

[21]See Bill Harper's contribution for a discussion of Kyburg's account of full risk-to-reward belief.

to an applied logic problem. The first two factors concern the properties of the formal framework and the properties of the problem to be modeled, and how each fit together. But judgments of fit are mitigated by a third factor, namely the purpose of the formal model. What we intend to *do* with a formal solution must be specified before we can evaluate its effectiveness. For example, a framework that is illuminating from a semantic point of view may be useless for computational modeling.

So, a complete standard for evaluating a proposed solution to the lottery paradox must identify the features of rational acceptance to be captured by the syntax and semantics of the representation language, and there must be a specification of what the formal framework is intended to do.[22]

In this section we will see how these three factors impact one another by discussing the relationship between a pair of properties that are often mentioned in informal treatments of the lottery paradox, *defeasible inference* and *conjunctive closure*, and their formal counterparts, namely *nonmonotonic consequence relations* and *aggregation properties* of a particular class of logics that we will specify later. Our treatment will be far from exhaustive. Rather, we will look at two recent impossibility results, one that targets nonmonotonic logics, the other which (in effect) targets logics with weakened aggregation properties. The goal is to pull together some formal results that are relevant to the lottery paradox, and to demonstrate the importance of weighing these three factors when evaluating formal solutions.

2.1 Logical Structure and Nonmonotonicity

The view that there are nonmonotonic argument structures goes at least as far back as R. A. Fisher's (1922; 1936) observation that statistical reduction should be viewed as a type of *logical*, non-demonstrative inference. Unlike demonstrative inference from true premises, the validity of a non-demonstrative, uncertain inference can be undermined by additional premises: a conclusion may be drawn from premises supported by the total evidence available now, but new premises may be added that remove any and all support for that conclusion.

This behavior of non-demonstrative inference does not appear in mathematical arguments, nor should it. If a statement is a logical consequence of a set of premises, that statement remains a consequence however we might choose to augment the premises. Once a theorem, always a theorem, which is why a theorem may be used as a lemma: even if the result of a lemma is misused in an invalid proof, the result of the lemma remains. We do not have to start the argument again from scratch.

Kyburg, following Fisher, has long stressed the nonmonotonic character-

---

[22]See (Wheeler 2005).

istics of non-demonstrative, uncertain inference. And evidential probability can be axiomatized within a specific class of nonmonotonic logics (Kyburg, Teng and Wheeler 2006). Still, some have challenged whether nonmonotonic logics are properly classified as logics. The aim of this section is to look carefully at an impossibility result offered against nonmonotonic logics and highlight the role that these three factors play in evaluating the normative force of this result. But first a remark about logical structure.

Logical structure

Formal approaches to the lottery paradox often mention the *structure* of rational acceptance, the *logic* of rational acceptance, or the *formal properties* of rational acceptance. But logical structure is better thought of as coming in degrees, and the level of structure that is appropriate depends upon what one expects the formal language to do.

Logical structure may mean what mathematical logicians mean by logical structure: a substitution function f on the free algebra of boolean formulas that is an endomorphism. Classical consequence has this degree of structure in the sense that consequence is preserved under uniform substitution of arbitrary sub-formulas for elementary formulas, within any formula in the language.

This sense of logical structure marks an important limit point. Let $\mathcal{L}$ be the set of Boolean formulas, $\Gamma \subseteq \mathcal{L}$, $\phi \in \mathcal{L}$ and '$\vdash$' denote the relation of logical consequence. Then define the operation of logical consequence Cn as $Cn(\Gamma) = \{\phi : \Gamma \vdash \phi\}$. A consequence relation $\mathrel{|\!\sim}$ is *supraclassical* if and only if $\vdash \subseteq \mathrel{|\!\sim} \subseteq 2^{\mathcal{L}} \times \mathcal{L}$, and a consequence operation C is supraclassical if and only if $Cn \leq C$. The reason that the uniform substitution function f marks an important sense of logical structure is that logical consequence is maximal with respect to f: there is no nontrivial supraclassical closure relation on a language $\mathcal{L}$ that expresses logical consequence that is closed under uniform substitution except for logical consequence.[23] If this degree of structure is necessary for a consequence relation, then any supraclassical relation would fail to be classified as a consequence relation.

But of course this isn't the only sense of logical structure. Arnold Koslow's (1992) sense of logical structure is considerably more general. Koslow considers logic to be concerned with the study of structures on arbitrary domains of elements that are generated by introducing a consequence relation of some kind or another over these elements. There is again a sense of closure under uniform substitution, but Koslow does not restrict the scope to only boolean languages, nor does he restrict the consequence relation to

---

[23]See Makinson (2005, p. 15) for discussion of this theorem.

classical consequence.[24] The advantage of Koslow's conception of logic is that he provides a general framework within which to compare different systems by which properties they share. We shall look at some weaker systems in section 2.2.

Nonmonotonicity and the Subset Principle

Are nonmonotonic logics real logics? Most nonmonotonic logics are supra-classical, and so fail to preserve uniform substitution of formulas appearing as premises. But this might be a feature of boolean languages and not a critical feature of consequence relations. Charles Morgan (1998, 2000) has recently addressed this question and argues that any consequence relation must be weakly, positively monotone,[25] regardless of the structural properties of the underlying language. But there are two reasons to doubt his claim that nonmonotonic consequence is a misnomer. The first is that the normative force of his impossibility result depends upon dubious assumptions about the structure of the epistemic notion of *joint acceptability*. The second reason is that, *formally*, monotonicity is not an essential property of a consequence relation in the sense that this property can be weakened without losing other key properties of consequence relations. If monotonicity were central to consequence relations, we would expect the removal of the monotonicity property to collapse the logic. But this in fact does not happen. We address the first point, the fit between the epistemic notion of *joint acceptability* and Morgan's formal model next. The second point is addressed in section 2.2. where some formal properties of nonmonotonic logics are discussed along with some proposals for weakening these systems.

Morgan appears to have a strong notion of logical structure in mind, although his aim is to produce a general impossibility result that is independent of the structure of the logical connectives of any particular language. He does this by building his main argument in terms of *belief structures*. A belief structure is a semi-ordered set whose elements are sets of sentences of a language $\mathcal{L}$. The focus of interest is the ordering relation LE defined over arbitrary sets, Γ and Δ. The idea is that a set of sentences is intended to represent "an instantaneous snapshot of the belief system of a rational agent" while Γ LE Δ expresses that "the degree of joint acceptability of the members [sentences] of Γ is less than or equal to the degree of joint acceptability of the members [sentences] of Δ" (2000, p. 328). On Morgan's view, a logic $\mathbf{L}$ is a set of arbitrary *rational* belief structures $\{LE_1, LE_2, ...\}$,

---

[24]See Levi, this volume, and Wheeler (2005, 2006) for examples.

[25]In Morgan (2000) there are actually 4 theorems given which aim to establish this impossibility result, viewed as an argument by four cases. The result we discuss here is the first of these theorems, and is first offered in (1998). It is the most general of his arguments.

where a rational belief structure is a subset of $\mathcal{P}(\mathcal{L}) \times \mathcal{P}(\mathcal{L})$ satisfying the structural properties of

Reflexivity: $\Gamma\ LE\ \Gamma$,

Transitivity: If $\Gamma\ LE\ \Gamma'$ and $\Gamma'\ LE\ \Gamma''$, then $\Gamma\ LE\ \Gamma''$, and

The Subset Principle: If $\Gamma \subseteq \Delta$, then $\Delta\ LE\ \Gamma$.

Soundness and completeness properties for **L** with respect to a provability relation $\vdash_b$ are defined as follows:

Soundness: If $\Gamma \vdash_b A$, then $\Gamma\ LE\ \{A\}$ for all (most) rational belief structures $LE \in$ **L**.

Completeness: If $\Gamma\ LE\ \{A\}$, then $\Gamma \vdash_b A$ for all (most) rational belief structures $LE \in$ **L**.[26]

Morgan's proposal, then, is for **L** to impose minimal prescriptive restrictions on belief structures, selecting the most general class that are "rational":

> Each distinct logic will pick out a different set of belief structures; those for classical logic will be different from those for intuitionism, and both will be different than those for Post logic. But the important point is that from the standpoint of Logic **L**, all and only the belief structures in **L** are rational (Morgan 2000: 329).

Hence, if every arbitrary set of rational belief structures is monotonic, then every logic must be monotonic as well—which is precisely the result of Morgan's Theorem 1.

THEOREM 1. *Let **L** be an arbitrary set of rational belief structures which are reflexive, transitive, and satisfy the subset principle. Further suppose that logical entailment $\vdash_b$ is sound and complete with respect to the set **L**. Then logical entailment is monotonic; that is, if $\Gamma \vdash_b A$, then $\Gamma \cup \Delta \vdash_b A$.*

Proof. (Morgan 2000):

1. $\Gamma \vdash_b A$	given.
2. $\Gamma\ LE\ \{A\}$ for all $LE \in L$.	1, soundness.
3. $\Gamma \cup \Delta\ LE\ \Gamma$ for all $LE \in L$.	subset principle
4. $\Gamma \cup \Delta\ LE\ \{A\}$ for all $LE \in L$.	2, 3 transitivity.
5. $\Gamma \cup \Delta \vdash_b A$	4, completeness.

∎

[26]The inclusion of 'most' in each construction appears only in (2000). Morgan states that a corollary to Theorem 1 may be established if 'most' means more than 50% (2000: 330).

The normative force of Theorem 1 rests upon the claim that *reflexivity*, *transitivity* and *the subset principle* are appropriate rationality constraints for belief structures.

Consider Morgan's case for the subset principle:

> [The subset principle] is motivated by simple relative frequency considerations....[A] theory which claims both *A and B* will be more difficult to support than a theory which claims just *A*. Looking at it from another point of view, there will be fewer (or no more) universe designs compatible with both *A* and *B* than there are compatible with just *A*. In general, if Γ makes no more claims about the universe than Δ, then Γ is at least as likely as Δ (2000: 329).

Morgan's argument appears to be that a set of sentences Δ has fewer compatible universe designs than any of its proper subsets, so Δ is less likely to hold than any of its proper subsets. Hence, Δ is harder to support than any of its subsets. Therefore, the degree of support for a set Δ is negatively correlated with the number of possible universe designs compatible with Δ.

There are, however, two points counting against this argument. First, it is misleading to suggest that the subset principle is motivated by relative frequency considerations. Morgan is not referring to repetitive events (such as outcomes of a gaming device) but rather to the likelihood of universe designs, for which there are no relative frequencies for an epistemic agent to consider. Second, even though there are fewer universe designs that satisfy a given set of sentences than with any of its proper subsets, this semantic feature of models bears no relationship to the degree of joint acceptability that may hold among members of an arbitrary set of sentences: "Harder to satisfy" does not entail "harder to support". Relations such as *prediction* and *justification* are classic examples of epistemic relations between sentences that bear directly on the joint epistemic acceptability of a collection of sentences but do not satisfy the subset principle.[27]

To see this last point, recall the notion of joint acceptability that underpins the interpretation of *LE*. The subset principle is equivalent to the following proposition:

PROPOSITION 2. *If a set of sentences Γ is a subset of the set of sentences Δ, then the degree of joint acceptability of the members of Δ is less than or equal to the degree of joint acceptability of Γ.*

Proposition 2, however, is false. Suppose that a hypothesis H predicts all and only observations $o_1, ..., o_n$ occur for some $n > 1$. Then H receives

---

[27] Note also that these particular notions are not outside the scope of Morgan's program as he conceives it: "[O]ur rational thought processes involve modeling and predicting aspects of our environment under conditions of uncertainty. Here we will assume that our formal language is adequate for such modeling and for the expression of claims about our world that are important to us" (2000, p. 324).

maximal evidential support just when o_1 occurs and ... and o_n occurs, which is represented by sentences $O_1, ..., O_n$, respectively. Hence, it is more rational to accept a set Δ consisting of $H, O_1, ..., O_n$ than Δ without some observation statement O_i, since the set $\{O_1, ..., O_n\}$ is better support for H than any of its proper subsets. Let $\Gamma = \{H, O_1, ..., O_{n-1}\}$. Then $\Gamma \subseteq \Delta$ but the joint degree of acceptability of Δ is not less than the joint degree of acceptability of Γ.

The behavior of joint acceptability of sentences evoked by this example is common and reasonable. Ångström measured the wavelengths of four lines appearing in the emission spectrum of a hydrogen atom (410 nm, 434 nm, 486 nm, and 656 nm), from which J.J. Balmer noticed a regularity that fit the equation

$$\frac{1}{\lambda} = R\left(\frac{1}{2^2} - \frac{1}{n^2}\right), \text{when } n = 3, 4, 5, 6,$$

where $R = 1.097 \times 10^7 m^{-1}$ is the *Rydberg* constant, and λ is the corresponding wavelength. From the observations that $\lambda = 410$ nm iff $n = 3$, $\lambda = 434$ nm iff $n = 4$, $\lambda = 434$ nm iff $n = 5$, and $\lambda = 656$ nm iff $n = 6$, Balmer's hypothesis—that this equation describes a series in emission spectra beyond these four visible values, that is for $n > 6$—is predicated on there being a series of measured wavelengths whose intervals are specified by $\frac{1}{\lambda}$. Later experimenters confirmed over time that the Balmer series holds for values $n > 6$ through the ultraviolet spectrum. (We now know that it does not hold for all of the non-visible spectrum, however.) The point behind this example is that the grounds for a Balmer series describing the emission spectrum of a hydrogen atom *increased* as the set of confirmed values beyond Ångström's initial four measurements increased.

Returning to Morgan, the normative force of his impossibility result rests on a particular view about the structure of joint acceptability. What we have demonstrated in the discussion is that it is unreasonable to interpret the formal constraints of a belief structure to be normative constraints on joint acceptability. In other words, belief structures are not a good formal model for joint acceptability. So, it is irrelevant that belief structures are necessarily monotone.

2.2   System P and Aggregation

Nevertheless we might wonder whether there is any interesting logical structure to nonmonotonic logics. Even if Morgan's views on joint acceptability are mistaken, perhaps he's right about the broader point that nonmonotonic logics fail to have enough structure to be properly classified as logics. To attack this question, first observe three standard properties that classical consequence $\vdash$ enjoys:

Reflexivity $(\alpha \vdash \alpha)$,

Transitivity (If $\Gamma \vdash \delta$ for all $\delta \in \Delta$ and $\Delta \vdash \alpha$, then $\Gamma \vdash \alpha$), and

Monotonicity (If $\Gamma \vdash \alpha$ and $\Gamma \subseteq \Delta$, then $\Delta \vdash \alpha$),

which were thinly disguised in Morgan's constraints on belief structures. What we want to investigate is whether these three properties are necessary to generate a non-trivial consequence relation. Or, more specifically, we wish to investigate whether there are any non-trivial consequence relations that do not satisfy the monotonicity condition.

Dov Gabbay (1985) noticed that a restricted form of transitivity was helpful in isolating a class of nonmonotonic consequence relations which nevertheless enjoy many properties of classical consequence. The result is important because it reveals that, *pace* Morgan, monotonicity isn't a fundamental property of consequence relations.

Gabbay observed that the general transitivity property of $\vdash$ is entailed by reflexivity, monotonicity, and a restricted version of transitivity he called *cumulative transitivity*:

Reflexivity $(\alpha \vdash \alpha)$,

Cumulative Transitivity (If $\Gamma \vdash \delta$ for all $\delta \in \Delta$ and $\Gamma \cup \Delta \vdash \alpha$, then $\Gamma \vdash \alpha$), and

Monotonicity (If $\Gamma \vdash \alpha$ and $\Gamma \subseteq \Delta$, then $\Delta \vdash \alpha$).

This insight opened a way to systematically weaken the monotonicity property by exploring relations constructed from reflexivity and cumulative transitivity, which yields the class of cumulative nonmonotonic logics.[28]

Since Gabbay's result, many semantics for nonmonotonic logics and conditional logics have been found to share a core set of properties identified by Kraus, Lehmann and Magidor (1990), which they named axiom System P. System P is defined here by six properties of the consequence relation $\mathrel|\joinrel\sim$:

Reflexivity $\qquad\qquad\qquad\qquad \alpha \mathrel|\joinrel\sim \alpha$

Left Logical Equivalence $\qquad \dfrac{\models \ulcorner \alpha \leftrightarrow \beta \urcorner; \ \alpha \mathrel|\joinrel\sim \gamma}{\beta \mathrel|\joinrel\sim \gamma}$

Right Weakening $\qquad\qquad \dfrac{\models \ulcorner \alpha \to \beta \urcorner; \ \gamma \mathrel|\joinrel\sim \alpha}{\gamma \mathrel|\joinrel\sim \beta}$

---

[28] See Makinson (2005) for discussion.

And
$$\frac{\alpha \mathrel{\vert\!\sim} \beta; \quad \alpha \mathrel{\vert\!\sim} \gamma}{\alpha \mathrel{\vert\!\sim} \ulcorner \beta \wedge \gamma \urcorner}$$

Or
$$\frac{\alpha \mathrel{\vert\!\sim} \gamma; \quad \beta \mathrel{\vert\!\sim} \gamma}{\ulcorner \alpha \vee \beta \urcorner \mathrel{\vert\!\sim} \gamma}$$

Cautious Monotonicity
$$\frac{\alpha \mathrel{\vert\!\sim} \beta; \quad \alpha \mathrel{\vert\!\sim} \gamma}{\ulcorner \alpha \wedge \beta \urcorner \mathrel{\vert\!\sim} \gamma}$$

System P is commonly regarded as the core set of properties that every nonmonotonic consequence relation should satisfy.[29] This assessment is based primarily on the observation that a wide range of nonmonotonic logics have been found to satisfy this set of axioms,[30] including probabilistic semantics for conditional logics.[31] So, there is very little formal basis for regarding belief structures to capture minimal constraints on consequence relations.

However, some have drawn a stronger conclusion from the System P axioms than that nonmonotonic logics are legitimate logical systems. Some authors have argued that System P marks minimal *normative* constraints on nonmonotonic inference.[32] For instance, an impossibility argument may be extracted from Douven and Williamson (2006) to the effect that no coherent probabilistic modeling of rational acceptance of a sentence p *can* be constructed on a logic satisfying axiom System P for values of $p < 1$, and that any "formal" solution to the lottery paradox must have at least this much structure.[33]

But it is one thing to assert that many nonmonotonic logics are cumulative, or that it is formally desirable for nonmonotonic logics to satisfy system P to preserve horn rules, say, or that there is no coherent and non-trivial probability logic satisfying System P, and it is quite another matter to say that nonmonotonic argument forms *should* satisfy System P, or to say that a logical account of rational acceptance *must* minimally satisfy the [**And**] rule of System P. One must have clear in mind what class of argument structures are to be modeled, what are the most important properties

---

[29] See for instance the review article by Makinson (1994) and textbook treatment in Halpern (2003).

[30] An important exception is Ray Reiter's default logic (1980). See Marek and Truszczynski (1991) and Makinson (2005) for textbook treatments of nonmonotonic logic.

[31] See Judea Pearl (1988, 1990) which is developed around Adams' infinitesimal ϵ semantics, and Lehman and Magidor (1990) which is built around non-standard probability.

[32] See Makinson (1994).

[33] Although Douven and Williamson do not mention System P nor Gabbay's result, the weakened form of transitivity they discuss in their footnote 2 is cumulative transitivity, and the generality they gesture toward here suggests that they are discussing the class of cumulative nonmonotonic logics.

of those structures to preserve in the system, whether those properties can be sensibly modeled, and whether those properties can be captured within the framework of choice.

Weakening System P

There are good reasons to think that statistical inference forms do not satisfy the [**And**], [**Or**] and [**Cautious Monotonicity**] axioms of System P.[34] And there are logically interesting probabilistic logics that are weaker than P. Among the weakest systems is System Y (Wheeler 2006), which preserves *greatest lower bound* on arbitrary joins and meets of probabilistic events, each of whose marginal lower probability is known but where nothing is known about the probabilistic relationship between the collection of events. This logic preserves *glb* by inference rules, called *absorption*, for combining conjunctive and disjunctive events, and it preserves this bound purely on the structural features of the measure. What is interesting about this system is that it behaves like a pure *progic* (probability logic), in this case demonstrating that there are non-trivial (even if extremely weak!) purely formal progics.

An interesting feature of System P is that it weakens the link between monotonicity and demonstrative inference: [**Cautious Monotonicity**] and the [**And**] rule together specify the restricted conditions under which nonmonotonically derived statements are aggregated.

Nevertheless, several authors reject monotone consequence operations *because* of the aggregation properties that are preserved under System P. This has led to several studies of non-aggregative logics that in effect reject or restrict the [**And**] rule.[35]

Rather than repeat the arguments for why a logic for rational acceptance *should* be weakly aggregative, let us instead highlight three issues that sub-P logics face. The first concerns the syntactic capabilities one would like the logic to have. Minimally we would like facilities in the object language for manipulating sentences that are rationally accepted. On this view rational acceptance is treated as a semantic value (or operator) assigned (attached) to sentences of the language, and our interest is to formulate rules for manipulating sentences purely on the basis of logical combinations of statements having this semantic value or being under the scope of such an operator.

---

[34]See (Kyburg, Teng and Wheeler 2006) for the KTW axioms for evidential probability, which includes counterexamples to these three properties of System P and discussion.

[35]The history of non-aggregative logics precedes and has developed independently from the study of nonmonotonic logics. See particularly (Jaskowski 1969), da Costa (1974), Schotch and Jennings (1989), Brown (1999) Arló-Costa (2002, 2005), Wheeler (2004; 2006), Kyburg, Teng and Wheeler (2006), Hawthorne (2006), Hawthorne and Makinson, forthcoming.

This is to be contrasted to a model theoretic approach under which one constructs all possible combinations of sentences that are probabilistically sound. One of the general challenges to formulating an adequate probabilistic logic is that attempts to introduce genuine logical connectives into the object language often erode the precision of the logic to effect proofs of all sound combinations of formulas. The short of it is that an inductive logic enjoying minimally interesting modularity properties in the object language may not be complete.[36]

Another issue facing sub-P logics is to specify the relationship that weakly aggregative logics have to the notion of rational acceptance we wish to model. For instance, the KTW axioms (Kyburg, Teng and Wheeler 2006) articulate a weakly aggregative logic that comes closest to axiomatizing evidential probability, although disjunction in this system is weaker than what one would expect from a standard probabilistic logic, and also different from non-aggregative modal logics. There are other conceptions of rational acceptance that authors have attempted to model as well.[37]

Even if we fix upon a structure for rational acceptance, it should be noted that while a sub-P logic may offer the most accurate representation of that structure from a knowledge representation point of view, this logic will be too weak on its own to be of much computational use. This is not an argument for interpreting System P as normative constraints on non-monotonic reasoning. But it is a recognition that the "given" structure of a problem domain, in this case the features of rational acceptance, may be considerably weaker than the minimal structure of an effective framework. Nevertheless, it is crucial to formally articulate what the given structure of a problem domain *is* in order to understand what structural properties must be added to, or assumed to hold within, a problem domain in order to generate an effective representation. Approaching the problem in this manner puts us in a position to articulate precisely what assumptions stand between an accurate formal representation of the problem and an effective model of that problem. And having the capacities to pinpoint these as-

---

[36]An example of a purely model theoretic approach to the paradox is (Bovens and Hawthorne 1999), whereas (Brown 1999) and (Wheeler 2006) are examples of syntactic approaches to the paradox. See (Wheeler 2005) for a discussion of Bovens and Hawthorne and arguments in favor of syntactic approaches.

[37]Erik Olsson (2004) views Hawthorne and Bovens's logic of belief (1999) as giving us a picture of Kyburg's theory in subjectivist terms, but Olsson remarks that the difficulty facing the logic of belief "is to explain why in the first place the agent should take the trouble of separating from other propositions those propositions in whose veracity she has a relatively high degree of confidence. What is the point of this separation business?" In reply, we've seen the motivation for rational acceptance in the first section. So, the question then is whether the logic of belief (or any of its competitors) represent the Kyburgian notion of rational acceptance? The answer here is clearly, No.

sumptions in this manner is necessary to evaluate when it is reasonable to make those assumptions.

The moral of this section is that formal solutions to applied logic problems are influenced both by the relationship between the formal language and the non-logical structure of the problem, and by the purpose of the formal model. Simply because a formal framework has a certain set of attractive mathematical or computational properties does not entail that the problem domain it is intended to model should have that structure. Assessing the basic structure of the problem domain is necessary before making a claim that a formal framework characterizes normative constraints for that problem domain. Furthermore, we may have good reason to select a less accurate formal representation of a problem domain, such as when that representation enables computation and we have a measure of the error that the framework introduces.

These remarks might suggest to some readers that formal approaches are prone to failure, particularly for philosophically contentious notions like rational acceptance, and so ought to be abandoned altogether. This would be rash. Failures often lie in under-specifying the problem to be solved. Whenever applying a formal model, it is very important to understand both the properties of that model and the properties of the problem. We construct formal solutions that may not directly match the concept or relation to be modeled in part to learn about the limitations of our models, and the nature of the problem we'd like to model. In this way we may understand the space of possible solutions offered by a particular framework, which is necessary in order to understand what can be done with that framework. A solution to this paradox, like many solutions to applied logic problems, will be in the form of an optimization problem. It will pick the salient properties of rational acceptance to model, and capture these features in a formal framework that balances precision against usefulness.

2.3   Applied Logic and Psychologism

The trade-offs involved in settling on an appropriate applied logic for a problem are influenced by what degree of structure one should demand, which is a question that is partly determined by the purpose one has in mind for the formal framework. But there is a general question of whether logic and probability are appropriate for representing features of "human reasoning" or for representing "inference". Gilbert Harman (1986, 2002) has long argued that there is no relationship between logical consequence and the psychological process of drawing an inference, and has pointed out that serious confusion results from thinking of logic as a theory of reasoning. Harman is right about the confusion that results from viewing logic as a theory of reasoning, but he is wrong to assume that formal epistemology in general, and a logic for rational acceptance in particular, must adopt a psychologistic view of logic.

Logic is a branch of mathematics that, like many other branches, lends itself to application. There is no conceptual confusion in applying logic to model features of a state transition system, a grammar, an automated theorem prover, or a formal ontology. Likewise, there is no conceptual confusion in proposing to use formal languages to model evidential relations, including those which hold between evidence and probability assignments. The results from attempting to achieve an appropriate model may be disappointing—either because the model does not enjoy a rich enough logical structure, or fails to capture all of the salient features of evidential relations, or results in a formal model that doesn't suit our particular needs. But, we can readily make sense of the aim of the project and evaluate the outcome along each of these lines. Even if there are systematic failures in this project, this does not entail that the cause is a "category mistake".

The problem Harman highlights is not a category mistake arising from failing to observe a distinction between the psychological process of *inference* and the logical relation of *implication*, but a methodological error traced to regarding the necessary properties of a logic to be necessary rules of thought, and viewing derivations within a system to provide an instruction set for the psychological process of drawing a reasonable inference. This methodological mistake is perpetuated by well-intended but fundamentally confused courses in critical thinking offered to high school and undergraduate students, which are a disservice to logic and have (to my knowledge) not a shred of empirical evidence showing their effectiveness. And the confusion shows up in professional philosophy and computer science, too. So, I sympathize with Harman's general point. But there is nothing about the application of logic to the study of epistemic relations that forces this methodological error upon us.

3   On Being Reasonable But Wrong

We started this essay listing a number of approaches to resolving the lottery paradox, but soon set them aside to review the original motivations for the paradox, what the theory of rational acceptance is designed to achieve, some facts about nonmonotonic logics that are relevant to formal work on the paradox, and how formal work on it should aim to provide some insight into uncertain inference. We also outlined a method for doing formal epistemology without a commitment to psychologism.

Let's now return to the literature and consider an approach that resolves the paradox by denying that it is rational to accept that a ticket will lose.

The particular approach that we'll consider is one that is based on a claim that statistical evidence is an insufficient basis for rational acceptance. Since there are additional conditions for rational acceptance besides statistical evidence, the argument runs, and the lottery thought experiment is missing those additional conditions, one should reject the set-up for the paradox by denying that it is rational to accept that a ticket loses.

But what are the grounds for adding additional, minimal constraints on rational acceptance? And is our understanding of uncertainty best advanced by hand-crafted solutions to the lottery ticket thought experiment? We address these two questions in the next two subsections.

3.1   Statistical evidence and rational belief

Within traditional epistemology it is commonly held that statistical evidence is an insufficient basis for knowledge because knowledge is *factive*— that is, it is thought that a known proposition must be a true proposition. (Kyburg and Teng's conception of "risky knowledge" (2002) denies that knowledge is factive, but we may sidestep this point for the moment.) Nevertheless, the conditions that make a belief *justified* may be, and often are, uncertain. In the lingo of epistemology, justified beliefs are often fallible, and so are not necessarily factive.

The lottery paradox takes root in epistemology because it is conceived to be about justified belief. One may accept the standard analysis of knowledge and deny that statistical evidence is a sufficient basis for *knowledge*, but nevertheless think that statistical evidence is (often) sufficient grounds for *justified belief*.

However, Dana Nelkin (Nelkin 2003) has challenged this conventional view by arguing that statistical evidence is both an insufficient basis for knowledge and an insufficient basis for rational (justified) belief. Her view is that statistical evidence alone doesn't yield sufficient grounds to reasonably believe a proposition, because an agent must also grasp a causal or explanatory relationship between his believing a proposition and the facts that make that belief true. Since these conditions extend beyond what is provided by statistical evidence alone, the lottery is thus "solved" by simply rejecting the conditions for rational acceptance that generate the paradox.

But Nelkin's argument is based upon a common-sense analysis of why it is not rational to believe that a ticket of the lottery loses, rather than upon a thorough analysis of uncertain inference.

> If I believe that p (say, my ticket will lose) on the basis of a high statistical probability for p, and I find out that not-p (I won!), then there is nothing at all in my reasons to reject. I still believe the same odds were in effect, and I still believe that they made my losing extremely probable. I have no reason to think that my evidence failed to bear a connection to my conclusion that I previously thought it

did. Learning that my belief is false puts no pressure on me to find some problem in my reasons. Thus, there is a way in which they are not "sensitive" to the truth, or at least to what I conceive of as the truth. It seems to me that, given the role of rationality as a guide toward the truth, this lack of sensitivity to the truth in the case of P-inferences [statistical inferences] might help explain why such inferences are not rational (Nelkin 2004, p. 401).

But the view of statistical inference sketched here is untenable.

Believing that evidence bears a connection to a conclusion is not to suppose that the evidence entails that conclusion. Sometimes perfectly good reasoning from strong evidence leads to a false conclusion, which is reflected in a distinction we may draw between *mistakes* and *errors*. Statistical methods are designed to control error, rather than eliminate it, but we scrutinize the tools and methods used to build a statistical model to determine whether the level of epistemic risk the model tells us we are exposed to is accurate. In other words, we look for mistakes in our reasoning that would expose us to more epistemic risk than we are prepared to accept.

Errors come in two forms. There are type (I) errors, which occur when one accepts a falsehood, and type (II) errors, which occur when one fails to accept a truth. The reason that errors are controlled rather than eliminated is that in practice the methods that reduce the frequency of one type of error typically increase the frequency of the other.

So when a false statement is accepted (or a true statement is rejected) on the basis of statistical evidence, it is perfectly sensible to ask whether it was the result of a mistake or whether it is within the boundaries of our accepted level of epistemic risk.

There are analogous but informal methods used for correcting mistakes of common-sense reasoning. After learning that a reasonably held belief is false, we may ask whether we overlooked a point in our evidence or misjudged the odds of being wrong, just as we investigate and probe for flaws in an experimental design. We may ask whether we correctly assessed the epistemic risk attending the evidence, just as we assess whether the statistical model correctly reflects the error probabilities we face. We may ask whether we correctly worked through the consequences of the evidence and its bearing on the conclusion we have drawn, just as we ask whether we correctly calculated values within our experimental model. In each of these cases we are investigating whether a mistake was made when we learn that an accepted conclusion from an uncertain inference is false.

Nelkin's argument relies upon a particular view of "rationality guiding us to the truth": in effect, she is presupposing an account of reasoning from evidence to accurate probability assessments that is demonstrative in character. But one difference between a false conclusion reached from a demonstrative inference, and a false conclusion reached from a non-demonstrative inference, is that there is no guarantee from a non-demonstrative inference that the cause of the false conclusion is to be found in your reasoning or your commitments. Nor should there be one. If the proportion of F's that are G's is between 0.99 and 1, a is judged to be an F and there is no evidence that says a isn't a G, it is practically certain that a is a G. Now a might in fact fail to be a G because of a mistaken classification of a among the F's, or by over-estimating the proportion of F's that are G's, or because we are missing a piece of evidence that identifies a to be a G. But, given cogent reasoning from what we reasonably accept, a might in fact fail to be a G simply by a bad luck of the draw.

3.2   Full but possibly false belief

The lottery paradox has been taken by some to be literally about what attitude to take toward lottery tickets in so-called "lottery contexts", rather than a short argument designed to fix in readers' minds that uncertain evidential relations are surprisingly tricky to model. But, designing a theory that addresses what one ought to do in lottery contexts is unlikely to yield much insight into the problem of uncertain evidential relations. This point is worth stressing, since the paradox has spawned a literature and sub-culture of its own—much of it based upon informal intuitions about buying lottery tickets in situations spelled out in the argument. But basing a theory of rational belief on the particulars of the lottery thought experiment risks confusing inessential features of the thought experiment for general features of rational acceptance. For instance, in describing the setup for the lottery paradox, some think that it doesn't make sense to ascribe rational acceptability to the claim that ticket i won't win *before* the drawing because the agent knows in advance that one and only one ticket will win. The advice here is to recognize that the agent is in a "lottery context" and that he should suspend judgment until the ticket is drawn.

But this would be a bad policy for a general theory of rational acceptance, since the fact that one and only one ticket will be drawn is inessential to a logic for rational belief.[38] When I book a flight from Philadelphia to Denver I judge it rational to accept that I will arrive without serious incident since I stand rough odds of 1 : 350,000 of dying on a plane trip in the U.S. any given year[39], and stand considerably better odds than this of arriving without incident if I book passage on a regularly scheduled U.S. commercial flight. I judge these odds to be those that characterize my flight from Philadelphia to Denver. Nevertheless, I believe that there will be a fatality from an airline accident on a U.S. carrier in the coming year even though

---

[38] Compare to (Hawthorne and Bovens 1999).

[39] According to the *National Safety Council*

I don't believe that there must be at least one fatal accident each year. Indeed, 2002 marked such an exception.[40]

Notice that I am not speaking of my decision to board the plane, but my belief that I stand better than $1 : 350,000$ odds of getting off alive. The reason why I believe I will land safely in Denver is that I view my trip to be a random member of the class of domestic trips on US domestic carriers. My epistemological stance toward the belief that I land safely is distinct from what other actions I may be willing to take given this stance, such as booking a car or hotel. In short, I believe that I will survive this flight to Denver. My belief that I will survive the trip is part of my background knowledge I have to base my decisions about actions on options I will, or should, take while in Denver. My belief that there will be an accidental fatality on US carriers in this coming year is unlikely to serve as a basis for actions I might take, but it is a belief that I nonetheless have. This belief is revealed to me by my surprise, which perhaps you share, to learn that there was a recent year in which there were no fatalities on US airline carriers.

4   Conclusion

One issue that distinguishes the Kyburgian view of uncertain inference from orthodox Bayesian approaches is a sharp distinction drawn between thought and action, and an insistence that each be treated independently. My airplane beliefs are one thing; my airplane bets, another. We can find an analogue of this distinction within statistics between *informative inference* and *decision*. We close with a quotation from "Two World Views" that remarks on the importance of this distinction.

> To have a coherent approach to decision and action in the world which is not self-defeating, is important, and the personalistic approach to statistics provides an enormous amount of clarification in this regard. But, we want to understand the world, as well as to act in it, and it is in connection with the mechanisms for rejecting and accepting hypotheses that other approaches to statistics are important. That we do not have to understand the world in order to act coherently in it is true but irrelevant. One might also claim that one need not act coherently in order to understand the world. In point of fact, we want both to act in the world and to understand it ... (Kyburg 1970b, p. 348).

It may be that we will be able to capture informative inference within a broadly decision theoretic framework, and to capture the features of evidential probability within a unified framework for uncertainty. In which case, the ordinary language I have used to describe my airplane beliefs may need to be changed. But we are not there yet, and this review along with the essays in this volume spell out some of the obstacles we face. Much of

---

[40] According to the 2002 *National Transportation Safety Board* there were 34 accidents on U.S. commercial airlines during 2002 but zero fatalities, a first in twenty years.

the literature on the lottery paradox rests atop these open problems in the foundations of statistics and decision; any proposal to resolve the paradox, in so far as it assumes an effective reduction of informative inference to decision, is failing to engage the problems of uncertain inference that rest at the core of this puzzle.[41]

BIBLIOGRAPHY

[Adams, 1966] E. Adams. Probability and the Logic of Conditionals. In J. Hintikka and P. Suppes, eds. *Aspects of Inductive Logic*, Amsterdam, North Holland Press, pp. 265–316, 1966.

[Adams, 1975] E. Adams. *The Logic of Conditionals*, Dordrecht: D. Reidel, 1975.

[Alchourrón et al., 1985] C. Alchourrón, P. Gärdenfors, and D. Makinson. On the Logic of Theory Change: Partial Meet Contraction Functions and their Associated Revision Functions, *Journal of Symbolic Logic*, 50:510–30, 1985.

[Arló-Costa, 2002] H. Arló-Costa. First order extensions of classical systems of modal logic; the role of the Barcan schemas, *Studia Logica*, 71(1): 87–118, 2002.

[Arló-Costa and Parikh, 2005] H. Arló-Costa and R. Parikh. Conditional Probability and Defeasible Inference, *Journal of Philosophical Logic*, 34: 97–119, 2005.

[Arló-Costa, 2005] H. Arló-Costa. Non-adjunctive Inference and Classical Modalities, *Journal of Philosophical Logic*, 34:581–605, 2005.

[Baumann, 2004] P. Baumann. Lotteries and Contexts, *Erkenntnis*, 61: 415–428, 2004.

[Bonjour, 1985] L. Bonjour. *The Structure of Empirical Knowledge*, Cambridge, MA: Harvard Press, 1985.

[Braithwaite, 1946] R. B. Braithwaite. Belief and Action, *Proceedings of the Aristotelian Society*, 20: 1–19, 1946.

[Brown, 1999] B. Brown. Adjunction and Aggregation, *Nous*, 33(2): 273–283, 1999.

[Carnap, 1962] R. Carnap. *Logical Foundations of Probability*, Chicago: University of Chicago Press, 1962.

[Carnap, 1968] R. Carnap. On Rules of Acceptance, in I. Lakatos, ed. *The Problem of Inductive Logic*, Amsterdam: North Holland, pp. 146–150, 1968.

[Christensen, 2004] D. Christensen. *Putting Logic in Its Place*, Oxford: Oxford University Press, 2004.

[Cohen, 1988] S. Cohen. How to Be a Fallibilist, *Philosophical Perspectives*, 2: 91–123, 1988.

[Cohen, 1998] S. Cohen. Contextualist Solutions to Epistemological Problems: Skepticism, Gettier and the Lottery, *Australasian Journal of Philosophy*, 76: 289–306, 1998.

[Conee, 1992] E. Conee. Preface Paradox, *A Companion to Epistemology*, ed. J. Dancy and E. Sosa. Oxford: Blackwell Publishers, 1992.

[da Costa, 1974] N. C. A. da Costa. On the theory of inconsistent formal systems, *Notre Dame Journal of Formal Logic* 15: 497–510, 1974.

[de Cooman, 2005] G. de Cooman. Belief Models: An order-theoretic investigation, *Annals of Mathematics and Artificial Intelligence*, 45:5–34., 2005.

[de Cooman and Zaffalon, 2004] G. de Cooman and M. Zaffalon. Updating Beliefs with Incomplete Observations, *Artificial Intelligence*, 159:74–125, 2004.

[Delgrande et al., 1994] J. P. Delgrande, T. Schaub and W. K. Jackson. Alternative approaches to default logic, *Artificial Intelligence* 70: 167–237, 1994.

[DeRose, 1996] K. DeRose. Knowledge, Assertion and Lotteries, *Australasian Journal of Philosophy*, 74: 568–80, 1996.

[41]This essay is based upon work supported by FCT award SFRH/BPD-13699-2003 and the *Leverhulme Trust*. Thanks to Bryson Brown, Marco Castellani, Stephen Fogdall, Jan-Willem Romeyn, Choh Man Teng, and Jon Williamson for their comments.

[Douven, 2002] I. Douven. A New Solution to the Paradoxes of Rational Acceptability, *The British Journal for the Philosophy of Science*, 53(3): 391–410, 2002.

[Douven and Williamson, 2006] I. Douven and T. Williamson. Generalizing the Lottery Paradox, 2006. *The British Journal for the Philosophy of Science*, in press.

[Dretske, 1971] F. Dretske. Conclusive Reasons, *Australasian Journal of Philosophy*, 49: 1–22, 1971.

[Evnine, 1999] S. Evnine. Believing Conjunctions, *Synthese*, 118: 201–227, 1999.

[de Finetti, 1974] B. de Finetti. *Theory of Probability*, vol. 1. London: Wiley, 1974.

[Fisher, 1922] R. A. Fisher. On the mathematical foundations of theoretical statistics, *Philosophical Transactions of the Royal Society of London*, 222: 309–68. 1922.

[Fisher, 1935] R. A. Fisher. The Fiducial Argument in Statistical Inference, *Annals of Eugenics*, 6: 391–398. 1935.

[Fisher, 1936] R. A. Fisher. Uncertain Inference, *Proceedings of the American Academy of Arts and Sciences*, 71: 245–258, 1936.

[Foley, 1992] R. Foley. The Epistemology of Belief and the Epistemology of Degrees of Belief, *American Philosophical Quarterly*, 29: 111–21, 1992.

[Gabbay, 1985] D. M. Gabbay. Theoretical foundations for nonmonotonic reasoning in expert systems, in *Logics and Models of Concurrent Systems*, K. Apt (ed.). Berlin: Springer-Verlag, 1985.

[Gabbay and Smets, 1998] D. M. Gabbay and P. Smets, eds. *Handbook on Defeasible Reasoning and Uncertainty Management Systems*, Dordrecht: Kluwer Academic Press, 1998.

[Girard, 1987] J.-Y. Girard. *Proof Theory and Logical Complexity*, vol. 1, Amsterdam: Elsevier Science, 1987.

[Goldman, 1986] A. Goldman. *Epistemology and Cognition*, Cambridge, MA: Harvard Press, 1986.

[Harman, 1967] G. Harman. Detachment, Probability and Maximum Likelihood, *Nous* 1: 401–11, 1967.

[Harman, 1968] G. Harman. Knowledge, Inference, and Explanation, *American Philosophical Quarterly*, 5: 164–73, 1968.

[Harman, 1986] G. Harman. *Change in View*, Cambridge, MA: MIT Press, 1986.

[Harman, 2002] G. Harman. Internal Critique: A Logic is Not a Theory of Reasoning and a Theory of Reasoning is Not a Logic, in Gabbay *et. al.* (eds.) *Studies in Practical Reasoning*, Amsterdam: Elsevier Science B. V., pp. 171–186, 2002.

[Hawthorne, 2007] J. Hawthorne. Nonmonotonic Conditionals that Behave Like Conditional Probabilities, *Journal of Applied Logic*, forthcoming.

[Hawthorne and Bovens, 1999] J. P. Hawthorne and L. Bovens. The Preface, the Lottery, and the Logic of Belief, *Mind* 108: 241–264, 1999.

[Hawthorne, 2004] J. P. Hawthorne. *Knowledge and Lotteries*, Oxford: Oxford University Press, 2004.

[Hempel, 1960] C. Hempel. Inductive Inconsistencies, *Synthese*: 12: 439–469, 1960.

[Herron *et al.*, 1997] T. Herron, T. Seidenfeld, and L. Wasserman. Divisive conditioning: further results on dilation, *Philosophy of Science*, 64: 411–444, 1997.

[Hintikka and Hilpinen, 1966] J. Hintikka and R. Hilpinen. Knowledge, Acceptance, and Inductive Logic, in J. Hintikka and P. Suppes (eds.) *Aspects of Inductive Logic*, Amsterdam: North Holland Press, 1966.

[Hunter, 1996] D. Hunter. On the Relation Between Categorical and Probabilistic Belief, *Nous* 30(1): 75–98, 1996.

[Jaskowski, 1969] S. Jaskowski. A Propositional Calculus for Contradictory Deductive Systems, *Studia Logica* 24: 143–157, 1960. (English translation of *Soc. Sci. Torunensis* A1(5):55–77, 1948.)

[Jeffrey, 1956] R. Jeffrey. Valuation and Acceptance of Scientific Hypotheses, *Philosophy of Science*, 23(3): 237–246, 1956.

[Jeffrey, 1965] R. Jeffrey. *Logic of Decision*, New York: McGraw Hill, 1965.

[Jeffrey, 1970] R. Jeffrey. Dracula Meets Wolfman: Acceptance vs. Partial Belief, in M. Swain (ed.) *Induction, Acceptance, and Rational Belief, Synthese Library* vol. 26. Dordrecht: Reidel, 1970.

[Kaplan, 1981] M. Kaplan. A Bayesian Theory of Rational Acceptance, *Journal of Philosophy*, 78(6): 305–330, 1981.

[Kitcher, 1992] P. Kitcher. The Naturalists Return, *Philosophical Review*, 101: 53–114, 1992.

[Klein, 1985] P. Klein. The Virtues of Inconsistency, *Monist*, 68: 103–135, 1985.

[Koslow, 1992] A. Koslow. *A Structural Theory of Logic*, Cambridge: Cambridge University Press, 1992.

[Kraus *et al.*, 1990] S. Kraus, D. Lehman, and M. Magidor. Nonmonotonic Reasoning, Preferential Models and Cumulative Logics, *Artificial Intelligence*, 44: 167–207, 1990.

[Kvanvig, 1998] J. Kvanvig. The Epistemic Paradoxes, in Edward Craig (ed.) *Routledge Encyclopedia of Philosophy*, London: Routledge Press, 1998.

[Kyburg, 1961] H. E. Kyburg, Jr. *Probability and the Logic of Rational Belief*, Middletown: Wesleyan University Press, 1961.

[Kyburg, 1970a] H. E. Kyburg, Jr. Conjunctivitis, Marshall Swain, (ed.) *Induction, Acceptance, and Rational Belief*, Dordrecht: Reidel, 1970.

[Kyburg, 1970b] H. E. Kyburg, Jr. Two World Views, *Nous*. Reprinted with alterations in *Epistemology and Inference*, 18–27, 1970.

[Kyburg, 1983] H. E. Kyburg, Jr. *Epistemology and Inference*, Minneapolis: Minnesota Press, 1983.

[Kyburg, 1984] H. E. Kyburg, Jr. *Theory and Measurement*, New York: Cambridge University Press, 1984.

[Kyburg, 1988] H. E. Kyburg, Jr. Full Belief, *Theory and Decision*, 25:137–162, 1988.

[Kyburg, 1990] H. E. Kyburg, Jr. *Science and Reason*, Oxford: Oxford University Press, 1990.

[Kyburg, 1997] H. E. Kyburg, Jr. The Rule of Adjunction and Rational Inference, *Journal of Philosophy* 94:109-25, 1997.

[Kyburg and Teng, 1999] H. E. Kyburg, Jr. and C. M. Teng. Statistical Inference as Default Logic, *International Journal of Pattern Recognition and Artificial Intelligence* 13(2) : 267-283, 1999.

[Kyburg and Teng, 2001] H. E. Kyburg, Jr. and C. M. Teng. *Uncertain Inference*, Cambridge: Cambridge University Press, 2001.

[Kyburg and Teng, 2002] H. E. Kyburg, Jr. and C. M. Teng. The logic of risky knowledge, in *Proceedings of WoLLIC*, São Paulo, Brazil, 2002.

[Kyburg *et al.*, 2006] H. E. Kyburg, Jr., C. M. Teng and G. Wheeler. Forthcoming. Conditionals and Consequences, *Journal of Applied Logic*.

[Lehman and Magidor, 1990] D. Lehman and M. Magidor. What does a conditional knowledge base entail?, *Artificial Intelligence*, 55: 1–60, 1990.

[Levi, 1967] I. Levi. *Gambling with Truth*, Cambridge, MA: MIT Press, 1967.

[Levi, 2004] I. Levi. *Mild Contractions*, Oxford: Oxford University Press, 2004.

[Lehrer, 1981] K. Lehrer. A Self Profile, in R. Boghan *Keith Lehrer*, Dordrecht: Reidel, 1981.

[Lewis, 1996] D. Lewis. Elusive Knowledge, *Australasian Journal of Philosophy*, 74:549–67, 1996.

[Makinson, 1965] D. Makinson. The Paradox of the Preface, *Analysis* 25:205–207, 1965.

[Makinson, 1994] D. Makinson. General Patterns in Nonmonotonic Reasoning, in Gabbay, Hogger and Robinson (eds.) *Handbook of Logic in Artificial Intelligence and Logic Programming. Vol. 3: Nonmonotonic Reasoning and Uncertain Reasoning.* Oxford: Clarendon Press, 1994.

[Makinson, 2005] D. Makinson. *Bridges From Classical to Nonmonotonic Logic*, London: King's College Publications, 2005.

[Marek and Truszczynski, 1991] V. W. Marek and M. Truszczyński. *Nonmonotonic Logic*, Berlin: Springer-Verlag, 1991.

[McGinn, 1984] C. McGinn. The Concept of Knowledge, in French, P., Uehling, T., and Wettstein, H. (eds.) *Midwest Studies in Philosophy*, vol. 9. Minneapolis: Minnesota Press, 1984.

[Morgan, 1998] C. G. Morgan. Nonmonotonic Logic is Impossible, *Canadian Artificial Intelligence* 42:19-25, 1998.

[Morgan, 2000] C. G. Morgan. The Nature of Nonmonotonic Reasoning, *Minds and Machines* 10:321-360, 2000.

[Nelkin, 2000] D. Nelkin. The Lottery Paradox, Knowledge, and Rationality, *Philosophical Review*, 109(3): 373–409, 2000.

[Neufeld and Goodwin, 1998] E. Neufeld and S. Goodwin. The 6-49 Lottery Paradox, *Computational Intelligence*, 14(3): 273–286, 1998.

[Neyman, 1935] J. Neyman. Statistical Problems in Agricultural Experimentation, *Journal of the Royal Statistical Society*, 2(Supplement): 107–80, 1935.

[Nilsson, 1986] N. J. Nilsson. Probabilistic Logic, *Artificial Intelligence*, 28(1): 71-87, 1986.

[Nozick, 1981] R. Nozick. *Philosophical Explanations*, Cambridge, MA: Harvard University Press, 1981.

[Olsson, 2006] E. Olsson. Levi and the Lottery, in E. Olsson (ed.) *Knowledge and Inquiry: The Pragmatism of Isaac Levi*, Cambridge University Press, 2006.

[Pearl, 1988] J. Pearl. *Probabilistic Reasoning in Intelligence Systems*, San Francisco, CA: Morgan Kaufmann, 1988.

[Pearl, 1990] J. Pearl. System Z: A Natural Ordering of Defaults with Tractable Applications to Default Reasoning, *Proc. of Theoretical Aspects of Reasoning about Knowledge (TARK)*, pp.121–135, 1990.

[Pollock, 1986] J. L. Pollock. The Preface Paradox, 53: 246–58, 1986.

[Pollock, 1993] J. L. Pollock. Justification and Defeat, *Artificial Intelligence*, 67: 377-407, 1993.

[Ramsey, 1931] F. P. Ramsey. *The Foundations of Mathematics and Other Essays, volume 1*. New York: Humanities Press, 1931.

[Reiter, 1980] R. Reiter. A Logic for Default Reasoning, *Artificial Intelligence*, 13(1): 81-132, 1980.

[Romeyn, 2005] J.-W. Romeyn. Bayesisan Inductive Logic, Ph.D. Thesis, Department of Philosophy, University of Groningen. (Published by Haveka BV, Amsterdam, ISBN: 90-367-2279-9, 2005.)

[Ryan, 1991] S. Ryan. The Preface Paradox, *Philosophical Studies*, 64: 293–307, 1991.

[Savage, 1972] L. Savage. *Foundations of Statistics*, New York: Dover, 1972.

[Schotch and Jennings, 1989] P. Schotch and R. Jennings. On Detonating, in Priest, Routley and Norman (eds.) *Paraconsistent Logic: Essays on the Inconsistent*, Munich: Philosophia Verlag, pp. 306–327, 1989.

[Seidenfeld, 1979] T. Seidenfeld. *Philosophical Problems of Statistical Inference: Learning from R. A. Fisher*, Dordrecht: D. Reidel, 1979.

[Seidenfeld and Wasserman, 1993] T. Seidenfeld and L. Wasserman. Dilation for Sets of Probabilities, *Annals of Statistics* 21: 1139-1154, 1993.

[Sheoman, 1987] F. Schoeman. Statistical vs. Direct Evidence, *Nous*, 21(2):179–198, 1987.

[Sorensen, 1988] R. Sorensen. Dogmatism, Junk Knowledge, and Conditionals, *Philosophical Quarterly* 38: 433–54, 1988.

[Stalnaker, 1987] R. Stalnaker. *Inquiry*. Cambridge, MA: MIT Press, 1987.

[Thomson, 1984] J. J. Thomson. Remarks on Causation and Liability, *Philosophy and Public Affairs*, 13: 101–133, 1984.

[Vogel, 1990] J. Vogel. Are there Counterexamples to the Closure Principle? in Roth, M. and Ross, G. (eds.) *Doubting: Contemporary Perspectives on Skepticism*, Dordrecht: Kluwer, 13–27, 1990.

[Walley, 1991] P. Walley. *Statistical Reasoning with Imprecise Probabilities*, London: Chapman and Hall, 1991.

[Walley, 2000] P. Walley. Towards a unified theory of imprecise probability, *International Journal of Approximate Reasoning*, 24: 125–148, 2000.

[Weatherson, 2005] B. Weatherson. Can we do without pragmatic encroachment?, *Philosophical Perspectives*, 19: 417–433, 2005.

[Wheeler, 2004] G. Wheeler. A resource bounded default logic, J. Delgrande and T. Schaub (eds.) *Proceedings of the 10th International Workshop on Non-Monotonic Reasoning (NMR 2004)*, Whistler, British Columbia, Canada, pp. 416-422, 2004.

[Wheeler, 2005] G. Wheeler. On the Structure of Rational Acceptance, *Synthese*, 144(2): 287–304, 2005.

[Wheeler, 2006] G. Wheeler. Rational Acceptance and Conjunctive/Disjunctive Absorption, *Journal of Logic, Language and Information*, 15(1-2): 49–63, 2006.

[Williamson, 1996] T. Williamson. Knowing and Asserting, *Philosophical Review*, 105(4): 489–523, 1996.

Acceptance and Scientific Method

WILLIAM HARPER

Gregory Wheeler closed his discussion of the lottery paradox by quoting a selection from a passage in Henry Kyburg's paper, "Two World Views." I want to open my discussion of acceptance in scientific method by quoting a similar selection from the revised version of that same interesting paper.

> To have a coherent approach to decision and action in the world which is not self-defeating, is important, and the personalistic approach to statistics provides an enormous amount of clarification in this regard. A strongly instrumentalist view of science is perfectly adequate to the design of experimental apparatus as well as to the creation of engineering wonders. But, we want to understand the world, as well as to act in it, and it is in connection with the mechanisms for rejecting and accepting hypotheses that other approaches to statistics and to scientific theory are important. [Kyburg, 1983, 27]

This quotation challenges the Bayesian approach to statistics by pointing out the importance of rejecting and accepting hypotheses for the practice of science as an enterprise for seeking to understand the world.

I will argue that attention to the details of Newton's argument for universal gravitation and its application to solar system motions reveals a methodology centered on inferences to provisionally accept theoretical claims, as guides to research, in an enterprise of seeking successively more accurate approximations. Newton's inferences from phenomena are theory-mediated measurements that are far more effective at making empirical phenomena inform our understanding of the world than any Bayesian approach that does not afford a method of theory acceptance. These inferences are also richer and more effective than the inferences to accept or reject hypotheses in classical statistics.

This richer and more effective Newtonian methodology continues to guide research in gravitation today. Among many examples are tests of the equivalence principle [Harper and DiSalle, 1996]. The development and application of the parameterized post-Newtonian formalism for assessing alternative relativistic theories of gravity is another a striking example of Newton's method at work [Harper, 1997]. It can be argued that an important part of what distinguishes what we now characterize as the natural sciences is the

method exemplified in the successful application of universal gravity to the
solar system.[1]

After giving some background details of Newton's method and its application to the solar system, I will explore some issues for interpreting scientific
acceptance that arise from recent work on relating knowledge to lotteries
and practical interests. I will explore the extent to which such models of
acceptance can usefully illuminate Newton's method.

1   Newton's method

1.1   Newton's basic argument and its illustration of the role of acceptance in his method.

Newton begins by introducing, quantity of matter, quantity of motion, iner-
tia, and impressed force [Cohen and Whitman, 1999, pp. 403–408].[2] These
are concepts basic to his fundamental theory of motion. He also introduces
his special theoretical concept of a centripetal force construed as a capacity
or power to induce impressed forces toward a center (405–408). His controversial scholium on time, space, place, and motion, introduces conventions,
which allow his laws of motion to be formulated and applied to argue for his
theory of gravity and its application to the motions of solar system bodies
(408–415).[3]

Newton characterizes his laws of motion as "accepted by mathematicians
and confirmed by experiments of many kinds" [Cohen and Whitman, 1999,
p. 424]. In a corollary he shows how the laws of motion are confirmed by
the detailed compositions and resolutions of impressed forces exhibited in
the actions of such fundamental machines as wheels, pulleys, levers, wedges,
stretched strings, as well as the forces of tendons to move the bones of ani-
mals (418–420). In his scholium to the laws, he points out that the first two
laws of motion are strikingly confirmed by experiments with falling bodies,
projectile motions, pendulums and daily experience with clocks (424). He
goes on to discuss the influential experiments with pendulums by Huygens,
Wren, and Wallis that confirm the application of the third law of motion to
collisions, and then to report on his own, very carefully carried out, repeti-
tions and refinements of such pendulum collision experiments (424–427). He

---

[1]See Harper and Smith 1995 and Harper 1997.

[2]The citation [Cohen and Whitman, 1999, pp. 403–408] is to the cited pages in the
translation of the *Principia* by I. B. Cohen and Anne Whitman, 1999.

[3]Howard Stein's classic paper, "Newtonian Space-Time," transformed the understanding available to philosophers of science of Newton treatment of space and time. Robert
DiSalle's recent book, [DiSalle, 2006], is a culmination of investigations initiated by Stein.
It is the most philosophically informative treatment I know of the complex methodological and evidential issues raised by Newton's conventions about space, time and motion
and their radical transformation by Einstein.

then addresses the application of the third law of motion to attractions between bodies. He describes empirical experiments with magnets and pieces
of iron floating on wooden blocks and supplements these with arguments
based on thought experiments [Cohen and Whitman, 1999, 427–430].

Newton's three laws together with propositions inferred from them in
books 1 and 2 of his *Principia* are theoretical claims that he accepts as
propositions that can be appealed to as resources to make motion phenomena measure centripetal forces.[4] His classic inferences from orbital phenomena to inverse-square centripetal forces are backed up by systematic
dependencies afforded by these accepted theoretical resources. For example, Propositions 1 and 2 of book 1, together, establish that Kepler's rule of
areas holds if and only if the force acting on the moving body is centripetal
[Cohen and Whitman, 1999, pp. 444–448]. A corollary then adds that the
rate at which areas are swept out by radii from the center is increasing just
in case the net force is off-center in the direction of motion, and decreases
just in case it is in the opposite direction (447). These systematic dependencies make the constancy of the rate at which areas are swept out measure
the centripetal direction of the force maintaining a body in an orbit.[5] Similar systematic dependencies from the accepted theoretical background back
up Newton's inferences to the inverse-square variation of orbital centripetal
forces from Kepler's 3/2 power rule for a system of orbits about a common
center (451)[6] and from the absence of orbital precession of any single orbit

---

[4]These are the mathematical principles referred to in Newton's full title of his *Principia*, "Mathematical Principles of Natural Philosophy."

[5][Harper, 1998] shows that these subjunctive supporting systematic dependencies
make inferences like Newton's avoid the counterexamples based of "unnatural" material conditionals constructed by Christensen, which led to the demise of Clark Glymour's
bootstrap confirmation as a serious candidate for explicating scientific inference.

[6]According to Kepler's harmonic law for a system of orbits about a single center, the
periods are as the $\frac{3}{2}$ power of the mean distances. That is the ratio $\frac{t^2}{R^3}$, where t is the
period and R is the mean distance is a constant for those orbits. Corollary 7, Proposition
4 book 1, is equivalent to the following universal systematic dependency $t \propto R^s$ iff
$f \propto R^{(1-2s)}$, where f is the force maintaining a body in an orbit with period t and
radius R. Corollary 6 follows when s equals $\frac{3}{2}$. For each of a whole range of alternative
power law proportions of periods to orbital radii, cor. 7 establishes the equivalent power
law proportion to radii for the centripetal forces that would maintain bodies in those
orbits.

To have the periods be as some power $s > \frac{3}{2}$ would be to have the centripetal forces
fall off faster than the -2 power of the radii, while having the periods be as some power
$s < \frac{3}{2}$ would be to have the centripetal forces fall off more slowly than the -2 power of
the radii. These systematic dependencies make the harmonic law phenomenon ($s = \frac{3}{2}$)
for a system of orbits measure the inverse-square (-2) power for the centripetal forces
maintaining bodies in those orbits. This constitutes a very strong sense in which the
harmonic law for a system of orbits carries the information that the forces maintaining
bodies in those orbits satisfy the inverse-square power law. Newton proves proposition 4

$(543).^7$

In addition to accepting these theoretical background assumptions, Newton also accepts the cited phenomena (C&W, 797-801). These phenomena are relative motions of orbiting bodies with respect to centers of bodies about which they orbit. They are patterns exhibited in sets of data, which can be expected to remain reasonably stable as new data are added. Given that these phenomena are expected to continue to hold, measurements of parameters by them can be expected to be backed up by increasingly large sets of data. For planetary motions, the data available to Newton are geocentric angular positions at specified times estimated from observations.

Newton's inferences to inverse-square centripetal forces are inferences to what Howard Stein has called inverse-square "fields of acceleration."[8] Each centripetal force is counted as a power or capacity characterized by assigning inverse-square varying centripetal accelerations to locations around the central body. The systematic dependencies backing up Newton's inferences allow the cited data from the several orbits fitting Kepler's 3/2 power rule to count as agreeing measurements of what the inverse-square adjusted acceleration toward the central body would be at any given distance.[9]

Newton's moon-test shows that measurements of the acceleration toward the center of the earth afforded by inverse-square adjusting the centripetal acceleration exhibited by the lunar orbit agree with the more precise measurements of the acceleration of gravity at the surface of the earth afforded by pendulum motion phenomena [Cohen and Whitman, 1999, pp. 803–805]. Newton's inference to identify the force maintaining the moon in its

---

and its corollaries for concentric circular orbits. As Newton was aware, the result goes over directly to the corresponding result for elliptical orbits with force toward a focus. To do this one takes the semi-major axis as the mean distance. See [Harper, 2002b] and especially [Harper, 1999].

[7]Newton cites corollary 1 of proposition 45 book 1, according to which Precession is p degrees per revolution if and only if the centripetal force f is as the $(\frac{360}{360+p})^2 - 3$ power of distance.

If a planet in going from aphelion to return to it, makes an angular motion against the fixed stars of $360 + p$ degrees, then the aphelion is precessing forward with p degrees per revolution. The aphelion is the farthest point from the sun. According to this corollary, zero precession is equivalent to having the centripetal force be as the -2 power of distance; forward precession is equivalent to having the centripetal force fall off faster than the inverse-square; and backward precession is equivalent to having the centripetal force fall off slower than the inverse-square power of distance. Newton's proposition 45, book 1, and its corollaries are proved for orbits that are very nearly circular. The results, however, can be extended to orbits of arbitrarily great eccentricity. Indeed, orbital eccentricity increases the sensitivity of absence of unaccounted-for precession as a null experiment measuring inverse-square variation of a centripetal force. See [Valluri et. al., 1997] and [Valluri et al., 1999].

[8]See [Stein, 1970; 1977; 2002].

[9]See [Harper, 2002a].

orbit with the earth's gravity makes these two distinct phenomena count as agreeing measurements of the strength a single inverse-square field of acceleration extending out from the surface of the earth.[10]

The moon-test inference, like the basic inferences to inverse-square centripetal forces, strikingly realizes an ideal of empirical success as convergent accurate measurement of theoretical parameters by the phenomena which they are taken to explain. This ideal of empirical success is stronger than the, more strictly empiricist, conception of empirical success as limited to prediction alone. To realize Newton's stronger ideal of empirical success a theory would need to do more than to accurately predict the phenomena it purported to explain; it would need, in addition, to have those phenomena accurately measure the theoretical parameters by which they are to be explained.

Newton goes on to identify the, previously inferred, inverse-square centripetal forces toward the sun, Jupiter and Saturn as inverse-square gravity towards those bodies [Cohen and Whitman, 1999, pp. 805–806]. He then extends the attribution of such inverse-square gravity to planets without known satellites to afford orbital phenomena to measure centripetal accelerations toward them (806).

The inference to extend the theory by attributing inverse-square gravity toward planets without satellite orbits to measure it is backed up by an explicit rule for doing natural philosophy that conveys central features of the role of acceptance in Newton's methodology.

Rule 4. *In experimental philosophy, propositions gathered from phenomena by induction should be considered either exactly or very nearly true notwithstanding any contrary hypotheses, until yet other phenomena make such propositions either more exact or liable to exceptions* [Cohen and Whitman, 1999, p. 796].

This rule instructs us to consider propositions gathered from phenomena by induction as "either exactly or very nearly true" and to maintain this in the face of any contrary hypotheses. Note that Newton explicitly allows for propositions considered to be approximations to be counted as accepted. We want to clarify what are to count as such propositions gathered from phenomena by induction and how they differ from what are to be dismissed as mere contrary hypotheses.

The classic inferences from phenomena which open the argument for Universal Gravitation are measurements of the centripetal direction and the inverse square accelerative quantity of gravitation maintaining moons and planets in their orbits. To extend the attribution of centripetally directed gravitation with inverse square accelerative quantity to planets with-

---

[10]See [Harper, 2002b].

out moons is to treat such orbital phenomena as measurements of these quantifiable features of gravity for planets generally.

What would it take for an alternative proposal to succeed in undermining this generalization of gravity to planets without moons? The inferences we have been examining suggest that Newton's rule 4 would have us treat such an alternative proposal as a mere "contrary hypothesis" unless it is sufficiently backed up by measurements from phenomena to count as a rival to be taken seriously as an alternative prospect for realizing Newton's ideal of empirical success.

Consider the skeptical challenge that the argument has not ruled out the claim that there is a better alternative theory in which these planets do not have gravity. Rule 4 will count the claim of such a skeptical challenge as a mere contrary hypothesis to be dismissed, unless such an alternative is given with details that actually deliver on measurement support sufficient to make it count as a serious rival. On Newton's method, it is not enough for a skeptic to show that such an alternative has not been logically ruled out by the explicitly cited premises of an argument.

Newton backs up his argument for assigning gravities as inverse square centripetal acceleration fields to all planets by appealing to diverse phenomena that provide agreeing measurements of the equality of accelerations toward planets for bodies at equal distances from them [Cohen and Whitman, 1999, pp. 806–810]. He begins with gravitation toward the earth. He describes pendulum experiments which measure the equality of the ratio of weight to inertial mass for pairs of samples of nine varied materials (806–807). The equality of the periods of such pairs of pendulums counts as a phenomenon which measures the equality of these ratios for laboratory sized bodies near the surface of the earth to considerable precision. Newton describes these experiments as sufficient to have revealed differences of even less than one part in a thousand (807).[11]

A second phenomenon is the outcome of the moon-test. The agreement between the acceleration of gravity at the surface of the earth and the inverse-square-adjusted centripetal acceleration exhibited by the lunar orbit measures the agreement between the ratio of the moon's weight toward the earth to its mass and the common ratio to their masses of the inverse-square-adjusted weights toward the earth that terrestrial bodies would have at the lunar distance. Given Newton's assumptions, inverse-square adjusting the centripetal acceleration of the lunar orbit for each of the six lunar distance estimates cited in the third edition of his *Principia* yields a 95% Student's t-

---

[11]Repetitions of these experiments, carried out at a conference at St. Johns College, Annapolis, in 1999, suggest that the bounds claimed by Newton are quite modest. See [Wilson, 1999].

confidence estimate of the acceleration of gravity at the surface of the earth that would bound Eötvös ratios expressing differences from the pendulum measurements to less than about 0.03.[12]

The equality of the periods of pairs of pendulums in Newton's experiments are phenomena which measure the equalities of ratios of weight to mass for terrestrial bodies. Similarly, the outcome of the moon-test counts as a rougher agreeing measurement. These phenomena count as agreeing measurements bounding toward zero an earth-parameter, Δ_e, representing differences between ratios of inverse-square adjusted weight toward the earth to mass for bodies at even very great distances above its surface.[13]

Newton follows up his argument for the earth with an appeal to the harmonic law for Jupiter's moons as a phenomenon which measures the equality of the ratio of mass to inverse-square-adjusted weight toward Jupiter for bodies at the distances of its moons. The data Newton cites measure the equality of these ratios to fair precision. They bound toward zero a Jupiter-parameter, Δ_j, representing differences between ratios of inverse-square adjusted weight toward Jupiter to mass for bodies. Similarly, the data Newton cites for the harmonic law for the primary planets bound toward zero a sun-parameter. To show the equality of ratios of mass to weight toward the sun at equal distances Newton also appeals to three additional phenomena — absence of polarization toward or away from the sun of orbits of respectively Jupiter's moons, Saturn's moons and the earth's moon. If the ratio of mass to weight toward the sun for a moon were greater or less than the corresponding ratio for the planet then the orbit of that moon would be shifted toward or away from the sun. Absence of such orbital polarization counts as a phenomenon measuring the equality of ratios of mass to weight toward the sun at equal distances.[14]

All these phenomena count as agreeing measurements bounding toward zero a single general parameter representing differences between bodies of the ratios of their inertial masses to their inverse-square-adjusted weights toward the sun or any planet. These measurements all establish empirical limits on violations of a fundamental property of gravity that counts as an important part of Einstein's more inclusive equivalence principle.[15]

---

[12]See [Harper and DiSalle, 1996].

[13]Ibid.

[14]See [Harper et al., 2002] for an account of the calculations and application to modern data.

[15]See [Harper and Disalle, op. cit]. For more detail on Einstein's equivalence principle and empirical tests of it see [Will, 1993, pp.13–66]. The $TH\epsilon\mu$ formalism [Will, 1993, pp. 45–58] for testing the equivalence principle is a background framework that realizes Newton's methodology of using theoretical background assumptions to make data afford empirical measurement bounds limiting expected violations.

The final steps of Newton's basic argument appeal to applications of his third law of motion of a sort which, for example, would specifically construe the equal and opposite reaction to the action of the sun's gravity on a planet to be the action of that planet's gravity on the sun. Such applications of his third law of motion are cleverly extended to parts of planets to arrive at his theory of gravity as a universal force of interaction between pairs of particles directly proportional to their masses and inversely proportional to the square of the distance between them. These applications of his third law of motion transform his conception of inverse-square centripetal forces into his conception of gravity as a universal *force of interaction* between bodies characterized by his *law of interaction*.[16]

1.2   The application to motions among solar system bodies.

The application of Newton's theory of gravity to the motions of solar system bodies illustrates a very important feature of the role of acceptance as a guide to research in a method of dealing with daunting complexity by successive approximations. Applications of this method to this problem by later researchers illustrate the extensive support afforded to back up measurements of parameters by phenomena from what can be very large open-ended sets of data fit by those phenomena.

One useful consequence of Newton's applications of the third law of motion is that his agreeing measurements of the relative strengths of the inverse-square acceleration fields toward the sun and planets with moons give agreeing measurements of the relative masses of those central bodies. These measurements of masses lead to his center of mass resolution of the Two Chief World Systems problem. This problem of deciding between geocentric and heliocentric world systems, the topic of Galileo's famously controversial dialogue, had been a dominant question of natural philosophy in the seventeenth century. Newton's measurements support his empirical resolution:

> The sun is engaged in continual motion but never recedes far from the common center of gravity of all the planets [Cohen and Whitman, 1999, p. 816].

Corollary 4 of Newton's laws of motion is central to this application of universal gravity to the solar system.

> The common center of gravity of two or more bodies does not change its state whether of motion or of rest as a result of the actions of the bodies upon one another; and therefore the common center of gravity of all bodies acting upon one

---

[16]See [Stein, 1991] and [Harper, 2002a] for treatments of the complex evidential issues raised by this special application of Newton's third law to treat the equal and opposite reaction to a planets attraction toward the sun to be an attraction of the sun toward that planet.

another (excluding external actions and impediments) either is at rest or moves uniformly forward [Cohen and Whitman, 1999, p. 421].

This allows the measurements of the relative masses of planets and the sun, afforded by his theory of gravity, to support Newton's surprising resolution of the two chief world systems problem.[17] Neither the sun nor the earth can count as the center for determining what count as the true motions of solar system bodies among themselves. The sun, however, can count as a good enough approximation from which to begin using deviations from Keplerian orbits to guide construction of successively more accurate approximations.

One striking feature of Newton's basic argumentindexNewton! theory of gravity! and Keplerian phenomena, that has been much remarked upon by philosophers, is that its conclusion, his theory of universal gravitation, requires revising the Keplerian phenomena that he accepted as empirical premises. For example, the application of the third law of motion to the sun and planets leads to two-body corrections to the harmonic law. Newton also expected that interactions between Jupiter and Saturn would sensibly disturb their orbital motions. The treatments of such deviations by Newton and later researchers exemplify a method of successive approximations that informs applications of universal gravitation to motions of solar system bodies. On this method deviations from the model developed so far count as new *theory-mediated* phenomena to be exploited as carrying information to aid in developing a more accurate successor. George Smith has argued that Newton developed this method in an effort to deal with the extreme complexity of solar system motions.[18] He points to a striking passage, which he calls "the Copernican scholium," that Newton added to an intermediate augmented version of his *De Motu* tract, before it grew into the *Principia*.

> By reason of the deviation of the Sun from the center of gravity, the centripetal force does not always tend to that immobile center, and hence the planets neither move exactly in ellipses nor revolve twice in the same orbit. There are as many orbits of a planet as it has revolutions, as in the motion of the Moon, and the orbit of any one planet depends on the combined motion of all the planets, not to mention the action of all these on each other. But to consider simultaneously all these causes of motion and to define these motions by exact laws admitting of easy calculation exceeds, if I am not mistaken, the force of any human mind [Wilson, 1989, 253].

It appears that shortly after articulating this daunting complexity problem, Newton was hard at work developing resources for responding to it with successive approximations.

The corrections by d'Alembert and Euler of Newton's treatment of the precession of the equinoxes, and their extension to account for the nutation

---

[17]See [Stein, 1967].
[18]See [Smith, 1999, pp. 50–51; Smith, 2002, pp. 153–167].

— an additional small wobble in the motion of the earth's axis discovered by Bradley, are among the many examples of this method of using deviations as theory mediated phenomena to inform the development of successively better approximations to account for the true motions.[19] Such developments through the work of Laplace at the turn of the 19th century and on through the work of Simon Newcomb at the turn of the 20th century yield, not just, increasingly accurate accounts of the motions that fit increasingly precise data from observations. They yield such accounts that are, also, empirically backed up by affording increasingly accurate convergent measurements of the masses of solar system bodies from those data. The work of Newcomb, which by the beginning of the twentieth century was leading to the preeminence of the U.S. naval observatory in predictive astronomy, is informed by his making the determination of a single consistent assignment of the relative masses an, explicitly stated, central focus of solar system research.[20]

R. E. Laubscher presents a Newtonian perturbation- theoretic account of the motion of Mars, which is fit to a body of over 5000 geocentric angular position data fixed from meridian observations of Mars taken between 1751 and 1969 as well as to 773 radar ranging data.[21] One result reported was a new correction to the estimated mass of Venus afforded by its perturbations of Mars revealed in details of Mars's motion.[22] A somewhat crude illustration of the support afforded to estimates of masses by this perturbation corrected account of the motion of Mars is given by agreeing estimates of the mass of the sun computed from 77 of Laubscher's cited meridian observation data.[23]

Contrary to Kuhn, the transition from Newton's theory to Einstein's general relativity theory is in accordance with Newton's method. Einstein showed that his theory could account for the 43 seconds per century of Mercury's perihelion motion not accounted for by Newtonian perturbations while at the same time preserving the Newtonian perturbation-theoretic account as an approximation good enough to continue to account for the other 531 seconds per century that had been accounted for by Newton's theory.[24] Without the Newtonian limit, which allowed recovery of the 531 seconds

---

[19]See [Wilson, 1995; 1987].

[20]See [Newcomb, 1895, 7, 10–13].

[21]See [Laubscher, 1981, pp. 374–375] for general remarks and pp. 415-483 for a list of the meridian observations with differences from the calculated values and pp. 484-490 for a list with residuals for the radar ranging data.

[22][Laubscher, 1981, pp. 388–392]. This fairly recent example of measurements of masses from orbital motions is now backed up by, more recent, measurements afforded by spacecraft orbits.

[23]See [Harper et al., 1994].

[24]See [Will, 1993, pp. 176–181].

per century perihelion motion accounted for by Newton's theory, Einstein's theory would not have been successful as an account of Mercury's perihelion motion. The Newtonian limit, built into Einstein's theory, insures that the perturbation-corrected motions and the corresponding measurements of masses can be counted as approximations good enough for general relativity to recover, for its parameters, the extensive empirical successes afforded to Newton's theory by its applications to solar system motions. Einstein's great excitement over his Mercury perihelion calculation is appropriate to the fact that it showed that his theory does better than Newton's theory by Newton's own standards.[25]

Newton's standard of what is to be counted as a more accurate successor applies even to theory change as fundamental as the revolutionary transition from Newton's to Einstein's. This shows that even such a radical theory change can be accommodated by Newton's methodology for seeking what can be counted as successively more accurate approximations.[26]

2 What is acceptance?

2.1 Acceptance as epistemic certainty.

In contrast to the ratios of betting odds appealed to by many philosophers, Howard Raiffa appeals to preferences between scaling and reference lotteries to represent degrees of belief.[27] Let A be the proposition that the American league team will win the next World Series. Now consider a lottery that would pay some substantial positive prize, say $100,000, if A turns out to be true and would pay nothing if not-A. This counts as a scaling lottery for proposition A. Let us call this the A-lottery. Now consider a fair 1000 ticket lottery with 600 red and 400 blue tickets, where you will receive the same positive prize, $100,000, if a red ticket is drawn and nothing if a blue ticket is drawn. Let us call this a reference lottery corresponding to the specified A-lottery. Suppose the reference lottery outcome will be known and the payoffs made at the same time when the outcome of that world series is known and the A-lottery is paid. On Raiffa's model your preference for the A-lottery over this reference lottery would show that your degree of belief in proposition A is higher than probability 0.6.

Now consider a proposition O for which you have the sort of epistemic access that one might expect you to have toward a relatively optimal observation claim. For example, let O be the claim that it is indeed a desk which you see and feel, as you look at and lean on your own desk while sitting in your chair before it. Consider a hypothetical scaling lottery for O:

---

[25]See [Pais, 1982, 253].
[26]See [Harper, forthcoming].
[27]See [Raiffa, 1970, pp. 108–110].

- Receive \$100,000 if O is really is true and status quo if not-O.

Now, consider the hypothetical choice between such a scaling lottery for O and a corresponding hypothetical fair 1000 ticket reference lottery with 999 good and one bad ticket. I have found that many people have quite clear preferences about this sort of abstract hypothetical choice. In this case almost everyone clearly prefers the O-lottery to this corresponding 999/1000 chance reference lottery. I found that quite a few people also shared, what was then, my clear preference for this sort of O-lottery over any corresponding fair finite reference lottery, however large, so long as it had at least one bad ticket in it. Such preferences correspond to a degree of belief that is closer to probability one than one minus any positive rational. To have such preferences is to exhibit a degree of belief in proposition O that differs at most infinitesimally from the complete certainty represented by the extreme subjective probability of 1.0.

According to a strict version of Leonard Savage's classic treatment, for an agent to treat proposition O as so certain as to assign it maximal epistemic probability $P(A) = 1$, that agent would have to be indifferent between the lotteries

- win \$10,000 if not-O, maintain status quo if O

- loose \$10,000 if not-O, maintain status quo if O

Such an agent would treat not-O as null.[28] They would not be at all guided by their conditional preferences on the hypothesis not-O. I found that only a very few of those asked were ready to so dismiss not-O.

There is a somewhat embarrassing tension between this result and Savage's commitment to the basic Bayesian updating model on which an agent shifts to conditional probability $P(\cdot|O)$ after a learning experience the total epistemic input of which can be represented as coming to accept O as new evidence. Richard Jeffrey's probability kinematics removes this tension by providing an alternative Bayesian updating model, which does not require assigning epistemic probability one to observations.[29] A more certainty-friendly alternative than Jeffrey's would be to drop the Archimedean axiom so as to allow for agents that can have full belief in O that is infinitesimally close to one, without treating not-O as null. This generates extended conditional probability belief systems and a non-monotonic logic of acceptance corresponding to the extended conditional belief systems I had developed to generate Popper functions from taking conditional acceptance as

---

[28][Savage, 1954, 24].
[29]See [Jeffrey, 1965, pp. 153–170].

primitive.[30] Such systems allow for a natural distinction between taking a proposition as a convention or conceptual commitment and taking it as a full belief without making it a conceptual commitment. They can also be naturally extended to allow for rational conceptual change.[31]

2.2   Myrvold's Example.

Wayne Myrvold was able to shake my faith in the applicability to scientific observation claims of this full-belief model of acceptance. You are in a closed room. Let M be the proposition that all the air molecules will collect into a single corner of that room some time within in the next hour. Consider now a scaling lottery for not-M:

- Receive $100,000 if not-M, status quo if M.

Compare this with the above corresponding scaling lottery for O.

- Receive $100,000 if O is really is true and status quo if not-O

I do not strictly prefer betting on it really being a desk in front of me now to betting that it will not be the case that all the air molecules in my room will collect into a single corner within the next hour. I am also not committed assigning at most infinitesimal belief to the claim that, as could be computed according to statistical mechanics, there is a positive, even if extremely small, finite objective chance that M will be true. I am just not that confident that statistical mechanics is wrong about this. Therefore, I cannot assign not-M an epistemic probability that differs at most infinitesimally from one. This raises problems for my earlier commitment to accept observation claims according to the model of acceptance as epistemic probability infinitesimally close to absolute certainty.

It raises some worries about my efforts to generate conditional assumption contexts from taking extended conditional epistemic probabilities, represented by Popper functions, as primitive.[32] This difficulty, however, does

---

[30]See [Harper, 1978; 1997]. I understand that [Pearl, 1988] has developed very interesting versions of such systems.

[31]In his *Principles of Mechanics* of 1900, Hertz explicitly took the invariance with respect to Galilean transformations characteristic of Newton's space-time conventions as an *a priori* conceptual commitment. In "Conceptual Change, Incommensurability and Special Relativity Kinematics", I showed how to apply an extended conditional belief system to model the transition from such a commitment to Newtonian kinematics to the kinematics of special relativity as a rational conceptual change that could be empirically guided by experimental evidence.

[32]See [Harper and Hajek, 1997]. The formal results of van Fraassen and McGee reported on pages 92-93 [with typo "$P^*(B)/P^*(AB)$" replaced by the correct "$P^*(AB)/P^*(A)$"] may be, I now think, less significant than I had taken them to be.

not undercut taking the acceptance context and system of associated assumption contexts as primitive, so that conditional epistemic probabilities relevant to a given scientific investigation are to be fixed relative to these more fundamental acceptance and assumption contexts. There are, perhaps, good reasons to suppose that preferences with respect to lotteries with very small or very large chances would not be reliable guides for assessing epistemic probabilities, even for otherwise quite rational agents. More importantly, I expect that an account of how to empirically assess alternative ways to revise the content of what is accepted is more informative for understanding the world than accounts limited to assessing how to revise epistemic probabilities of hypotheses which are neither accepted or rejected.

2.3 Acceptance with finite probability of error.

One way to extend the Bayesian approach to include acceptance is to model acceptance decision- theoretically by introducing epistemic utilities. Isaac Levi proposed such an epistemic utility based decision-theoretic Bayesian model of acceptance in his 1967 book *Gambling with Truth* and extended this approach in his 1980 book *The Enterprise of Knowledge*, which allowed for considerable flexibility for assessment of epistemic utility. I expect that an account of Newton's rich notion of empirical success as an epistemic utility might be a very informative way to further refine a Bayesian approach to scientific acceptance.

Henry Kyburg has long defended the importance of having an account of acceptance compatible with finite probability of error. His 1990 account of (r/s) full belief is motivated by the idea that what is crucial in whether a person acts as if a proposition is true is the ratio of the amount risked to the amount gained. Let us attempt to inform our intuitions about assessing Newton's empirical success as an epistemic utility by considering what might count as the epistemic risks and gains appropriate to the decision to accept or reject one of Newton's inferences, by a natural philosopher trying to understand the world with the sort of background and assumptions available in Newton's day.

Huygens, who had accepted Newton's inferences to inverse-square gravities toward the sun and planets, refused to accept Newton's applications of the third law of motion to count the gravity toward the earth that maintains the moon in its orbit as a two-body interaction.[33] He had developed a quite sophisticated causal hypothesis that would predict the centripetal

---

[33] In 1690 Huygens published his *Discourse on the Cause of Gravity* as an addition to *Treatise on Light*. His *Discourse* included specific comments on Newton's argument that he added to it after reading the *Principia*. George Smith has made available to me a manuscript of a translation by Karen Bailey. One nice feature of this translation is that it cites corresponding page numbers from Huygens 1690 publication.

accelerations of bodies as resulting from collisions on their internal particles by rapidly swirling spherical shells of very tiny vortical particles surrounding the center of the earth. His hypothesis can recover the phenomenon of weight toward a center by having many different very small layers or shells of vortical particles swirling in all different directions. The transverse tendencies imparted by the actions of these very tiny layers of vortical particles will cancel out, while the centripetal tendencies add up to produce the weight of a body. On his theory the correct application of Newton's third law would count the sum of the centrifugal actions of the particles making up the moon on the tiny vortical particles colliding with them as the equal and opposite reaction of the moon's gravitation toward the earth. His alternative theory had what he, and most natural philosophers of his age, regarded as the enormous advantage of showing how the action of gravity could be made intelligible by hypothesizing a mechanical cause acting by contact.

By refusing to accept this application of Law three, Huygens had to do without Newton's convergent measurements of relative masses of the sun, Jupiter, Saturn and the earth. He, therefore, had to do without Newton's center of mass resolution of the two chief worlds systems problem, with its acceptance of Keplerian orbits as provisional starting points to be corrected by successively better approximations, as perturbations are sought for and exploited. Instead, Huygens counted Newton's inferences to the sun's inverse-square gravity as grounds for counting Kepler's laws to be exact.

Somewhat crudely, we could represent the epistemic value afforded by Huygens' rejection of Newton's application of law three as those of accepting

H Kepler's laws are exact and are made mechanically intelligible by his account of the cause of gravity.

The epistemic value of the alternative if Newton's inference were to be correct might be that of

N Kepler's laws are good approximations and the strengths of the centripetal forces are proportional to the masses of the central bodies.

Huygens' commitment to the importance of making gravity mechanically intelligible would allow him to assign considerably higher epistemic value to **H** than to **N**. One way to represent Huygens' act to reject Newton's inference is as a gamble that would yield a gain of h = the epistemic value he assigns to **H** if he is correct and loss of $-n$, where n is the epistemic value of the alternative **N** he would miss out on by accepting **H** instead of Newton's inference. Let us assign $(n/h) = 5/10$ as the ratio of these epistemic values.

Suppose that Huygens had accepted **H** with full (20/1) belief according to Kyburg's model of (r/s) full belief. Then since (5/10) ≤ (20/1) he would in this situation simply act as if **H** were true and reject Newton's inference.

As our version of Huygens read on, he would have been confronted with the convergent agreeing measurements of the masses of the planets afforded from the orbits by Newton's inference. This would result in a revised understanding of the Newtonian alternative.

N1 Kepler's laws are good approximations and the strengths of the centripetal forces are proportional to the masses of the central bodies and lead to convergent accurate measurements of the relative masses of those central bodies.

Let us suppose our Huygens would have assigned significantly higher epistemic value to **N1** than to **N** but would not yet assign it an epistemic value as high as **H**. Let us suppose the epistemic value he assigned to this new richer Newtonian alternative **N1** would be n1 = 8. He would still reject Newton's inference.

After reading to the end of the *Principia*, our Huygens might, on our crude story, describe the Newtonian alternative as

N2 Kepler's laws are good approximations and the strengths of the centripetal forces are proportional to the masses of the central bodies and lead to convergent accurate measurements of the relative masses of those central bodies. In addition there is an account of tidal phenomena that affords estimates of the mass of the moon from tide ratio phenomena, etc..., including the solution of the variational inequality of the lunar orbit which afford corrections to the lunar areal rate corresponding to the perturbation of the earth-moon system by the sun. There are prospects for more general improvement of fit to data by corrections of Keplerian orbits to account for perturbations

For purposes of our story let us suppose that our Huygens assigned an epistemic value to this alternative that was even higher than that he assigned to **H**. Even if this value n2 = 30, our Huygens, like the real Huygens, would still have rejected Newton's inference. Unless he had significantly lowered his epistemic probability of **H** and no longer had (20/1) full belief in **H**, the fact that his risk to gain (30/10) ≤ (20/1) would lead him to continue to act as if **H** were true and reject Newton's inference.

In the first couple of decades after the *Principia*, there had not yet been developments of perturbations and observations to give Newtonian corrections that significantly improved on basic Keplerian orbits. But, by the 1750's the Newtonian corrections were becoming well established. Indeed, long before Laplace's successful treatment of the long recalcitrant Jupiter-Saturn perturbation, the evidence against **H** and the evidence for **N2** were so strong that the position of our Huygens would no longer have been tenable.

We have been allowing considerable weight to Huygens' commitment to making motion phenomena intelligible by conjecturing hypothetical mechanical causes. On Newton's methodology the viability of the Huygens alternative to count as a serious rival would depend on its realization of accurate measurement of its parameters by the phenomena it purports to explain. Unlike Descartes' much cruder vortex theory, Huygens theory does realize some Newtonian empirical success. Huygens was able to use a very clever calculation, from empirical pendulum results, of the enormous velocity his vortical shells would need to recover the measured gravity at the earth's surface. By the stage indicated as **N1**, we have, on Newton's method, the measurements afforded by this calculation to set against the convergent accurate measurements of the relative masses of the sun, Jupiter, Saturn and the earth afforded by Newton's inference. If we leave out counting any epistemic value to the prospect for making Kepler's laws intelligible by a merely possible hypothetical mechanical explanation then, perhaps, it would not have been inappropriate for Newton to have dismissed Huygens' alternative as a mere contrary hypothesis even at this stage. By the stage indicated by **N2**, with all the rest of the results in the Principia available, it would more certainly have been legitimate to so dismiss Huygens' theory. I suggest that the empirical success realized by the convergent accurate agreeing measurements afforded for the parameters of Newton's theory from the phenomena it was applied to explain were clearly greater than those empirical measurement success afforded to Huygens' theory by his estimate of vortical velocities from pendulum calculations.

I believe Newton was as sensitive to the oddness of the idea of action at a distance as Huygens and the other mechanical philosophers. His genius was not that he did not share their metaphysical qualms, but, rather, that he did not let them undercut his acceptance of inferences supported by empirical measurement successes.

It is, perhaps, not surprising that the demand for a mechanical explanation by contact, and an avoidance of apparent commitment to action at a distance, continued to be counted as an important epistemic utility until well after the *Principia* had become available. It may well have not been very inappropriate to have required the additional results afforded by perturbation theory by successors to override a commitment as fundamental as that to avoid apparent commitment to action at a distance.[34] I do,

---

[34]Like Newton earlier, Euler continued to hope for a mechanical explanation for gravity

however, want to defend the appropriateness of having these extensive empirical successes override even this powerfully intuitive commitment, as they eventually did do. It is an important part of what makes science such an effective way of empirically informing our understanding of the world that the empirical successes of Newton's theory were able to lead to acceptance of action at a distance. It is also an important part of what makes science so effective at empirically informing our understanding of the world that it required empirical successes of a developed rival theory by Einstein, rather than a priori objections, to overturn this acceptance of action at a distance.

BIBLIOGRAPHY

[Cohen and Smith, 2002] I. B. Cohen and G. E. Smith, eds. *The Cambridge Companion to Newton*, Cambridge: Cambridge University Press, 2002.

[Cohen and Whitman, 1999] I. B. Cohen and A. Whitman, trans. *Isaac Newton: The Principia, Mathematical Principles of Natural Philosophy: a new translation*, Los Angeles: University of California Press (C & W), 1999.

[DiSalle, 2006] R. DiSalle. *Understanding Space-time: The Philosophical Development of Physics from Newton to Einstein*, Cambridge: Cambridge University Press, 2006.

[Euler, 1843] L. Euler. *Letters to a German Princess on Different Subjects in Natural Philosophy: 1768–1772.*, Hunter, H. (trans.), with notes by Brewster, D. and Griscom, J. New York: Harper and Brothers, 1843.

[Galilei, 1632] G. Galilei. *Dialogue Concerning the Two Chief World Systems — Ptolemaic & Copernican*, Drake, S. (trans.), 1970. Los Angeles: University of California Press.

[Harper, 1991] W. Harper. Newton's Classic Deductions from Phenomena, *PSA1990*, vol. 2, 183–196, 1991.

[Harper, 1997] W. Harper. Isaac Newton on Empirical Success and Scientific Method, in Earman, J. and Norton, J. D. (eds.), *The Cosmos of Science: Essays of Exploration*, Pittsburgh: University of Pittsburgh Press, 55–86, 1997.

[Harper, 1998] W. Harper. Measurement and Approximation: Newton's Inferences from Phenomena versus Glymour's Bootstrap Confirmation, in Weingartner, P., Schurz, G. and Dorn, G. (eds.), *The Role of Pragmatics in Contemporary Philosophy*. Vienna: Hölder-Pichler-Tempsky, 265–287, 1998.

[Harper, 1999] W. Harper. The First Six Propositions in Newton's Argument for Universal Gravitation, *The St. John's Review*, vol. XLV(2): 74–93, 1999.

[Harper, 2002a] W. Harper. Howard Stein on Isaac Newton: Beyond Hypotheses? In D. B. Malament, (ed.), 71–112, 2002.

[Harper, 2002b] W. Harper. Newton's argument for universal gravitation, in I. B. Cohen and G. E. Smith, (eds.), 174–201, 2002.

[Harper, forthcoming] W. Harper. Newton's Methodology and Mercury's Perihelion before and after Einstein, forthcoming in *(PSA2006)*, 2006.

[Harper et al., 1994] W. Harper, B. H. Bennett, and S. R. Valluri. Unification and Support: Harmonic Law Ratios Measure the Mass of the Sun, in Prawitz, D. and Westertahl, (eds.) *Logic and Philosophy of Science in Uppsalla*, Dordrecht: Kluwer Academic Publishers, 131–146, 1994.

[Harper andDiSalle, 1996] W. Harper and R. DiSalle. Inferences from Phenomena in Gravitational Physics, *Philosophy of Science*, 63: S46-S54, 1996.

into the late 1760's, but, also like Newton before him, he did not take this to undercut applications of universal gravity to solar system motions. See his *Letters to a German Princess*.

[Harper *et al.*, 2002] W. Harper, S. R. Valluri, and R. B. Mann. Jupiter's Moons As a Test of the Equivalence Principle. In Gurzadyan, R. V. G., Jantzen, R.T., Ruffini, R. (eds.) *Proceedings of the Ninth Marcel Grossman Meeting on General Relativity*, Singapore: World Scientific, 2002.

[Hertz, 1990] H. Hertz. *The principles of Mechanics.* Reprinted, 1956. New York: Dover, 1990.

[Jeffrey, 1965] R. C. Jeffrey. *The Logic of Decision.* New York: McGraw-Hill Book Company, 1965.

[Kyburg, 1970] H. E. Kyburg, Jr. Two World Views, *Nous* 4: 337–348, 1970. Reprinted with modifications and additions in [Kyburg, 1983, 18–27].

[Kyburg, 1983] H. E. Kyburg, Jr. *Epistemology and Inference.* Minneapolis: University of Minnesota Press, 1983.

[Kyburg, 1990] H. E. Kyburg, Jr. *Science and Reason*, Oxford: Oxford University Press, 1990.

[Laubscher, 1981] R. E. Laubscher. *The Motion of Mars 1751-1969.* Washington: Nautical Almanac Office, 1981.

[Levi, 1967] I. Levi. *Gambling with Truth.* New York: Alfred Knopf, 1967.

[Levi, 1980] I. Levi. *The Enterprise of Knowledge: An Essay on Knowledge, Credal Probability and Chance.* Cambridge Mass.: MIT Press, 1980.

[Huygens, 1690] C. Huygens. *Discourse on the Cause of Gravity*, originally published as as an addition to his *Treatise on Light*, 1690. Translation by Karen Bailey with annotations by Karen Baily and George Smith, manuscript.

[Malament, 2002] D. E. Malament, ed. *Reading Natural Philosophy: Essays in the History and Philosophy of Science and Mathematics.* Chicago : Open Court Publishing Company, 2002.

[Newcomb, 1895] S. Newcomb. *The Elements of the Four Inner Planets and the Fundamental Constants of Astronomy.* Washington: government printing office, 1895.

[Pais, 1982] A. Pais. *Subtle is the Lord: The Science and the Life of Albert Einstein.* New York: Oxford University Press, 1982.

[Palter, 1970] R. Palter, ed. *The* Annus Mirabilis *of Sir Isaac Newton 1666-1966.* Cambridge Mass.: The MIT Press, 1970.

[Pearl, 1988] J. Pearl. *Probabilistic Reasoning in Intelligent Systems.* San Francisco: Morgan Kaufmann, 1988.

[Raiffa, 1968] H. Raiffa. *Decision Analysis: Introductory Lectures on Choices under Uncertainty.* Menlo Park : Addison-Wesley Publishing Company, 1968.

[Savage, 1954] L. J. Savage. *The Foundations of Statistics.* New York: John Wiley and Sons, 1954.

[Smith, 1999] G. E. Smith. How Did Newton Discover Universal Gravity, *The St. John's Review*, XLV(2), 32–63, 1999.

[Smith, 2002] G. E. Smith. The methodology of the *Principia*, In Cohen, I. B. and Smith, G. E. (eds.), pp. 138–173, 2002.

[Stein, 1967] H. Stein. Newtonian Space-Time, *The Texas Quarterly* Vol. X, No.3, 174-200, 1967. Reprinted with revision in R. Palter, (ed.), 1970, 258-284.

[Stein, 1970] H. Stein. On the Notion of Field in Newton, Maxwell, and Beyond, In Stuewer, R.H. (ed.), *Historical and Philosophical Perspectives of Science.* Minneapolis: University of Minnesota Press, 264–287, 1970.

[Stein, 1977] H. Stein. Some Philosophical Prehistory of General Relativity. In Earman, J., Glymour, C. and Stachel, J. (eds.), *Minnesota Studies*, vol. 8. Minneapolis: University of Minnesota Press, 3–49, 1977.

[Stein, 1991] H. Stein. 'From the Phenomena of Motions to the Forces of Nature': Hypothesis or Deduction?, in *PSA1990*, vol. 2, 209–222, 1991.

[Stein, 2002] H. Stein. Newton's metaphysics. In Cohen, I. B. and Smith, G. (eds.), 256-307, 2002.

[Taton and Wilson, 1989] R. Taton and C. Wilson, eds. *The General History of Astronomy, vol. 2, Planetary astronomy from the Renaissance to the rise of astrophysics, Part A: Tycho Brahe to Newton.* Cambridge: Cambridge University Press, 1989.

[Taton and Wilson, 1995] R. Taton and C. Wilson, eds. *The General History of Astronomy, vol. 2, Planetary astronomy from the Renaissance to the rise of astrophysics, Part B: The eighteenth and nineteenth centuries.* Cambridge: Cambridge University Press, 1995.

[Valluri *et al.*, 1997] S. R. Valluri, C. Wilson, and W. Harper. Newton's Apsidal Precession theorem and Eccentric Orbits, *Journal for the History of Astronomy*, xxvii: 13–27, 1997.

[Valluri *et al.*, 1999] S. R. Valluri, W. Harper, and R. Biggs. Newton's Precession Theorem, Eccentric Orbits and Mercury's Orbit, *Proceedings of the Eighth Marcel Grossman Meeting on General Relativity*, Piran, T. and Ruffini, R. (eds.), Singapore: World Scientific, 485–487, 1999.

[Will, 1993] C. M. Will. *Theory and experiment in gravitational physics* 2nd revised ed. Cambridge: Cambridge University Press, 1993.

[Wilson, 1987] C. A. Wilson. D'Alembert versus Euler on the Precession of the Equinoxes and the Mechanics of Rigid Bodies, *Archive for History of Exact Sciences* 233-273, 1987.

[Wilson, 1989] C. A. Wilson. The Newtonian achievement in astronomy. In Taton and Wilson (eds), 233–274, 1989.

[Wilson, 1995] C. A. Wilson. The precession of the equinoxes from Newton to d'Alembert and Euler. In Taton and Wilson (eds.), 47–54, 1995.

[Wilson, 1999] C. A. Wilson. Re-doing Newton's Experiment for Establishing the Proportionality of Mass and Weight, *The St. John's Review*, vol. XLV(2): 64–73, 1999.

Conflict and Consistency

CHOH MAN TENG

Henry Kyburg has introduced me to a world of ideas and fun. Over the years whatever I have to say about evidential probability I have said to him in person. Whatever made sense was put to some use and whatever didn't was unceremoniously discarded but occasionally resurrected. Here are more used and unused thoughts on Kyburg's work.

ABSTRACT. Data give rise to relative frequencies, and relative frequencies give rise to probability intervals. A key notion in evidential probability is the *conflict* of statistical intervals and its resolution. Two intervals are in conflict if neither is included in the other. Conflicting statistical intervals can be assimilated into one single probability interval based on the relationship between the intervals themselves and their corresponding supporting reference classes. We examine conflict in relation to consistency in evidential probability, noting how the conflict resolution principles are formulated such that consistency can be preserved among conflicting intervals. In view of these considerations we appraise the three principles for resolving conflict and explore some weak constraints for characterizing and relating evidential update.

1 Objectivity in Conflict and Consistency

In evidential probability, which is first elucidated in [Kyburg, 1961], the probability of a statement is derived from a collection of statistical statements, which themselves are construed to be based on objective relative frequencies compiled from a corpus of data. Such statements are naturally approximate and uncertain, and they can be represented as intervals in the knowledge base.

Although the starting point of evidential probability is usually a set of statistical statements together with other accompanying background knowledge, it is implicitly understood that these statistical statements are distilled from a collection of data, not necessarily all from the same source. The data sources are often not assumed to be available; otherwise we would not have to counter the difficulty of having statistics for two reference classes but none, except the trivial, for their intersection.

The grounding in objective data provides justified backing for the principles advanced for adjudicating differences between competing statistical statements. For instance, a reference set that is a subset of another more general reference set can be expected to include only a subset of the data points supporting the latter, and the statistical interval derived typically is correspondingly wider than the one for the latter. This lends support to the principle of specificity, reminiscent of Reichenbach's principle for preferring the narrowest class [Reichenbach, 1949]. This principle selects the interval of the more specific class only when the information captured by the additional specificity is of relevance to the target probability; otherwise it selects the interval of the more general class which can be expected to be non-conflicting and narrower by virtue of a larger collection of supporting data points.

Key to the integrity of evidential probability is the notion of conflict and the principles devised to adjudicate between conflicting intervals. These principles, among them the principle of specificity described above, are motivated by their grounding in objective but approximate relative frequency data. Applied haphazardly, however, they may lead to non-unique and inconsistent probability intervals in some situations.

We will discuss how some of the measures adopted for handling multiple, potentially conflicting, statistical statements help preserve consistency in evidential probability. The constraints imposed by the framework sometimes result in probability intervals that are not very informative, and in many cases make evidential updating inconvenient. We consider some weak characterizations of evidential update, including a partially incremental version incorporating practical certainty.

2   Background

First let us recall some of the basic elements of evidential probability. A statistical statement

$$\%\bar{x}(\tau(\bar{x}), \rho(\bar{x}), p, q)$$

loosely denotes that, of the objects that are ρ's, a proportion between p and q are also τ's. For example, a fair coin might give rise to the following statistical statement:

$$\%x(tails(x), coin\_toss(x), 0.49, 0.51).$$

These statistical statements are candidates for determining the probability of a target statement.[1] For instance, based on the above statistical statement on coin tosses we may say that the probability that the next toss of

---

[1] There are constraints on the terms that can take the role of each of the arguments in a statistical statement, as well as conditions specifying what constitutes a candidate

this coin will land tails is [0.49, 0.51]. However, in general we can expect that our knowledge base contains many statistical statements that may be applicable to our target situation. For example, we may also have in our knowledge base

$$\%x(tails(x), my\_toss(x), 0.32, 0.71),$$
$$\%x(tails(x), toss\_on\_sunday(x), 0.24, 0.43),$$
$$\%x(tails(x), left\_hand\_toss(x), 0.68, 0.84),$$

and even

$$\%x(tails(x), next\_toss(x), 0.00, 1.00),$$

this last statistical statement reflecting our absolute ignorance with regard to how the next toss will land.

Three principles are used to vet and combine such candidate statistical statements to obtain a single probability interval for the target statement. These principles make use of the following definitions.

[**Conflict**] Two intervals $[l, u]$ and $[l', u']$ *conflict* iff neither is included in the other. Two statistical statements $\%\bar{x}(\tau(\bar{x}), \rho(\bar{x}), l, u)$ and $\%\bar{x}'(\tau'(\bar{x}'), \rho'(\bar{x}'), l', u')$ *conflict* iff their associated intervals $[l, u]$ and $[l', u']$ conflict.

Note that conflicting intervals are not necessarily disjoint. They may overlap as long as neither is included entirely in the other.

[**Cover**] The *cover* of a set of intervals is the shortest interval including all the intervals. The cover of a set of statistical statements is the cover of their associated intervals.

[**Closure under Difference**] Given a set of intervals I, a set of intervals I' is *closed under difference* with respect to I iff $I' \subseteq I$ and I' contains every interval in I that conflicts with an interval in I'.

Given a set of statistical statements K, a set of statistical statements K' is *closed under difference* with respect to K iff the intervals associated with K' are closed under difference with respect to the intervals associated with K.

The three principles for resolving conflict between candidate statistical statements pertaining to a target statement S may be stated as follows.

---

statement for a given target probability. We will omit the details here but they can be found in [Kyburg and Teng, 2001].

1. [**Richness**] If two statistical statements conflict and the first is based on a marginal distribution while the second is based on the full joint distribution, disregard the first.

2. [**Specificity**] If two conflicting statistical statements both survive the principle of richness, and the second employs a reference class that is known to be included in the first, disregard the first.

3. [**Strength**] Those statistical statements not disregarded by the principles of richness and specificity are called *relevant*. The probability of S is the cover that is the shortest among all covers of non-empty sets of statistical statements closed under difference with respect to the set of relevant statistical statements; alternatively it is the intersection of all such covers.

The probability for a statement S derived according to the above three principles is unique and consistent. Key to this claim is the way these principles are constructed and applied.

3 Preserving Consistency

The consistency of evidential probability is safeguarded by several measures. Starting from a set of candidate statistical statements, inconsistency obtains when the three principles described in the previous section can be applied in different ways to arrive at different final probability intervals for a target statement.

Note that even if the two final probability intervals are not in conflict, they are inconsistent in the sense that each can be construed as a non-unique probability for the target statement. The principle of strength does not apply to probability intervals. It applies to candidate statistical statements, from which a final probability interval is derived.

Let us examine how inconsistency is prevented by the way evidential probability is formulated, in particular the way the three principles are used to handle conflicting information.

3.1 Order

The order of application of the three principles is important; so is the order in which pairs of statistical statements are considered for each principle. First we apply the principle of richness. Only when all possible pairs of statistical statements from corresponding marginal and joint distributions have been vetted do we apply the principle of specificity. Again we deal with all general-specific pairs of statistical statements before the principle of strength is invoked.

We may consider other orderings of the principles, some making more sense than others. (For example, it seems reasonable that the principle of strength should be applied last.) However, the applications of the principles must be exhaustive at each stage and not interleave.

Similarly, pairs of candidate statistical statements need to be considered in a specific order for each principle. Violating the prescribed order can result in inconsistent probability intervals.

An example will help illustrate this point. Suppose we have the following three statistical statements.

$$(s_1) \quad \%x(d(x), a(x) \wedge b(x) \wedge c(x), 0.2, 0.6),$$
$$(s_2) \quad \%x(d(x), a(x) \wedge b(x), 0.1, 0.4),$$
$$(s_3) \quad \%x(d(x), a(x), 0.3, 0.5).$$

Given that we know object o satisfies $a(o)$, $b(o)$ and $c(o)$, all three of the above statistical statements are candidates for the probability of $d(o)$. Consider two scenarios with different orderings of the pairs of statements to be vetted.

1. [**Correct**] (s_1) defeats (s_2) by specificity. Both (s_1) and (s_3) survive, and by the principle of strength, the probability of $d(o)$ is taken to be $[0.3, 0.5]$.

2. [**Incorrect**] (s_2) and (s_3) are compared first and (s_3) is eliminated by specificity. (s_1) and (s_2) are then compared and (s_2) is in turn eliminated by specificity. The probability of $d(o)$ then is taken to be $[0.2, 0.6]$.

Considering the pairs of candidate statements in different orders gives rise to inconsistent results. For the principle of specificity, pairs of more specific statements should be compared before pairs of more general statements, and statements that have been eliminated by a previous application of a principle cannot be used to eliminate further statements. Thus, in the above example, the probability of $d(o)$ should be as derived in Case (1), that is, $[0.3, 0.5]$.

3.2   Assimilate

Another potential source of inconsistency is multiple sources of data, or, multiple statistical statements arising from separate mechanisms. Each of these statistical statements or data items may support a different candidate interval, and we need to be able to assimilate the information in an objective and consistent manner.

An example of this is the repeated measurement of a quantity. Using a ruler whose measurement errors were distributed normally with mean 0 and variance σ^2, we measured the length of a stick once and got l_1, and then we measured it again (independently) and got l_2. Among others we have the following two statistical statements:

$$\%x(length(x) \in l_1 \pm \sigma, measured(x) = l_1, \ 0.68, \ 0.68),$$
$$\%x(length(x) \in l_2 \pm \sigma, measured(x) = l_2, \ 0.68, \ 0.68).$$

Where l_1 and l_2 differ, these two statements appear to give contradictory values for the length of the measured stick. Intuitively, however, we should consider the two observations combined together rather than separately. We then have a third statistical statement regarding the average of the two measurements:

$$\%x(length(x) \in \frac{l_1 + l_2}{2} \pm \frac{\sigma}{\sqrt{2}}, measured_1(x) = l_1 \wedge$$
$$measured_2(x) = l_2, \ 0.68, \ 0.68).$$

Now we have three statements, all applicable to our stick and each supporting a different length interval.

The apparent inconsistency can be averted by noting that the target formulas, that is, the formulas serving as the first arguments in the statistical statements, need to be logically equivalent to be comparable. The three target formulas in question, $length(x) \in l_1 \pm \sigma$, $length(x) \in l_2 \pm \sigma$ and $length(x) \in \frac{l_1+l_2}{2} \pm \frac{\sigma}{\sqrt{2}}$, are all similar but not equivalent (assuming $l_1 \neq l_2$) and thus cannot be compared directly.

Although the above three statistical statements concerning the length of the stick are incompatible in this sense, straightforward transformation will allow us to derive alternative, standardized, statements with equivalent target formulas based on these same measurements. The generic forms of the candidate statements are

$$\%x \left(length(x) \in [m, n], measured(x) = l, \int_m^n N(l, \sigma^2)dz, \int_m^n N(l, \sigma^2)dz \right)$$

for the individual measurements and

$$\%x \left(length(x) \in [m, n], measured_1(x) = l \ \wedge \ measured_2(x) = l', \right.$$
$$\left. \int_m^n N(\frac{l+l'}{2}, \frac{\sigma^2}{2})dz, \int_m^n N(\frac{l+l'}{2}, \frac{\sigma^2}{2})dz \right)$$

for the averaged measurement.

We can generate any number of statements adhering to these generic forms, but for the statements to be comparable, the values of m and n need to be held constant across the statements. We can then apply the three principles of conflict resolution to determine the probability that the true length of the stick falls within $[m, n]$. Since the standard deviation for the averaged measurement is smaller than that for an individual measurement, for a fixed $[m, n]$ the statement for the averaged measurement can be expected to give rise to a statistical interval with higher values.

For example, let $\sigma = 0.05$, $l_1 = 31.70$ and $l_2 = 31.60$. The generic statements may be instantiated as follows.

$\%x(length(x) \in [31.62, 31.67], measured(x) = 31.70, \ 0.22, \ 0.22),$

$\%x(length(x) \in [31.62, 31.67], measured(x) = 31.60, \ 0.26, \ 0.26),$

$\%x(length(x) \in [31.62, 31.67], measured_1(x) = 31.70 \ \wedge$

$$measured_2(x) = 31.60, \ 0.52, \ 0.52).$$

By specificity (or richness) the first two statements, based on individual measurements, are eliminated by the third statement which is based on the combined measurements. The probability that the length of the stick falls within $[31.62, 31.67]$ is thus $[0.52, 0.52]$.

As we make more measurements, these measurements can be combined to support an increasing probability that the true length of the stick falls within a fixed interval (provided that the fixed interval includes the average of the measurements). Repeated measurements thus can be leveraged to strengthen the evidential import of the combined observations.

3.3   Take Cover

Methods such as that described in the previous section for assimilating information are not always applicable. Sometimes the data cannot be combined and the conflict among candidate statistical statements cannot be resolved completely. We are then left with several statistical statements each of which is supportive of a different interval. These intervals may be overlapping, disjoint, or nested.

Examples abound for this situation. Here is one. Suppose we have the following statistical statements:

$$\%x(sour(x), red\_jelly\_bean(x), 0.72, 0.89),$$

$$\%x(sour(x), round\_jelly\_bean(x), 0.38, 0.44).$$

We have a red and round jelly bean. Before we devour it we would like to know the probability of its being sour. No other relevant information is available, in particular, there is no statistics available concerning the sub-population of jelly beans that are both red and round. Thus neither of the two statistical statements above can be eliminated on the grounds of richness or specificity.

We have two applicable intervals, [0.72, 0.89] and [0.38, 0.44], but this does not automatically constitute an inconsistency. The principle of strength ensures that in such a situation a single interval denoting the target probability can be composed from the uneliminated intervals. Loosely stated this is achieved by removing candidate intervals that include all others and then taking the cover of the remaining intervals. In our example neither of the two candidate intervals can be removed based on strength, and the probability of our red and round jelly bean being sour is therefore their cover [0.38, 0.89].

The result is consistent but it does illustrate one worry. Applying the principle of strength in some cases may give rise to a final probability interval that is wider, sometimes much wider, than the constituent intervals, even when the constituent intervals themselves are narrow but far apart. Arguably some information has been lost during the process.[2] However, by taking the cover of the surviving intervals, consistency is preserved even when the conflict between candidate statistical statements cannot be resolved in other ways. In the extreme case we obtain the vacuous probability interval [0, 1], not very interesting but nonetheless consistent.

4   Evidential Update

Evidential probability admits broader, interval-valued, ranges than traditional point based formulations. The three principles for resolving conflict, in particular the principle of strength, render evidential probability less tightly constrained than classical probability. Intervals for statements that are related are only loosely coupled. Derivative relationships between distributions in classical probability do not necessarily hold in evidential probability.

One area of contention is the handling of update in evidential probability. Much has been said about the relationship (or the absence thereof) between evidential probability intervals and their revision by conditioning: [Levi, 1980; Kyburg, 1983], for example, and also contributions to this volume by Kyburg, Levi, and Seidenfeld. We will examine the interaction between conflict resolution and conditioning and advance some reasons why

---

[2]We can also argue for the reverse, that the resulting interval is not wide enough, since it is possible that the probability in question in the jelly bean example lies outside of the evidential probability interval. We will leave this aside for the moment, except to note that to be *completely* certain the only admissable interval is [0, 1].

conditioning is not unquestionably accepted as the method for update. We will consider some circumstances under which conditioning is sanctioned and also other ways to characterize evidential update in special cases.

4.1   Conditioning

Evidential probability can be said to be always conditional in the sense that it is defined relative to a given knowledge base. Let us write $Pr(s \mid K)$ for the evidential probability interval of a statement s based on a body of knowledge K.

Kyburg allows for conditioning under some restricted circumstances, for instance those prescribed by the principle of richness. However, in general adding a piece of information to the knowledge base does not amount to updating the original probability by conditioning on that piece of information. The probability distribution $Pr(\cdot \mid K \cup \{a\})$ cannot be derived directly from the probability distribution $Pr(\cdot \mid K)$.

For classical probability, $Pr(s \mid K \cup \{a\}) = \frac{Pr(a \mid K\{s\}) \cdot Pr(s \mid K)}{Pr(a \mid K)}$ where their values exist. Let us set aside for the moment the question of how interval probabilities are to be multiplied and divided. We may apply similar calculations to the relative frequency bounds of statistical statements to obtain bounds that take into account an additional piece of information.

For example, suppose we have two types of coins with different biases for yielding tails, and we have the following statistical information regarding the coins in our knowledge base K.

(c_1) $\%x(tails(x), coin_1(x) \vee coin_2(x), 0.4, 0.6)$,

(c_2) $\%x(tails(x), coin_1(x), 0.3, 0.3)$,

(c_3) $\%x(coin_1(x), coin_1(x) \vee coin_2(x), 0.2, 0.2)$.

(For simplicity, degenerate point-valued intervals are used in some statements.) Regarding a generic coin o, we have $Pr(coin_1(o) \mid K) = [0.2, 0.2]$ (from (c_3)). We then flip the coin o and obtain tails. Applying conditioning to the frequency bounds, we can calculate

$$\%x(coin_1(x), (coin_1(x) \vee coin_2(x)) \wedge tails(x), p, q).$$

The interval $[p, q]$ is bounded by $[0.10, 0.15]$, which conflicts with the interval $[0.2, 0.2]$ obtained from the more general statistical statement (c_3) regarding $coin_1(o) \vee coin_2(o)$. The probability $Pr(coin_1(o) \mid K \cup \{tails(o)\})$ is therefore $[0.10, 0.15]$.

Now consider a similar situation but the statistical statement (c_1) for obtaining tails in general is replaced by the following.

(c_1^*) $\%x(tails(x), coin_1(x) \vee coin_2(x), 0.2, 0.4)$.

The information conveyed by this statement is not any more precise than that from (c_1) but now tails occur less often than in the previous case. Conditioning on $tails(x)$ again, we obtain

$$\%x(coin_1(x), (coin_1(x) \vee coin_2(x)) \wedge tails(x), 0.15, 0.30).$$

Even though this statement refers to a more specific reference class, its statistical interval is wider and includes the interval $[0.2, 0.2]$ of the more general statement (c_3). The principle of strength in this case directs us to retain the narrower interval $[0.2, 0.2]$ as the updated probability $Pr(coin_1(o) \mid K \cup \{tails(o)\})$.

In the first case, the updated probability is obtained by a process akin to conditioning, whereas in the second case the non-conflicting, narrower, interval of the more general statement takes precedence. The rationale underlying the differential treatment of the two cases is that conditioning should be allowed to update a probability interval only when it yields non-trivial information.

Intuitively, when the general statistical bounds for tails is $[0.4, 0.6]$ (from (c_1)) and the more specific $coin_1$ bounds for tails is the lower $[0.3, 0.3]$ (from (c_2)), observing tails should justifiably lower the probability that the coin is of type $coin_1$. On the other hand, when the general bounds for tails, $[0.2, 0.4]$ from (c_1^*), includes the more specific $coin_1$ bounds for tails, $[0.3, 0.3]$, observing tails does not provide much extra information regarding the probability that the coin in question is of type $coin_1$. The principles for resolving conflict see to such distinctions between these two situations.

4.2   Related Updates

We will consider another way of characterizing update, namely the relationship between probabilities derived from a knowledge base updated with related information. We will focus on the distributions

$$Pr(\cdot \mid K \cup \{a(o)\}), \; Pr(\cdot \mid K \cup \{b(o)\});$$
$$Pr(\cdot \mid K \cup \{a(o) \wedge b(o)\}).$$

In other words, what can we deduce about the probability of a statement $s(o)$ relative to $a(o) \wedge b(o)$ (and K) given the probabilities of $s(o)$ relative to $a(o)$ (and K) and relative to $b(o)$ (and K)?

For classical probability, in general $Pr(s(o) \mid K \cup \{a(o) \wedge b(o)\})$ does not bear any relation to either $Pr(s(o) \mid K \cup \{a(o)\})$ or $Pr(s(o) \mid K \cup \{b(o)\})$. This is the same for evidential probability. If there is relative frequency information regarding the proportion of $s(x)$ in the joint reference set $a(x) \wedge b(x)$, and this information is relatively precise (according to the principle of strength), then the statistical interval with the joint reference set is taken to be the probability of $s(o)$.

If on the other hand the only relevant information we have concerns the proportions of $s(x)$ among separately $a(x)$ and $b(x)$ and their supersets, and we have only vague or no relative frequency information regarding the proportion of $s(x)$ among the joint reference set $a(x) \wedge b(x)$, the evidential probability of $s(o)$, where the object o also satisfies $a(o)$ and $b(o)$ (and consequently also any of their supersets), is constructed from the available intervals concerning the individual conjuncts and their supersets. In this situation we have the following relationship.

The probability $Pr(s(o) \mid K \cup \{a(o) \wedge b(o)\})$ is a subset of the cover of $Pr(s(o) \mid K \cup \{a(o)\})$ and $Pr(s(o) \mid K \cup \{b(o)\})$.

This rather loose relationship may be generalized to more than two conjuncts. It holds even when some or all of the conjuncts are not supported by statistical statements of their own in the knowledge base, and their probabilities in turn have to be derived from, for instance, statistical statements concerning some even more general reference sets.

In the simplified case illustrated in the jelly bean example in Section 3.3, we have no information other than the statistical statements for the two separate conjuncts, red jelly beans and round jelly beans. The target probability of being sour relative to the joint reference set red and round jelly beans is then taken to be the cover of the intervals relative to the two individual conjuncts.

5 Update with Practical Certainty

Many constraints that are in place for classical, point-valued, probability do not hold for evidential probability. Updating by conditioning in general is not valid, and updates obtained from related information are only tangentially connected. Although less appealing and less convenient, to perform an update we need to add the new piece of information to the knowledge base and recompute the probability intervals from scratch.

Evidential probability is relativized to a knowledge base. This emphasis is particularly relevant here, since unlike classical probability, the updated probability distribution cannot readily be derived from the original distribution without referring to the underlying knowledge base.

We may consider a partially incremental approach to updating, extended from work variously in [Teng, 1997; Kyburg and Teng, 2002]. We will accept as certain those statements whose lower evidential probability values are above a particular threshold $1 - \epsilon$. This threshold denotes what counts as "practical certainty". Statements with high enough probability are simply accepted and regarded as true for all practical purposes and for further reasoning and update.

The threshold parameter ϵ is a fixed (but maybe context dependent) numeric value, not an infinitesimal arbitrarily close to 0 as advocated in some other approaches [Adams, 1975; Pearl, 1990; Bacchus *et al.*, 1993, for example]. This choice is in line with evidential probability's program to provide an objective account of probability. An infinitesimal cannot serve as an objective, empirical, criterion since we cannot be expected to be able to verify that a probability value is arbitrarily close to unity. Rather, the threshold is based on a fixed finite value, such that the condition for acceptance can be easily tested.

This formulation is in the same tradition as statistical hypothesis testing: A hypothesis is rejected with respect to a fixed chance of error, not with respect to an unverifiable infinitesimal one. It also reflects our everyday pattern of reasoning: except in the most synthetic environment, few non-trivial events are certain and we frequently simply accept events deemed to be of high enough probability. In the extreme (but not necessarily infinitesimal) case, high thresholds exceed our resolution limit, which render very high probability indistinguishable from certainty.

This acceptance of practical certainty is nonmonotonic. Further evidence to the contrary of what we have accepted could undermine our body of practical certainty and lead us to retract some of the statements previously accepted as true due to their high, above threshold, but less than unity probability values.

The body of accepted knowledge therefore grows in a stepwise manner (except for nonmonotonic retractions), with an increasing tolerance of error. Updates are carried out with respect to the corrigible knowledge base previously accepted at a particular error threshold.

This approach is only partially incremental because, apart from being nonmonotonic, the probabilities of those statements exceeding the tolerance of error still need to be recomputed with each update. Furthermore, this process of thresholding must not be allowed to carry on beyond the point where the strength of the accepted statements is deemed too weak to lend support to pragmatic reasoning.

The body of accepted knowledge is regarded as a body of practical certainty. We should not accept any uncertain statement whose chance of error exceeds our tolerance of error, unless we first re-evaluate the risk conferred by the body of corrigible knowledge we have accepted.

6 Concluding Remarks

Evidential probability is grounded in objective data, and as such admits of a certain level of imprecision which manifests as intervals that in many cases do not agree. The conflict resolution principles prescribed for adjudicating between competing statistical intervals have drawn some criticisms. They sometimes result in broad probability intervals that are not very informative, and they invalidate update by conditioning.

We argued that these conflict resolution principles, even if perhaps not strictly indispensable, play a significant role in maintaining consistency in evidential probability. The tension between conflict and consistency is mitigated by careful formulation and judicious application of these principles. We surveyed a number of potentially problematic situations and showed how the principles were instrumental in preserving overall consistency in each case.

These principles for resolving conflict however make it difficult to characterize evidential probability analytically. There is only a weak connection between probability distributions that we usually consider as related. In general updated probabilities need to be derived from the updated knowledge base of statistical statements, rather than more succinctly from the original probabilities. Although not as convenient, the formulation of evidential probability strikes a balance between conflict and consistency. We considered some conditions under which weak constraints can be established. In addition we considered an approach to practical certainty that allows for partially incremental update.

BIBLIOGRAPHY

[Adams, 1975] Ernest W. Adams. *The Logic of Conditionals: An Application of Probability to Deductive Logic*. Dordrecht, 1975.

[Bacchus *et al.*, 1993] Fahiem Bacchus, Adam J. Grove, Joesph Y. Halpern, and Daphne Koller. Statistical foundations for default reasoning. In *Proceedings of the Thirteenth International Joint Conference on Artificial Intelligence*, pages 563–569, 1993.

[Kyburg and Teng, 2001] Henry E. Kyburg, Jr. and Choh Man Teng. *Uncertain Inference*. Cambridge University Press, 2001.

[Kyburg and Teng, 2002] Henry E. Kyburg, Jr. and Choh Man Teng. The logic of risky knowledge. In *Electronic Notes in Theoretical Computer Science*, volume 67. Elsevier Science, 2002.

[Kyburg, 1961] Henry E. Kyburg, Jr. *Probability and the Logic of Rational Belief*. Wesleyan University Press, 1961.

[Kyburg, 1983] Henry E. Kyburg, Jr. The reference class. *Philosophy of Science*, 50:374–397., 1983.

[Levi, 1980] Isaac Levi. *The Enterprise of Knowledge,*. MIT Press, Cambridge, 1980.

[Pearl, 1990] Judea Pearl. System Z: A natural ordering of defaults with tractable applications to default reasoning. In *Theoretical Aspects of Reasoning about Knowledge*, pages 121–135, 1990.

[Reichenbach, 1949] Hans Reichenbach. *The Theory of Probability. An Inquiry into the Logical and Mathematical Foundations of the Calculus of Probability.* University of California Press, second edition, 1949.

[Teng, 1997] Choh Man Teng. Sequential thresholds: Context sensitive default extensions. In *Proceedings of the Thirteen Conference of Uncertainty in Artificial Intelligence*, pages 437–444, 1997.

Symmetry of Models versus Models of symmetry

GERT DE COOMAN AND ENRIQUE MIRANDA

ABSTRACT. A model for a subject's beliefs about a phenomenon may exhibit symmetry, in the sense that it is invariant under certain transformations. On the other hand, such a belief model may be intended to represent that the subject believes or knows that the phenomenon under study exhibits symmetry. We defend the view that these are fundamentally different things, even though the difference cannot be captured by Bayesian belief models. In fact, the failure to distinguish between both situations leads to Laplace's so-called Principle of Insufficient Reason, which has been criticized extensively in the literature.

We show that there are belief models (imprecise probability models, coherent lower previsions) that generalize and include the more traditional Bayesian belief models, but where this fundamental difference can be captured. This leads to two notions of symmetry for such belief models: weak invariance (representing symmetry of beliefs) and strong invariance (modeling beliefs of symmetry). We discuss various mathematical as well as more philosophical aspects of these notions. We also discuss a few examples to show the relevance of our findings both to probabilistic modeling and to statistical inference, and to the notion of exchangeability in particular.

1 Introduction

This paper deals with symmetry in relation to models of beliefs. Consider a model for a subject's beliefs about a certain phenomenon. Such a *belief model* may be *symmetrical*, in the sense that it is invariant under certain transformations. On the other hand, a belief model may try to capture that the subject believes that the phenomenon under study exhibits symmetry, and we then say that the belief model *models symmetry*. We defend the view that there is an important conceptual difference between the two cases: symmetry of beliefs should not be confused with beliefs of symmetry.[1]

---

[1] This echoes Walley's [1991, Section 9.5.6, p. 466] view that 'symmetry of evidence' is not the same thing as 'evidence of symmetry'.

Does this view need defending at all? That there is a difference may strike you as obvious, and yet we shall argue that Bayesian belief models, which are certainly the most popular belief models in the literature, are unable to capture this difference.

To make this clearer, consider a simple example. Suppose I will toss a coin, and you are ignorant about its relevant properties: it might be fair but on the other hand it might be heavily loaded, or it might even have two heads, or two tails (situation A). To you the outcomes of the toss that are practically possible are h (for heads) and t (for tails). Since you are ignorant about the properties of the coin, any model for your beliefs should not change if heads and tails are permuted, so the model that 'faithfully' captures your beliefs about the outcome of the toss should be symmetrical too, i.e., invariant under this permutation of heads and tails.

Suppose on the other hand that you know that the coin (and the tossing mechanism) I shall use is completely symmetrical (situation B). Your belief model about the outcome of the toss should capture this knowledge, i.e., it should model your beliefs about the symmetry of the coin.

Our point is that belief models should be able to catch the important difference between your beliefs in the two situations. Bayesian belief models cannot do this. Indeed—the argument is well-known—the only symmetrical probability model, which is in other words invariant under permutations of heads and tails, assigns equal probability $1/2$ to heads and tails. But this is automatically also the model that captures your beliefs that the coin is actually symmetrical, so heads and tails should be equally likely.

The real reason why Bayesian belief models cannot capture the difference between symmetry of models and modeling symmetry, is that they do not allow for *indecision*. Suppose that I ask you to express your preferences between two gambles, whose reward depends on the outcome of the toss. For the first one, a, you will win one euro if the outcome is heads, and lose one if it is tails. The second one, b, gives the same rewards, but with heads and tails swapped.

In situation B, because you believe the coin to be symmetrical, it does not matter to you which gamble you get, and you are *indifferent* in your choice between the two.

But in situation A, on the other hand, because you are completely ignorant about the coin, the available information gives you *no reason to (strictly) prefer a over b or b over a*. You are therefore *undecided* about which of the two gambles to choose.

Because decision based on Bayesian belief models leaves you no alternative but to either strictly prefer one action over the other, or to be indifferent between them, the symmetry of the model leaves you *no choice but to act as if you were indifferent* between a and b. We strongly believe that it is wrong to confuse indecision with indifference in this example (and elsewhere of course), but Bayesian belief models leave you no choice but to do so, unless you want to let go of the principle that if your evidence or your beliefs are symmetrical, your belief model should be symmetrical as well. The problem with Laplace's Principle of Insufficient Reason is precisely this: if you use a Bayesian probability model then the symmetry present in ignorance forces you to treat indecision (or insufficient reason to decide) between a and b as if it were indifference.[2] Or in other words, it forces you to treat symmetry of beliefs as if there were beliefs of symmetry.

If on the other hand, we consider belief models that allow for indecision, we can sever the unholy link between indecision and indifference, because in a state of complete ignorance, we are then allowed to remain undecided about which of the two actions to choose: in the language of preference relations, they simply become *incomparable*, and you need not be indifferent between them. As we shall see further on, similar arguments show that such belief models also allow us to distinguish between 'symmetry of models' and 'models of symmetry' in those more general situations where the symmetry involved is not necessarily that which goes along with complete ignorance.

So, it appears that in order to better understand the interplay between modeling beliefs and issues of symmetry, which is the main aim of this paper, we shall need to work with a language, or indeed, with a type of belief models that, unlike Bayesian belief models, take indecision seriously. For this purpose, we shall use the language of the so-called *imprecise probability models* [Walley, 1991], and in particular coherent lower previsions, which have the same behavioral pedigree as the more common Bayesian belief models (*in casu* coherent previsions, see de Finetti [1974–1975]), and which contain these models as a special case. We give a somewhat unusual introduction to such models in Section 2.[3] In Section 3, we provide the necessary mathematical background for discussing symmetry: we discuss monoids of transformations, and invariance under such monoids. After these introductory sections, we start addressing the issue of symmetry in relation to belief

---

[2]This may be a good explanation for why Keynes [1921, p. 83] renamed the 'Principle of Insufficient Reason' the 'Principle of Indifference'. He (and others, see Zabell [1989a]) also suggested that the principle should not be applied in a state of complete ignorance, but only if there is good reason to justify the indifference (such as when there is evidence of symmetry). By the way, Keynes was also among the first to consider what we shall call imprecise probability models, as his comparative probability relations were not required to be complete.

[3]For other brief and perhaps more conventional introductions to the topic, we refer to [Walley, 1996b; De Cooman and Zaffalon, 2004; De Cooman and Troffaes, 2004; De Cooman and Miranda, 2006]. A much more detailed account of the behavioral theory of imprecise probabilities can be found in Walley [1991].

models in Section 4. We introduce two notions of invariance for the imprecise probability models introduced in Section 2: *weak invariance*, which captures symmetry of belief models, and *strong invariance*, which captures that a model represents the belief that there is symmetry. We study relevant mathematical properties of these invariance notions, and argue that the distinction between them is very relevant when dealing with symmetry in general, and in particular (Section 5) for modeling complete ignorance. Further interesting properties of weak and strong invariance, related to inference, are the subject of Sections 6 and 7, respectively. We show among other things that a weakly invariant coherent lower prevision can always be extended to a larger domain, in a way that is as conservative as possible. This implies that, for any given monoid of transformations, there always are weakly invariant coherent lower previsions. This is not generally the case for strong invariance, however, and we give and discuss sufficient conditions such that for a given monoid of transformations, there would be strongly invariant coherent (lower) previsions. We also give various expressions for the smallest strongly invariant coherent lower prevision that dominates a given weakly invariant one (if it exists). In Section 8, we turn to the important example of coherent (lower) previsions on the set of natural numbers, that are shift-invariant, and we use them to characterize the strongly invariant coherent (lower) previsions on a general space provided with a single transformation. Further examples are discussed in Section 9, where we characterize weak and strong invariance with respect to finite groups of permutations. In particular, we discuss Walley's [1991] generalization to lower previsions of de Finetti's [1937] notion of exchangeability, and we use our characterization of strong permutation invariance to prove a generalization to lower previsions of de Finetti's representation results for finite sequences of exchangeable random variables. Conclusions are gathered in Section 10.

We want to make it clear at this point that this paper owes a significant intellectual debt to Peter Walley. First of all, we use his behavioral imprecise probability models [Walley, 1991] to try and clarify the distinction between symmetry of beliefs and beliefs of symmetry. Moreover, although we like to believe that much of what we do here is new, we are also aware that in many cases we take to their logical conclusion a number of ideas about symmetry that are clearly present in his work (mainly Walley [1991, Sections 3.5, 9.4 and 9.5] and Pericchi and Walley [1991]), sometimes in embryonic form, and often more fully worked out.

2   Imprecise probability models

Consider a very general situation in which uncertainty occurs: a subject is uncertain about the value that a variable X assumes in a set of possible values $\mathcal{X}$. Because the subject is uncertain, we shall call X an *uncertain*, or *random*, variable.

The central concept we shall use in order to model our subject's uncertainty about X, is that of a *gamble* (on X, or on $\mathcal{X}$), which is a bounded real-valued function f on $\mathcal{X}$. In other words, a gamble f is a map from $\mathcal{X}$ to the set of real numbers $\mathbb{R}$ such that

$$\sup f := \sup\{f(x)\colon x \in \mathcal{X}\} \text{ and } \inf f := \inf\{f(x)\colon x \in \mathcal{X}\}$$

are (finite) real numbers. It is interpreted as the reward function for a transaction which may yield a different (and possibly negative) reward $f(x)$, measured in units (called *utiles*) of a pre-determined linear utility,[4] for each of the different values x that the random variable X may assume in $\mathcal{X}$.

We denote the set of all gambles on X by $\mathcal{L}(\mathcal{X})$. For any two gambles f and g, we denote their point-wise sum by $f+g$, and we denote the point-wise (scalar) multiplication of f with a real number λ by λf. $\mathcal{L}(\mathcal{X})$ is a real linear space under these operations. We shall always endow this space with the *supremum norm*, i.e., $\|f\| = \sup|f| = \sup\{|f(x)|\colon x \in \mathcal{X}\}$, or equivalently, with the topology of uniform convergence, which turns $\mathcal{L}(\mathcal{X})$ into a Banach space.

An *event* A is a subset of $\mathcal{X}$. If $X \in A$ then we say that the event *occurs*, and if $X \notin A$ then we say that A *doesn't occur*, or equivalently, that the *complement(ary event)* $A^c = \{x \in \mathcal{X}\colon x \notin A\}$ occurs. We shall identify an event with a special $\{0,1\}$-valued gamble I_A, called its *indicator*, and defined by $I_A(x) = 1$ if $x \in A$ and $I_A(x) = 0$ elsewhere. We shall often write A for I_A, whenever there is no possibility of confusion.

2.1 Coherent sets of really desirable gambles

Given the information that the subject has about X, she will be disposed to accept certain gambles, and to reject others. The idea is that we model a subject's beliefs about X by looking at which gambles she accepts, and to collect these into a *set of really desirable gambles* $\mathcal{R}$.

EXAMPLE 1 (ROLLING A DIE). Assume that our subject is uncertain about the outcome X of my tossing a die. In this case $\mathcal{X} = \mathcal{X}_6 := \{1,2,3,4,5,6\}$ is the set of possible values for X. If the subject is rational, she will accept the gamble which yields a positive reward whatever the value of X, because she is certain to improve her 'fortune' by doing so. On the other hand, she will not accept a non-positive gamble that is negative somewhere, because

---

[4]This utility can be regarded as amounts of money, as is the case for instance in de Finetti [1974–1975]. It is perhaps more realistic, in the sense that the linearity of the scale is better justified, to interpret it in terms of probability currency: we win or lose lottery tickets depending on the outcome of the gamble; see Walley [1991, Section 2.2].

by accepting such a gamble she can only lose utility (we then say she *incurs a partial loss*). She will not accept the gamble which makes her win one utile if the outcome X is 1, and makes her lose five utiles otherwise, unless she knows for instance that the die is loaded very heavily in such a way that the outcome 1 is almost certain to come up.

Real desirability can also be interpreted in terms of the betting behavior of our subject. Suppose she wants to bet on the occurrence of some event, such as my throwing 1 (so that she receives 1 utile if the event happens and 0 utiles otherwise). If she thinks that the die is fair, she should be disposed to bet on this event at any rate r strictly smaller than $\frac{1}{6}$. This means that the gamble $I_{\{1\}} - r$ representing this transaction (winning $1 - r$ if the outcome of X is 1 and losing r otherwise) will be really desirable to her for $r < \frac{1}{6}$. ◆

Now, accepting certain gambles has certain consequences, and has certain implications for accepting other gambles, and if our subject is rational, which we shall assume her to be, she should take these consequences and implications into account. To give but one example, if our subject accepts a certain gamble f she should also accept any other gamble g such that $g \geq f$, i.e., such that g *point-wise dominates* f, because accepting g is certain to bring her a reward that is at least as high as accepting f does.

Actually, this requirement is a consequence [combine (D2) with (D3)] of the following four basic rationality axioms for real desirability, which we shall assume any rational subject's set of really desirable gambles $\mathcal{R}$ to satisfy:

(D1) if $f < 0$ then $f \notin \mathcal{R}$ [avoiding partial loss];

(D2) if $f \geq 0$ then $f \in \mathcal{R}$ [accepting sure gains];

(D3) if $f \in \mathcal{R}$ and $g \in \mathcal{R}$ then $f + g \in \mathcal{R}$ [accepting combined gambles]

(D4) if $f \in \mathcal{R}$ and $\lambda > 0$ then $\lambda f \in \mathcal{R}$ [scale invariance].

where $f < g$ is shorthand for $f \leq g$ and $f \neq g$.[5] We call any subset $\mathcal{R}$ of $\mathcal{L}(\mathcal{X})$ that satisfies these axioms a *coherent* set of really desirable gambles.

It is easy to see that these axioms reflect the behavioral rationality of our subject: (D1) means that she should not be disposed to accept a gamble which makes her lose utiles, no matter the outcome; (D2) means that she should accept a gamble which never makes her lose utiles; on the other hand, if she is disposed to accept two gambles f and g, she should also accept the combination of the two gambles, which leads to a reward $f + g$; this is an

---

[5]So, here and in what follows, we shall write '$f < 0$' to mean '$f \leq 0$ and not $f = 0$', and '$f > 0$' to mean '$f \geq 0$ and not $f = 0$'.

immediate consequence of the linearity of the utility scale. This justifies (D3). And finally, if she is disposed to accept a gamble f, she should be disposed to accept the scaled gamble λf for any $\lambda > 0$, because this just reflects a change in the linear utility scale. This is the idea behind condition (D4).

Walley [1991; 2000] has a further coherence axiom that sets of really desirable gambles should satisfy, which turns out to be quite important for conditioning, namely

(D5) if $\mathcal{B}$ is a partition of $\mathcal{X}$ and if $I_B f \in \mathcal{R}$ for all B in $\mathcal{B}$, then $f \in \mathcal{R}$ [full conglomerability].

Since this axiom is automatically satisfied whenever $\mathcal{X}$ is finite [it is then an immediate consequence of (D3)], and since we shall not be concerned with conditioning unless when $\mathcal{X}$ is finite (see Section 9), we shall ignore this additional axiom in the present discussion.

A coherent set of really desirable gambles is a convex cone [axioms (D3)–(D4)] that includes the 'non-negative orthant' $\mathcal{C}_+ := \{f \in \mathcal{L}(\mathcal{X}) : f \geq 0\}$ [axiom (D2)] and has no gamble in common with the 'negative orthant' $\mathcal{C}_- := \{f \in \mathcal{L}(\mathcal{X}) : f < 0\}$ [axiom (D1)].[6] If we have two coherent sets of really desirable gambles $\mathcal{R}_1$ and $\mathcal{R}_2$, such that $\mathcal{R}_1 \subseteq \mathcal{R}_2$, then we say that $\mathcal{R}_1$ is less committal, or more conservative, than $\mathcal{R}_2$, because a subject whose set of really desirable gambles is $\mathcal{R}_2$ accepts at least all the gambles in $\mathcal{R}_1$. The least-committal (most conservative, smallest) coherent set of really desirable gambles is $\mathcal{C}_+$. Within this theory, it seems to be the appropriate model for *complete ignorance*: if our subject has no information at all about the value of X, she should be disposed to accept only those gambles which cannot lead to a loss of utiles (see also the discussion in Section 5).

Now suppose that our subject has specified a set $\mathcal{R}$ of gambles that she accepts. In an elicitation procedure, for instance, this would typically be a finite set of gambles, so we cannot expect this set to be coherent. We are then faced with the problem of enlarging this $\mathcal{R}$ to a coherent set of really desirable gambles that is as small as possible: we want to find out what are the (behavioral) consequences of the subject's accepting the gambles in $\mathcal{R}$, taking into account *only* the requirements of coherence. This inference problem is (also formally) similar to the problem of inference (logical closure) in classical propositional logic, where we want to find out what are

---

[6]This means that the zero gamble 0 belongs to the set of really desirable gambles. This is more a mathematical convention than a behavioral requirement, since this gamble has no effect whatsoever in the amount of utiles of our subject. See more details in Walley [1991].

the consequences of accepting certain propositions.[7]

The smallest convex cone including $\mathcal{C}_+$ and $\mathcal{R}$, or in other words, the smallest subset of $\mathcal{L}(\mathcal{X})$ that includes $\mathcal{R}$ and satisfies (D2)–(D4), is given by

$$\mathcal{E}_{\mathcal{R}}^r := \left\{ g \in \mathcal{L}(\mathcal{X}) : g \geq \sum_{k=1}^{n} \lambda_k f_k \text{ for some } n \geq 0, \ \lambda_k \in \mathbb{R}^+ \text{ and } f_k \in \mathcal{R} \right\},$$

where $\mathbb{R}^+$ denotes the set of non-negative real numbers. If this convex cone $\mathcal{E}_{\mathcal{R}}^r$ intersects $\mathcal{C}_-$ then it is easy to see that actually $\mathcal{E}_{\mathcal{R}}^r = \mathcal{L}(\mathcal{X})$, and then it is impossible to extend $\mathcal{R}$ to a coherent set of really desirable gambles [because (D1) cannot be satisfied]. Observe that $\mathcal{E}_{\mathcal{R}}^r \cap \mathcal{C}_-' = \emptyset$ if and only if

$$\text{there are no } n \geq 0, \ \lambda_k \in \mathbb{R}^+ \text{ and } f_k \in \mathcal{R} \text{ such that } \sum_{k=1}^{n} \lambda_k f_k < 0,$$

and we then say that the set $\mathcal{R}$ *avoids partial loss*. Let us interpret this condition. Assume that it doesn't hold (so we say that $\mathcal{R}$ *incurs partial loss*). Then there are really desirable gambles $f_1, \ldots, f_n$ and positive $\lambda_1, \ldots, \lambda_n$ such that $\sum_{k=1}^{n} \lambda_k f_k < 0$. But if our subject is disposed to accept the gamble f_k then by coherence [axioms(D2) and (D4)] she should also be disposed to accept the gamble $\lambda_k f_k$ for all $\lambda_k \geq 0$. Similarly, by coherence [axiom (D3)] she should also be disposed to accept the sum $\sum_{k=1}^{n} \lambda_k f_k$. Since this sum is non-positive, and strictly negative in at least some elements of $\mathcal{X}$, we see that the subject can be made subject to a partial loss, by suitably combining gambles which she accepts. This is unreasonable.

When the class $\mathcal{R}$ avoids partial loss, and only then, we are able to extend $\mathcal{R}$ to a coherent set of really desirable gambles, and the smallest such set is precisely $\mathcal{E}_{\mathcal{R}}^r$, which is called the *natural extension* of $\mathcal{R}$ to a set of really desirable gambles. This set reflects only the behavioral consequences of the assessments present in $\mathcal{R}$: the acceptance of a gamble f not in $\mathcal{E}_{\mathcal{R}}^r$ (or, equivalently, a set of really desirable gambles strictly including $\mathcal{E}_{\mathcal{R}}^r$) is not implied by the information present in $\mathcal{R}$, and therefore represents stronger implications that those of coherence alone.

2.2 Coherent sets of almost-desirable gambles

Coherent sets of really desirable gambles constitute a very general and powerful class of models for a subject's beliefs (see Walley [1991, Appendix F] and Walley [2000] for more details and discussion). We could already discuss

---

[7]See Moral and Wilson [1995] and De Cooman [2000; 2005] for more details on this connection between natural extension and inference in classical propositional logic.

symmetry aspects for such coherent sets of really desirable gambles, but we shall instead concentrate on a slightly less general and powerful type of belief models, namely coherent lower and upper previsions. Our main reason for doing so is that this will allow us to make a more direct comparison to the more familiar Bayesian belief models, and in particular to de Finetti's [1974–1975] coherent previsions, or fair prices.

Consider a gamble f. Then our subject's *lower prevision*, or supremum acceptable buying price, $\underline{P}(f)$ for f is defined as the largest real number s such that she accepts the gamble $f - t$ for any price $t < s$, or in other words accepts to buy f for any such price t. Similarly, her *upper prevision*, or infimum acceptable selling price, $\overline{P}(f)$ for the gamble f is the smallest real number s such that she accepts the gamble $t - f$ for any price $t > s$, or in other words accepts to sell f for any such price t.

For an event A, the lower prevision $\underline{P}(I_A)$ of its indicator is also called the *lower probability* of A, and denoted by $\underline{P}(A)$. It can be interpreted as the supremum rate for betting on the event A. Similarly, $\overline{P}(I_A)$ is called the *upper probability* of A, and also denoted by $\overline{P}(A)$.

Since selling a gamble f for price s is the same thing as buying $-f$ for price $-s$, we have the following *conjugacy* relationship between an upper and a lower prevision:

$$\overline{P}(f) = -\underline{P}(-f).$$

This implies that from a given lower prevision $\underline{P}$, we can always construct the conjugate upper prevision $\overline{P}$, so they are mathematically equivalent belief models. In what follows, we shall mainly concentrate on lower previsions.

Now assume that our subject has a coherent set of really desirable gambles $\mathcal{R}$, then it is clear from the definition of lower and upper prevision that we can use $\mathcal{R}$ to define a lower prevision

(2.1) $\underline{P}_{\mathcal{R}}(f) = \sup\{s \in \mathbb{R}\colon f - s \in \mathcal{R}\}$ D-LPR

and an upper prevision $\overline{P}_{\mathcal{R}}(f) = \inf\{s \in \mathbb{R}\colon s - f \in \mathcal{R}\}$ for every gamble f on $\mathcal{X}$. So, given $\mathcal{R}$ we can construct two real-valued functionals, $\underline{P}_{\mathcal{R}}$ and $\overline{P}_{\mathcal{R}}$, whose interpretation is that of a supremum acceptable buying price, and an infimum acceptable selling price, respectively, and whose domain is $\mathcal{L}(\mathcal{X})$. We shall call these functionals *lower* and *upper previsions*.

We call a *coherent lower prevision* on $\mathcal{L}(\mathcal{X})$ any real-valued functional on $\mathcal{L}(\mathcal{X})$ satisfying the following three axioms:

(P1) $\underline{P}(f) \geq \inf f$ [accepting sure gains];

(P2) $\underline{P}(f + g) \geq \underline{P}(f) + \underline{P}(g)$ [super-additivity];

(P3) $\underline{P}(\lambda f) = \lambda \underline{P}(f)$ [non-negative homogeneity].

for all gambles f and g on $\mathcal{X}$, and all non-negative real λ.

It follows from the coherence axioms (D1)–(D4) for $\mathcal{R}$ that the lower prevision $\underline{P}_{\mathcal{R}}$ that corresponds to a coherent set of really desirable gambles $\mathcal{R}$ is coherent.[8]

So we see that with a coherent set of really desirable gambles $\mathcal{R}$, we can define a coherent lower prevision on $\mathcal{L}(\mathcal{X})$, using (2.1). We shall see further on that, conversely, given a coherent lower prevision $\underline{P}$ on $\mathcal{L}(\mathcal{X})$, we can always find a coherent set of really desirable gambles $\mathcal{R}$ such that $\underline{P}$ and $\mathcal{R}$ are related through (2.1). But unfortunately, the relationship between the two types of belief models is many-to-one: there are usually many coherent sets of really desirable gambles that lead to the same coherent lower prevision. This is why we said before that coherent sets of really desirable gambles are a more general and powerful belief model than coherent lower previsions. The ultimate reason for this is the following: suppose that a subject specifies her supremum buying price $\underline{P}(f)$ for a gamble f. This implies that she accepts all the gambles $f - \underline{P}(f) + \delta$, where $\delta > 0$. But the specification of $\underline{P}(f)$ says nothing about the gamble $f - \underline{P}(f)$ (where $\delta = 0$) itself: she might accept it, but then again she might not. And precisely because specifying a coherent lower prevision says nothing about this border behavior, it leads to a belief model that is less powerful than coherent sets of really desirable gambles, where this border behavior would be determined.

EXAMPLE 2 (ROLLING A DIE (CONT.)). Let us go back to the die example. Consider, for any x in $\mathcal{X}_6 = \{1, \ldots, 6\}$, the event $\{x\}$ that the outcome X of rolling the die is x. If, for some real number r, our subject accepts the gamble $I_{\{x\}} - r$, she is willing to pay r utiles in return for the uncertain reward $I_{\{1\}}$, or in other words to bet on the event $\{1\}$ at rate r. So her lower probability $\underline{P}(\{x\})$ for $\{x\}$, or equivalently, her lower prevision $\underline{P}(I_{\{x\}})$ for $I_{\{x\}}$, is the supremum rate at which she is willing to bet on $\{x\}$. This means that she accepts the gamble $I_{\{x\}} - s$ for any $s < \underline{P}(\{x\})$. But it doesn't imply that she actually accepts the gamble $I_{\{x\}} - \underline{P}(\{x\})$: this gamble is only claimed to be almost-desirable, as we shall see further on.

If she is completely ignorant about the properties of the die, her evidence about the die is symmetrical, i.e., doesn't change when the possible outcomes are permuted. A belief model that 'faithfully' captures the available evidence should therefore be symmetrical with respect to such permutations as well, so we infer that in particular $\underline{P}(\{1\})$, $\ldots$, $\underline{P}(\{6\})$ are all

---

[8]To prove (P1), use (D2); for (P2) use (D3); and for (P3) use (D4) for $\lambda > 0$ and (D1) and (D2) for $\lambda = 0$.

equal to some number p. Coherence [use (P1) and (P2)] then requires that $0 \leq p \leq \frac{1}{6}$. Any such p leads to a symmetrical lower probability defined on the singletons, and therefore reflects 'symmetry of beliefs'. As we have indicated above, the model corresponding to $p = 0$ is the one that reflects complete ignorance. We shall see further on (see Sections 4.2 and 9) that the choice $p = \frac{1}{6}$ leads to the only model that captures the belief that the die is fair, i.e., that reflects 'beliefs of symmetry'. ♦

In order to better understand the relationship between coherent lower previsions and coherent sets of really desirable gambles, we need to introduce, besides *real* desirability, a new and weaker notion, called *almost-desirability*, which will also play an important part in our discussion of symmetry further on. This notion is inspired by the ideas in the discussion above: we say that a gamble f is *almost-desirable* to a subject, or that she *almost-accepts* f, whenever she accepts $f + \delta$, or in other words $f + \delta$ is really desirable to her, for any strictly positive amount of utility $\delta > 0$. By stating that f is almost-desirable to her, nothing is specified about whether the subject accepts f itself: she might, but then again she also might not. If we generically denote by $\mathcal{D}$ a set of gambles that are almost-desirable to our subject, we see that the set $\mathcal{D}_{\mathcal{R}}$ of almost-desirable gambles that corresponds to a coherent set $\mathcal{R}$ of really desirable gambles, is given by

$$(2.2) \quad \mathcal{D}_{\mathcal{R}} = \{f \in \mathcal{L}(\mathcal{X}) \colon (\forall \delta > 0) f + \delta \in \mathcal{R}\} = \bigcap_{\delta > 0} [\mathcal{R} - \delta] \qquad \text{D-M}$$

so $\mathcal{D}_{\mathcal{R}}$ is the closure (in the topology of uniform convergence on $\mathcal{L}(\mathcal{X})$) of the convex cone $\mathcal{R}$.

We call any set of gambles $\mathcal{D}$ that satisfies the following five axioms a *coherent set of almost-desirable gambles*:

(M1) if $\sup f < 0$ then $f \notin \mathcal{D}$ [avoiding sure loss];

(M2) if $\inf f \geq 0$ then $f \in \mathcal{D}$ [accepting sure gains];

(M3) if $f \in \mathcal{D}$ and $g \in \mathcal{D}$ then $f + g \in \mathcal{D}$ [accepting combined gambles];

(M4) if $f \in \mathcal{D}$ and $\lambda > 0$ then $\lambda f \in \mathcal{D}$ [scale invariance];

(M5) if $f + \delta \in \mathcal{D}$ for all $\delta > 0$ then $f \in \mathcal{D}$ [closure].

It is a closed and convex cone in $\mathcal{L}(\mathcal{X})$ that includes the non-negative orthant $\mathcal{C}_+$ and does not intersect with the set $\mathcal{C}'_- = \{f \in \mathcal{L}(\mathcal{X}) \colon \sup f < 0\} \subset \mathcal{C}_-$.

It is easy to see that the set of almost-desirable gambles $\mathcal{D_R}$ that corresponds to a coherent set of really desirable gambles $\mathcal{R}$ is actually also coherent.[9]

It should at this point come as no surprise that coherent lower previsions and coherent sets of almost-desirable gambles are actually equivalent belief models. Indeed, consider a coherent set of almost-desirable gambles $\mathcal{D}$, i.e., $\mathcal{D}$ satisfies (M1)–(M5). Then the real-valued functional $\underline{P}_\mathcal{D}$ defined on $\mathcal{L}(\mathcal{X})$ by[10]

(2.3) $\quad \underline{P}_\mathcal{D}(f) := \max\{s \in \mathbb{R}\colon f - s \in \mathcal{D}\}$ M-LPR

satisfies (P1)–(P3) and therefore is a coherent lower prevision on $\mathcal{L}(\mathcal{X})$.[11]

Conversely, if we consider a coherent lower prevision $\underline{P}$ on $\mathcal{L}(\mathcal{X})$, i.e., $\underline{P}$ satisfies (P1)–(P3), then the set of gambles

(2.4) $\quad \mathcal{D}_{\underline{P}} := \{f \in \mathcal{L}(\mathcal{X})\colon \underline{P}(f) \geq 0\}$ LPR-M

satisfies (M1)–(M5) and is therefore a coherent set of almost-desirable gambles.[12] Moreover, the relationships (2.3) and (2.4) are bijective (one-to-one and onto), and they are each other's inverses.[13]

Finally, consider a coherent lower prevision $\underline{P}$ on $\mathcal{L}(\mathcal{X})$, and define the set of gambles $\mathcal{D}_{\underline{P}}^+ := \{f \in \mathcal{L}(\mathcal{X})\colon \underline{P}(f) > 0 \text{ or } f > 0\}$. Then $\mathcal{D}_{\underline{P}}^+ \cup \{0\}$ is a coherent set of really desirable gambles, i.e., it satisfies (D1)–(D4).[14] Moreover, any coherent set of really desirable gambles $\mathcal{R}$ that satisfies $\mathcal{D}_{\underline{P}}^+ \cup \{0\} \subseteq \mathcal{R} \subseteq \mathcal{D}_{\underline{P}}$, i.e., the union of whose (relative) topological interior with $\mathcal{C}^+$ is $\mathcal{D}_{\underline{P}}^+ \cup \{0\}$ and whose topological closure is $\mathcal{D}_{\underline{P}}$, has $\underline{P}$ as its associated lower prevision, through (2.1). This confirms what we claimed before: coherent lower previsions, or equivalently, coherent sets of almost-desirable gambles , are less powerful belief models than coherent sets of

---

[9]To prove (M1), use (D1) with $\delta = -\frac{\sup f}{2}$; to prove (M2), use (D2); to prove (M3), use (D3); to prove (M4), use (D4); and to prove (M5), use $\epsilon = \frac{\delta}{2}$ and the definition of $\mathcal{D_R}$ to prove that $f + \delta \in \mathcal{R}$ for all $\delta > 0$.

[10]The supremum in Eq. (2.1) now becomes a maximum, simply because the set $\mathcal{D}$ is closed.

[11](P1) follows from (M2), (P2) from (M3) and (P3) is a consequence of (M4).

[12]First, conditions (P1) and (P2) imply that $\underline{P}$ is monotone. Now, (P2) and (P3) imply that $0 = \underline{P}(0) \geq \underline{P}(f) + \underline{P}(-f) \geq \underline{P}(f) + \inf(-f)$, whence $\underline{P}(f) \leq \sup f$. From these two facts we deduce (M1). (M2) is a consequence of (P1), (M3) of (P2) and (M4) of (P3). Finally, the monotonicity of $\underline{P}$ implies that $\underline{P}(\mu) = \mu$ for any constant value μ, and from this we deduce that $\underline{P}(f + \delta) = \underline{P}(f) + \delta$ for any $\delta > 0$. This implies (M5).

[13]To see that they are each other inverses, it suffices to use that a coherent lower prevision satisfies $\underline{P}(f - s) = \underline{P}(f) - s$ for any gamble f and any real number s, and, conversely, that $f \in \mathcal{D}_{\underline{P}}$ if and only if $\underline{P}(f) \geq 0$; this implies also that both transformations are bijective.

[14]For (D1), use that a coherent lower prevision $\underline{P}$ satisfies $\underline{P}(f) \leq \sup f$ for any gamble f; for (D2), that $f \geq 0$ satisfies either $f > 0$ or $f = 0$; for (D3), use (P2) and the monotonicity of the coherent $\underline{P}$, and for (D4) use (P3).

really desirable gambles. If a subject specifies a coherent lower prevision $\underline{P}$, then she actually states that all gambles in the union $\mathcal{D}_{\underline{P}}^+ \cup \{0\}$ of $\mathcal{C}_+$ with the relative topological interior of $\mathcal{D}_{\underline{P}}$ are really desirable, but she doesn't specify whether the gambles in the topological boundary $\mathcal{D}_{\underline{P}} \setminus \mathcal{D}_{\underline{P}}^+$ of $\mathcal{D}_{\underline{P}}$ are: we only know that they are almost-desirable to her.

2.3   Natural extension for coherent lower previsions

There is one important problem that we skipped over in the discussion above, namely that of inference. Suppose a subject specifies a set $\mathcal{D}$ of gambles that are almost-desirable to her. In an elicitation procedure, for instance, this would typically be a finite set of gambles, so we cannot expect this set to be coherent. We are then, as before for really desirable gambles, faced with the problem of enlarging this $\mathcal{D}$ into a coherent set of almost-desirable gambles that is as small as possible: we want to find out what are the (behavioral) consequences of the subject's almost-accepting the gambles in $\mathcal{D}$, taking into account *only* the requirements of coherence.

The smallest closed convex cone including $\mathcal{C}_+$ and $\mathcal{D}$, or in other words, the smallest subset of $\mathcal{L}(\mathcal{X})$ that includes $\mathcal{D}$ and satisfies (M2)–(M5), is given by

$$(2.5) \quad \begin{aligned} \mathcal{E}_{\mathcal{D}}^m := \ & \{g \in \mathcal{L}(\mathcal{X})(\forall \delta > 0)(\exists n \geq 0, \lambda_k \in \mathbb{R}^+, \\ & f_k \in \mathcal{D})g \geq \sum_{k=1}^n \lambda_k f_k - \delta\}. \end{aligned} \qquad \text{M-NE}$$

This is the topological closure of the set $\mathcal{E}_{\mathcal{D}}^r$. If this convex cone $\mathcal{E}_{\mathcal{D}}^m$ intersects $\mathcal{C}_-' = \{f \in \mathcal{L}(\mathcal{X}): \sup f < 0\}$ then it is easy to see that actually $\mathcal{E}_{\mathcal{D}}^m = \mathcal{L}(\mathcal{X})$, and then it is impossible to extend $\mathcal{D}$ to a coherent set of almost-desirable gambles [because (M1) cannot be satisfied]. Observe that $\mathcal{E}_{\mathcal{D}}^m \cap \mathcal{C}_-' = \emptyset$ if and only if[15]

$$(2.6) \quad \sup\left[\sum_{k=1}^n \lambda_k f_k\right] \geq 0 \text{ for some } n \geq 0, \lambda_k \in \mathbb{R}^+ \text{ and } f_k \in \mathcal{D}, \qquad \text{M-ASL}$$

and we then say that the set $\mathcal{D}$ of almost-desirable gambles *avoids sure loss*. In that case, and only then, we are able to extend $\mathcal{D}$ to a coherent set of almost-desirable gambles, and the smallest such set is precisely $\mathcal{E}_{\mathcal{D}}^m$, which is called the *natural extension* of $\mathcal{D}$ to a set of almost-desirable gambles.

What does natural extension mean for the equivalent belief model of coherent lower previsions? Suppose our subject specifies a supremum acceptable buying price, or lower prevision, $\underline{P}(f)$ for each gamble f in some set of gambles $\mathcal{K} \subseteq \mathcal{L}(\mathcal{X})$.[16] We can then interpret $\underline{P}$ as a real-valued map

---

[15] Actually, this condition is equivalent to the one where we always choose $\lambda_k = 1$.

[16] This set of gambles $\mathcal{K}$ need not have any predefined structure; in particular, it does not have to be a linear space.

on $\mathcal{K}$, and we call $\underline{P}$ a *lower prevision on* $\mathcal{K}$, and say that $\mathcal{K}$ is the *domain* of $\underline{P}$.

To study the problem of natural extension for this lower prevision, we shall use what we already know about natural extension in the context of almost-desirable gambles. Recall that specifying $\underline{P}$ on $\mathcal{K}$ is tantamount to stating that the gambles in the set $\mathcal{D} := \{f - \underline{P}(f): f \in \mathcal{K}\}$ are almost-desirable. We now look at the natural extension of this $\mathcal{D}$. Using (2.6), we know that such a natural extension exists if and only if[17]

$$(2.7) \quad \begin{array}{l} \sup \left[\sum_{k=1}^{n} \lambda_k \left[f_k - \underline{P}(f_k)\right]\right] \geq 0 \text{ for all } n \geq 0,\ \lambda_k \in \mathbb{R}^+ \\ \text{and } f_k \in \mathcal{K}, \end{array} \qquad \text{LPR-ASL}$$

and we then say that the lower prevision $\underline{P}$ on $\mathcal{K}$ *avoids sure loss*. In this case, the natural extension $\mathcal{E}_{\mathcal{D}}^m$ is the smallest coherent set of almost-desirable gambles that includes $\mathcal{D}$, and consequently the coherent lower prevision $\underline{P}_{\mathcal{E}_{\mathcal{D}}^m}$ associated with $\mathcal{E}_{\mathcal{D}}^m$ through

$$\underline{P}_{\mathcal{E}_{\mathcal{D}}^m}(g) := \max\left\{s\colon g - s \in \mathcal{E}_{\mathcal{D}}^m\right\}$$

is the point-wise smallest coherent lower prevision on $\mathcal{L}(\mathcal{X})$ that dominates $\underline{P}$ on $\mathcal{K}$. We call this coherent lower prevision the *natural extension* of $\underline{P}$ and we denote it by $\underline{E}_{\underline{P}}$. We deduce from (2.5) that for all gambles g on $\mathcal{X}$:

$$(2.8) \quad \begin{aligned} \underline{E}_{\underline{P}}(g) &= \sup_{\substack{\lambda_k \geq 0, g_k \in \mathcal{D} \\ k=1\ldots,n, n \geq 0}} \inf \left[g - \sum_{k=1}^{n} \lambda_k g_k\right] \\ &= \sup_{\substack{\lambda_k \geq 0, f_k \in \mathcal{K} \\ k=1\ldots,n, n \geq 0}} \inf \left[g - \sum_{k=1}^{n} \lambda_k \left[f_k - \underline{P}(f_k)\right]\right]. \quad \text{LPR-NE} \end{aligned}$$

If $\underline{P}$ *incurs sure loss*, i.e., (2.7) is not satisfied, then $\mathcal{E}_{\mathcal{D}}^m = \mathcal{L}(\mathcal{X})$ and consequently $\underline{E}_{\underline{P}}$ assumes the value $+\infty$ in every gamble.

We shall call the lower prevision $\underline{P}$ on $\mathcal{K}$ *coherent*, whenever it can be extended to a coherent lower prevision on $\mathcal{L}(\mathcal{X})$, or in other words, whenever it coincides with its natural extension $\underline{E}_{\underline{P}}$ on every gamble in its domain $\mathcal{K}$. Taking into account (2.8), we see that this happens exactly when

$$(2.9) \quad \begin{array}{l} \sup \left[\sum_{k=1}^{n} \lambda_k \left[f_k - \underline{P}(f_k)\right] - \lambda_0 \left[f_0 - \underline{P}(f_0)\right]\right] \geq 0 \\ \text{for all } n \geq 0,\ \lambda_k \in \mathbb{R}^+ \text{ and } f_k \in \mathcal{K}, \end{array} \qquad \text{LPR-COH}$$

This coherence condition implies that $\underline{P}$ avoids sure loss.

Let us see if, for lower previsions, we can give a more immediate behavioral interpretation for avoiding sure loss, coherence, and natural extension. This should allow us to develop more intuition, as the approach we have

---

[17]Here too, this condition is equivalent to the one where we always choose $\lambda_k = 1$.

followed so far, which motivates these notions through the coherence axioms for real and almost-desirable gambles, is admittedly quite abstract. We begin with avoiding sure loss. Suppose that condition (2.7) is not satisfied. Then there are $n \geq 0$, $\lambda_1, \ldots, \lambda_n$ in $\mathbb{R}^+$ and $f_1, \ldots f_n$ in $\mathcal{K}$ such that $\sup\left[\sum_{k=1}^{n} \lambda_k [f_k - \underline{P}(f_k)]\right] < 0$, which implies that there is some $\delta > 0$ for which

$$\sum_{k=1}^{n} \lambda_k [f_k - \underline{P}(f_k) + \delta] \leq -\delta.$$

Now, by the definition of $\underline{P}(f_k)$, our subject accepts each of the gambles $f_k - \underline{P}(f_k) + \delta$, so she should also accept the combined gamble $\sum_{k=1}^{n} \lambda_k [f_k - \underline{P}(f_k) + \delta]$ [use axioms (D3) and (D4) for real desirability]. But this gamble leads to a sure loss of at least δ. In other words, if condition (2.7) doesn't hold, there are gambles which the subject accepts and which, if properly combined, make her subject to a sure loss.

Next, assume that condition (2.9) fails to hold. Then there are $n \geq 0$, $\lambda_0, \ldots, \lambda_n$ in $\mathbb{R}^+$ and $f_0, \ldots f_n$ in $\mathcal{K}$ such that $\sup[\sum_{k=1}^{n} \lambda_k[f_k - \underline{P}(f_k)] - \lambda_0[f_0 - \underline{P}(f_0)]] < 0$. Assume that $\lambda_0 > 0$, as we have already considered the case $\lambda_0 = 0$ in our discussion of avoiding sure loss. Then there is some $\delta > 0$ such that

$$\sum_{k=1}^{n} \frac{\lambda_k}{\lambda_0} [f_k - \underline{P}(f_k) + \delta] \leq f_0 - (\underline{P}(f_0) + \delta).$$

As before, the gamble on the left-hand side is a gamble that our subject accepts. But then she should also accept the gamble $f_0 - (\underline{P}(f_0) + \delta)$ since it point-wise dominates a gamble she accepts [use (D2) and (D3)]. This implies that she should be willing to pay a price $\underline{P}(f_0) + \delta$ for f_0, which is strictly higher than the supremum price $\underline{P}(f_0)$ she has specified for it. Coherence avoids this kind of inconsistency.

Finally, we turn to natural extension. Consider a gamble g on $\mathcal{X}$, then we infer from (2.8) that $\underline{E}_P(g)$ is the supremum s such that there are $n \geq 0$, $\lambda_1, \ldots, \lambda_n$ in $\mathbb{R}^+$ and $f_1, \ldots f_n$ in $\mathcal{K}$ for which

$$g - s \geq \sum_{k=1}^{n} \lambda_k [f_k - \underline{P}(f_k)]$$

Now the expression on the right-hand side is almost-desirable, because it is a non-negative linear combination of almost-desirable gambles [apply the axioms (M3) and (M4)]. So $g - s$ should be almost-desirable as well [apply the axioms (M2) and (M3)], and therefore our subject should be willing to buy g for any price $t < s$. So we deduce that $\underline{E}_P(g)$ is the supremum price for g that the subject can be forced to pay for the gamble g, by suitably combining transactions that she is committed to accept by her specifying the lower prevision $\underline{P}$ on $\mathcal{K}$. In other words, $\underline{E}_{\underline{P}}(g)$ is the lower prevision for g that is implied by the assessments in $\underline{P}$ and coherence *alone*.

2.4   Linear previsions: the Bayesian belief models

When a lower prevision $\underline{P}$ on $\mathcal{K}$ is *self-conjugate*, that is, when $\underline{P}(f) = \overline{P}(f)$ for any gamble f in $\mathcal{K}$, it is called a *prevision*. The common value $P(f)$ is then called the *prevision* of f; it is a *fair price* for the gamble f in the sense of de Finetti [1974–1975]. Formally, a real-valued function P on a class of gambles $\mathcal{K}$ is called a *linear, or coherent, prevision* whenever

$$\text{(2.10)} \qquad \sup \left[\sum_{k=1}^{n} [f_k - P(f_k)] - \sum_{j=1}^{m} [g_j - P(g_j)] \right] \geq 0$$

$$\text{for all } n, m \geq 0 \text{ and } f_k, g_j \in \mathcal{K}, \qquad\qquad \text{PR-COH}$$

A linear prevision is coherent, both as a lower and as an upper prevision. Moreover, if its domain is the class of all gambles, $\mathcal{L}(\mathcal{X})$, then condition (2.10) simplifies to

(PR1) $P(f + g) = P(f) + P(g)$ for any f and g in $\mathcal{L}(\mathcal{X})$ [additivity].

(PR2) $P(f) \geq \inf f$ for any f in $\mathcal{L}(\mathcal{X})$ [accepting sure gains].

Linear previsions are the familiar Bayesian belief models: any linear prevision on all gambles is indeed a coherent prevision in the sense of de Finetti [1974–1975]; and a prevision defined on an arbitrary set of gambles is coherent exactly when it is the restriction of some coherent prevision on all gambles. The restriction to (indicators of) events of a coherent prevision on all gambles is a finitely additive probability. We shall denote by $\mathbb{P}(\mathcal{X})$ the set of all coherent previsions on $\mathcal{L}(\mathcal{X})$.

There is an interesting relationship between coherent previsions and coherent lower previsions. Let $\underline{P}$ be a lower prevision with domain $\mathcal{K}$, and let us denote by

$$\mathcal{M}(\underline{P}) := \{ P \in \mathbb{P}(\mathcal{X}) \colon (\forall f \in \mathcal{K}) P(f) \geq \underline{P}(f) \}$$

the set of all coherent previsions on $\mathcal{L}(\mathcal{X})$ that *dominate* $\underline{P}$ on its domain. Then it can be checked[18] that $\underline{P}$ avoids sure loss if and only if $\mathcal{M}(\underline{P})$ is non-empty, that is, if and only if there is some coherent prevision on $\mathcal{L}(\mathcal{X})$ that dominates $\underline{P}$ on $\mathcal{K}$, and $\underline{P}$ is coherent if and only if it is the *lower envelope* of $\mathcal{M}(\underline{P})$, meaning that for all $\underline{P}$ in $\mathcal{K}$,

$$\underline{P}(f) = \min \{ P(f) \colon P \in \mathcal{M}(\underline{P}) \}.$$

---

[18] See Walley [1991, Sections 3.3–3.4] for proofs for these statements.

Also, any lower envelope of a set of coherent previsions is a coherent lower prevision. Moreover, the natural extension $\underline{E}_{\underline{P}}$ of $\underline{P}$ to all gambles can be calculated using the set $\mathcal{M}(\underline{P})$ of coherent previsions: for any gamble f on $\mathcal{X}$, we have

$$\underline{E}_{\underline{P}}(f) = \min\left\{P(f) \colon P \in \mathcal{M}(\underline{P})\right\}.$$

This means that from a *mathematical* point of view, a coherent lower prevision $\underline{P}$ and its set of dominating coherent lower previsions $\mathcal{M}(\underline{P})$, are equivalent belief models. It can be checked that this set is convex and closed in the weak* topology.[19] Moreover, there is a bijective relationship between weak*-closed convex sets of coherent previsions and coherent lower previsions (their lower envelopes). This fact can (but need not) be used to give coherent lower previsions a *Bayesian sensitivity analysis interpretation*, besides the direct behavioral interpretation given in Section 2.2: we might assume the existence of a precise but unknown coherent prevision P expressing a subject's behavioral dispositions, and we might model the information about P by means of a weak*-closed convex set of coherent previsions $\mathcal{M}$ (the set of possible candidates). Then, this set is *mathematically* equivalent to its lower envelope $\underline{P}$, which is a coherent lower prevision. We shall come back to the difference between the direct behavioral and the Bayesian sensitivity analysis interpretation of a lower prevision in Section 4.2, when we discuss the interplay between these interpretations and the notion of symmetry.

Taking into account the bijective relationship that exists between coherent lower previsions and sets of almost-desirable gambles, we may also establish a bijective relationship between sets of coherent previsions and sets of almost-desirable gambles: given a weak*-closed convex set $\mathcal{M}$ of coherent previsions on $\mathcal{L}(\mathcal{X})$, the class

$$\mathcal{D}_{\mathcal{M}} := \{f \in \mathcal{L}(\mathcal{X}) \colon (\forall P \in \mathcal{M})P(f) \geq 0\}$$

is a coherent set of almost-desirable gambles, that is, it satisfies the coherence conditions (M1)–(M5). Conversely, given a coherent set of almost-desirable gambles $\mathcal{D}$, the corresponding set of coherent previsions

$$\mathcal{M}(\mathcal{D}) := \{P \in \mathbb{P}(\mathcal{L}) \colon (\forall f \in \mathcal{D})P(f) \geq 0\}$$

is a weak*-closed convex set of coherent previsions.

Hence, there are at least three mathematically equivalent representations for the behavioral dispositions of our subject: coherent sets of almost-desirable gambles, coherent lower previsions, and weak*-closed convex sets

---

[19]The weak* topology on the set of all continuous linear functionals on $\mathcal{L}(\mathcal{X})$ is the topology of point-wise convergence. For more details, see Walley [1991, Appendix D].

of coherent previsions. The bijective relationships between them are summarized in Table 1.

$\diagup$	$\mathcal{D}$	$\underline{P}(\cdot)$	$\mathcal{M}$
$\mathcal{D}$		$\{f\colon \underline{P}(f)\geq 0\}$	$\{f\colon (\forall P\in\mathcal{M})P(f)\geq 0\}$
$\underline{P}(\cdot)$	$\max\{s\colon \cdot-s\in\mathcal{D}\}$		$\min\{P(\cdot)\colon P\in\mathcal{M}\}$
$\mathcal{M}$	$\{P\colon (\forall f\in\mathcal{D})P(f)\geq 0\}$	$\{P\colon (\forall f)P(f)\geq \underline{P}(f)\}$	

Table 1. Bijective relationships between the equivalent belief models: coherent sets of almost-desirable gambles $\mathcal{D}$, coherent lower previsions $\underline{P}$ on $\mathcal{L}(\mathcal{X})$, and weak*-closed convex sets $\mathcal{M}$ of coherent previsions on $\mathcal{L}(\mathcal{X})$

We now briefly discuss a number of belief models that constitute particular instances of coherent lower previsions. First, we consider n-monotone lower previsions, where $n \geq 1$. A lower prevision $\underline{P}$ is called n-monotone[20] when the following inequality holds for all $p \in \mathbb{N}$, $p \leq n$, and all f, f_1, …, f_p in $\mathcal{L}(\mathcal{X})$:

$$\sum_{I\subseteq\{1,\ldots,p\}} (-1)^{|I|}\underline{P}\left(f \wedge \bigwedge_{i\in I} f_i\right) \geq 0,$$

where, here and further on, $|I|$ denotes the number of elements in a finite set I. A similar definition can be given if the domain of $\underline{P}$ is only a *lattice of gambles*, i.e., a set of gambles closed under point-wise minimum $\wedge$ and point-wise maximum $\vee$. Such n-monotone lower previsions are particular instances of exact functionals [Maaß, 2003], i.e., they are scalar multiples of some coherent lower prevision. In particular, an n-monotone lower probability defined on a lattice of events $\mathcal{S}$ that contains $\emptyset$ and $\mathcal{X}$ is coherent if and only if $\underline{P}(\emptyset) = 0$ and $\underline{P}(\mathcal{X}) = 1$.

A *completely monotone* lower prevision is simply one that is n-monotone for any natural number $n \geq 1$. When it is defined on indicators of events, it is called a completely monotone lower probability. When $\mathcal{X}$ is finite, this leads to *belief functions* in the terminology of Shafer [1976].

Two particular cases of belief functions and their conjugate upper probabilities are *probability charges*, or finitely additive probabilities defined on a field of events [Bhaskara Rao and Bhaskara Rao, 1983] and *possibility measures*. The latter [De Cooman, 2001; Zadeh, 1978] are set functions Π satisfying $\Pi\left(\bigcup_{i\in I} A_i\right) = \sup_{i\in I} \Pi(A_i)$ for any family $(A_i)_{i\in I}$ of subsets of $\mathcal{X}$. Π is a coherent *upper* probability if and only if $\Pi(\mathcal{X}) = 1$.

---

[20]See De Cooman *et al.* [2006; 2005a; 2005b] for a detailed discussion of n- and complete monotonicity for lower previsions.

Finally, we can consider a particular instance of a completely monotone coherent lower prevision that allows us to model complete ignorance, the so-called *vacuous lower prevision*. It is given by

$$\underline{P}_{\mathcal{X}}(f) = \inf_{x \in \mathcal{X}} f(x),$$

for all gambles f on $\mathcal{X}$. It corresponds to the set of almost-desirable gambles $\mathcal{D} = \mathcal{C}_+ = \{f \colon f \geq 0\}$, and to the set $\mathcal{M} = \mathbb{P}(\mathcal{L})$ of all coherent previsions on $\mathcal{L}$. If we have no information at all about the values that X takes in $\mathcal{X}$, we have no reason to reject any coherent prevision P, and this leads to the vacuous lower prevision as a belief model. More generally, we can consider a vacuous lower prevision relative to some subset A of $\mathcal{X}$, which is given by

$$\underline{P}_A(f) = \inf_{x \in A} f(x).$$

A vacuous lower prevision relative to a set A is the adequate belief model when we know that the random variable X assumes values in A, and nothing else. The restriction to events of a vacuous upper prevision is a (zero-one-valued) possibility measure.

2.5 Incomparability and indifference

We claimed in the Introduction that Bayesian belief models do not take indecision seriously, and that we therefore need to look at a larger class of belief models that do not have this defect. Here, we present a better motivation for this claim.

Consider two gambles f and g on $\mathcal{X}$. We say that a subject *almost-prefers* f to g, and denote this as $f \succeq g$, whenever she accepts to exchange g for f in return for any (strictly) positive amount of utility. Given this definition, it is straightforward to check that we can express this in terms of the three equivalent belief models $\mathcal{D}$, $\underline{P}$ and $\mathcal{M}$ of the previous sections by

$$\begin{aligned} f \succeq g &\Leftrightarrow f - g \in \mathcal{D} \\ &\Leftrightarrow \underline{P}(f - g) \geq 0 \\ &\Leftrightarrow (\forall P \in \mathcal{M}) P(f) \geq P(g). \end{aligned}$$

The binary relation $\succeq$ is a partial pre-order on $\mathcal{L}(\mathcal{X})$, i.e., it is reflexive and transitive.[21] Observe also that $f \succeq g \Leftrightarrow f - g \succeq 0$ and that $f \succeq 0 \Leftrightarrow f \in \mathcal{D}$, so f is almost-preferred to g if and only if $f - g$ is almost-preferred to

---

[21] The binary relation $\succeq$ is actually a *vector ordering* on the linear space $\mathcal{L}(\mathcal{X})$, because it is compatible with the addition of gambles, and the scalar multiplication of gambles with non-negative real numbers.

the zero gamble, which in turn is equivalent to the fact that our subject *almost-accepts* $f - g$, i.e., that $f - g$ is almost-desirable to her.

Unless our subject's lower prevision $\underline{P}$ is actually a (precise) prevision P (meaning that $\mathcal{D}$ is the semi-space $\{f \colon P(f) \geq 0\}$, and that $\mathcal{M} = \{P\}$), this ordering is not linear, or total: it does not hold for all gambles f and g that $f \succeq g$ or $g \succeq f$. When, therefore, both $f \not\succeq g$ and $g \not\succeq f$, we say that both gambles are *incomparable*, or that the subject is undecided about choosing between f and g, and we write this as $f \parallel g$.

It is instructive to see why the relation $\parallel$ is non-empty unless $\underline{P}$ is a precise prevision P. If $\underline{P}$ is not precise (but coherent), there is some gamble h such $\underline{P}(h) < \overline{P}(h)$. Let x be any real number such that $\underline{P}(h) < x < \overline{P}(h)$. In this case, the subject does not express a willingness to buy h for the price x, because x is strictly greater than her supremum acceptable price $\underline{P}(h)$ for buying h. Nor does she express a willingness to sell h for a price x, because x is strictly smaller than her infimum acceptable price $\overline{P}(h)$ for selling h. But there is more. Consider the gambles $f := h - x$ (buying h for a price x) and $g := x - h$ (selling h for a price x). Then it follows from the coherence of $\underline{P}$ that

$$\underline{P}(f - g) = 2\underline{P}(h - x) = 2[\underline{P}(h) - x] < 0 \text{ and}$$
$$\underline{P}(g - f) = 2\underline{P}(x - h) = 2[x - \overline{P}(h)] < 0,$$

so $f \parallel g$: our subject is also undecided in the choice between buying h for x or selling h for that price.

We say that our subject is *indifferent* between f and g, and denote this as $f \approx g$ whenever both $f \succeq g$ and $g \succeq f$. This means that $\underline{P}(f - g) = \underline{P}(g - f) = 0$, or equivalently, $P(f) = P(g)$ for all P in $\mathcal{M}$. Clearly, $\approx$ is an equivalence relation (a reflexive, symmetrical and transitive binary relation) on $\mathcal{L}(\mathcal{X})$. It is important to distinguish between incomparability and indifference. Indifference between gambles f and g represents strong behavioral dispositions: it means that our subject almost-accepts to exchange f for g and *vice versa*; on the other hand, incomparability has no behavioral implications, it merely records the absence of a(n expressed) behavioral disposition to choose between f and g.

3   Monoids of transformations

Symmetry is generally characterized mathematically as invariance under certain transformations. In this section, we provide the necessary mathematical apparatus that will allow us to describe and characterize symmetry for the belief models we are interested in.

3.1   Transformations and lifting

We are interested in models for beliefs that concern a random variable X. So let us begin by concentrating on transformations of the set of possible values $\mathcal{X}$ for X. A *transformation* of $\mathcal{X}$ is defined mathematically as a map $T : \mathcal{X} \to \mathcal{X} : x \mapsto Tx$. At this point, we do not require that such a map T should be *onto* (or surjective), i.e., that $T(\mathcal{X}) := \{Tx : x \in \mathcal{X}\}$ should be equal to $\mathcal{X}$. Neither do we require that T should be *one-to-one* (or injective), meaning that $Tx = Ty$ implies $x = y$ for all x and y in $\mathcal{X}$. A transformation of $\mathcal{X}$ that is both onto and one-to-one will be called a *permutation* of $\mathcal{X}$, but we shall in the sequel also need to consider transformations of $\mathcal{X}$ that are not permutations.

Suppose we have two transformations, T and S, of $\mathcal{X}$ that are of interest. Then there is no real reason why we shouldn't also consider the combined action of T and S on $\mathcal{X}$, leading to new transformations $ST := S \circ T$ and $TS := T \circ S$, defined by $(ST)x := S(Tx)$ and similarly $TSx := T(Sx)$ for all x in $\mathcal{X}$. And of course, we could also consider in a similar way TST and STS, or for that matter $TTTSST$, which we shall also write as $T^3 S^2 T$. So it is natural in this context to consider a set $\mathcal{T}$ of transformations of $\mathcal{X}$ that is closed under composition, i.e.,

(3.1) $(\forall T, S \in \mathcal{T})(TS \in \mathcal{T})$ SG

Such a set is called a *semigroup of transformations*.[22]   If moreover the semigroup $\mathcal{T}$ contains the identity map $\mathrm{id}_\mathcal{X}$, defined by $\mathrm{id}_\mathcal{X}\, x := x$ for all x in $\mathcal{X}$, it is called a *monoid*. As the identity map leaves all elements of $\mathcal{X}$ unchanged, it has no implications as far as symmetry and invariance are concerned, and we can therefore in what follows assume without loss of generality that any $\mathcal{T}$ we consider actually contains $\mathrm{id}_\mathcal{X}$ (is a monoid).

A monoid $\mathcal{T}$ is *Abelian* if $ST = TS$ for all T and S in $\mathcal{T}$. An important example of an Abelian monoid is the following. Consider a single transformation T of $\mathcal{X}$, and the Abelian monoid $\mathcal{T}_T$ generated by T, given by

$$\mathcal{T}_T := \{T^n : n \geq 0\}\,,$$

where $T^0 := \mathrm{id}_\mathcal{X}$ is the identity map on $\mathcal{X}$, $T^1 := T$ and for $n \geq 2$,

$$T^n := \underbrace{T \circ T \circ \cdots \circ T}_{n \text{ times}}.$$

A monoid $\mathcal{T}$ of transformations is *left-* (respectively *right-*)*cancelable* when for every transformation T in $\mathcal{T}$ there is some S in $\mathcal{T}$ such that

---

[22] A semigroup is defined as a set with a binary operation that is internal and associative. Composition of maps is always an associative binary operation, and (3.1) guarantees that it is internal in $\mathcal{T}$.

$ST = \mathrm{id}_{\mathcal{X}}$ (respectively $TS = \mathrm{id}_{\mathcal{X}}$). This transformation S is then called a *left-* (respectively *right-*)*inverse* of T. If T is both left- and right-cancelable, then the left-and right-inverses of T are unique and coincide for any T in $\mathcal{T}$, and $\mathcal{T}$ is called a *group*. Any element of $\mathcal{T}$ is then a permutation of $\mathcal{X}$.

For our purposes here, we generally only need to assume that $\mathcal{T}$ is a monoid, because there interesting (and relevant) situations where $\mathcal{T}$ is not a group; this is for instance the case for the Abelian monoid of the *shift* transformations of the set of natural numbers $\mathbb{N}$:

(3.2) $\mathcal{T}_\theta := \{\theta^n : n \geq 0\}$,

where $\theta(m) = m + 1$, and $\theta^n(m) = m + n$ for all natural numbers m and n. Another important example is the monoid $\mathcal{T}_{\mathcal{X}}$ of all transformations of $\mathcal{X}$, which is generally not Abelian, nor a group.

Since we are also concerned with gambles f on $\mathcal{X}$, we need a way to turn a transformation of $\mathcal{X}$ into a transformation of $\mathcal{L}(\mathcal{X})$. This is done by the procedure of *lifting*: given any gamble f on $\mathcal{X}$, we shall denote by $T^t f$ the gamble $f \circ T$, i.e.,

$$T^t f(x) := f(Tx),$$

for all x in $\mathcal{X}$. For an event A, we have that $T^t I_A = I_{T^{-1}(A)}$, where $T^{-1}(A) := \{x \in \mathcal{X} : Tx \in A\}$ is the so-called *inverse image* of A under T. On the other hand, given a constant μ, we have $T^t \mu = \mu$ for any transformation T.

The following observation is quite important. Consider two transformations T and S on $\mathcal{X}$. Then for any gamble f on $\mathcal{X}$ we see that

$$(ST)^t f = f \circ (S \circ T) = (f \circ S) \circ T = (S^t f) \circ T = T^t(S^t f),$$

so $(ST)^t = T^t S^t$, and lifting reverses the order of application of the transformations: for x in $\mathcal{X}$, STx means that T is applied first to x, and then S to Tx. For f in $\mathcal{L}(\mathcal{X})$, $(ST)^t f$ means that S^t is applied first to f and then T^t to $S^t f$.

Any transformation T of $\mathcal{X}$ can therefore be lifted to a transformation T^t of $\mathcal{L}(\mathcal{X})$, and we denote the corresponding set of liftings by $\mathcal{T}^t$. $\mathcal{T}^t$ is then a monoid of transformations of $\mathcal{L}(\mathcal{X})$. Lifting preserves the most common properties of semigroups, taking into account the above-mentioned order-inversion: being a monoid, being Abelian, and being a group are preserved under lifting. But being left-cancelable is turned into being right-cancelable, and *vice versa*. Lifting also has the interesting property that it turns a transformation T on $\mathcal{X}$ into a *linear* transformation T^t of the linear space $\mathcal{L}(\mathcal{X})$: for any pair of gambles f and g on $\mathcal{X}$ and any real numbers λ and μ, we have

$$T^t(\lambda f + \mu g) = \lambda T^t f + \mu T^t g.$$

3.2 Invariant (sets of) gambles

We now turn to the important notions of invariance under transformations. We start with the invariance of a set of gambles, because that is the most general notion, from which all other notions of invariance can be derived. If $\mathcal{K}$ is a set of gambles on $\mathcal{X}$, and T any transformation of $\mathcal{T}$, then we denote by

$$T^t \mathcal{K} := \{ T^t f \colon f \in \mathcal{K} \}$$

the direct image of the set $\mathcal{K}$ under T^t, and we say that $\mathcal{K}$ *is $\mathcal{T}$-invariant* if

$$(\forall T^t \in \mathcal{T}^t)(T^t \mathcal{K} \subseteq \mathcal{K}),$$

i.e., if all transformations in $\mathcal{T}^t$ are *internal* in $\mathcal{K}$.[23]

A gamble f on $\mathcal{X}$ is called *$\mathcal{T}$-invariant* if the singleton $\{f\}$ is, i.e., if $T^t f = f$ for all transformations T in the monoid $\mathcal{T}$. We call an event A *$\mathcal{T}$-invariant* if its indicator I_A is, i.e., if $T^{-1}(A) = A$ for all T in $\mathcal{T}$.

Let us denote by $\mathcal{I}_\mathcal{T}$ the set of all $\mathcal{T}$-invariant events. It is easy to check that $\mathcal{I}_\mathcal{T}$ is an *ample field*, i.e., it contains $\emptyset$ and $\mathcal{X}$, and it is closed under arbitrary unions and complementation, and therefore also under arbitrary intersections. For any x in $\mathcal{X}$, we shall call

$$[x]_\mathcal{T} := \bigcap \{ A \colon A \in \mathcal{I}_\mathcal{T} \text{ and } x \in A \}$$

the *$\mathcal{T}$-invariant atom containing x*. It is the smallest $\mathcal{T}$-invariant event that contains x. Any $\mathcal{T}$-invariant event A is a union of $\mathcal{T}$-invariant atoms: $A = \bigcup_{x \in A} [x]_\mathcal{T}$. We shall denote by $\mathcal{A}_\mathcal{T}$ the set of all invariant atoms: $\mathcal{A}_\mathcal{T} := \{ [x]_\mathcal{T} \colon x \in \mathcal{X} \}$. It is a partition of $\mathcal{X}$. A gamble f on $\mathcal{X}$ is $\mathcal{T}$-invariant if and only if it is constant on the $\mathcal{T}$-invariant atoms of $\mathcal{X}$.

Of course, the bigger the set of transformations $\mathcal{T}$, the smaller the number of $\mathcal{T}$-invariant events (or, equivalently, the bigger the atoms $[x]_\mathcal{T}$). The following proposition relates the $\mathcal{T}$-invariant atoms $[x]_\mathcal{T}$ to the images of x under the transformations in $\mathcal{T}$.

PROPOSITION 1. *Let $\mathcal{T}$ be a monoid of transformations of $\mathcal{X}$, and let x be any element of $\mathcal{X}$. In general we have that $\{ Tx \colon T \in \mathcal{T} \} \subseteq [x]_\mathcal{T}$. If $\mathcal{T}$ is left-cancelable, then $[x]_\mathcal{T} = \{ Tx \colon T \in \mathcal{T} \}$.*

Proof. Fix x in $\mathcal{X}$. Let $\mathcal{T}(x) := \{ Tx \colon T \in \mathcal{T} \}$ for brevity of notation. Consider any T in $\mathcal{T}$. Since $[x]_\mathcal{T}$ is $\mathcal{T}$-invariant, we have that $T^{-1}([x]_\mathcal{T}) = [x]_\mathcal{T}$. Since $x \in [x]_\mathcal{T}$ because $\mathcal{T}$ is a monoid, we infer from this equality that $Tx \in [x]_\mathcal{T}$. Hence indeed $\mathcal{T}(x) \subseteq [x]_\mathcal{T}$.

---

[23] So $\mathcal{T}^t$ is a monoid of transformations of $\mathcal{K}$.

To prove the converse inequality, assume that $\mathcal{T}$ is left-cancelable. Consider any S in $\mathcal{T}$. If we can prove that $\mathcal{T}(x)$ is S-invariant, meaning that $S^{-1}(\mathcal{T}(x)) = \mathcal{T}(x)$, then the proof is complete, since then $\mathcal{T}(x)$ will be $\mathcal{T}$-invariant, and since this set contains x [because $\mathrm{id}_{\mathcal{X}} \in \mathcal{T}$], it must include the smallest $\mathcal{T}$-invariant set $[x]_{\mathcal{T}}$ that contains x. So we set out to prove that $S^{-1}(\mathcal{T}(x)) = \mathcal{T}(x)$. Consider any y in $\mathcal{X}$. First assume that $y \in \mathcal{T}(x)$. Then there is some T in $\mathcal{T}$ such that $y = Tx$, whence $Sy = STx \in \mathcal{T}(x)$, since $ST \in \mathcal{T}$. Conversely, assume that $y \in S^{-1}(\mathcal{T}(x))$, or equivalently, that $Sy \in \mathcal{T}(x)$, then there is some T in $\mathcal{T}$ such that $Sy = Tx$, and since $\mathcal{T}$ is assumed to be left-cancelable, there is some S' in $\mathcal{T}$ such that $S'S = \mathrm{id}_{\mathcal{X}}$, whence $\mathcal{T}(x) \ni S'Tx = S'Sy = y$, since $S'T \in \mathcal{T}$.   ∎

An important special case is the following. Consider a transformation T of $\mathcal{X}$, and the Abelian monoid $\mathcal{T}_T = \{T^n : n \geq 0\}$ generated by T. Then a set of gambles $\mathcal{K}$ is $\mathcal{T}_T$-invariant if and only if $T^t \mathcal{K} \subseteq \mathcal{K}$, and we simply say that $\mathcal{K}$ is T-*invariant*. Similarly, a gamble f is $\mathcal{T}_T$-invariant if and only if $T^t f = f$, and we say that f is T-*invariant*. In what follows, we shall always use the phrase 'T-invariant' for '$\mathcal{T}_T$-invariant'. Also $\mathcal{I}_T$ is the set of T-invariant events, and it is an ample field whose atoms are denoted by $[x]_T$. With this notation, we have for an arbitrary monoid $\mathcal{T}$ that $\mathcal{I}_{\mathcal{T}} = \bigcap_{T \in \mathcal{T}} \mathcal{I}_T$.

For instance, the particular case of the shift transformations of $\mathbb{N}$ given by Eq. (3.2) concerns the Abelian monoid generated by θ. Here, the only θ- (or shift-)invariant events are $\emptyset$ and $\mathbb{N}$, and consequently a gamble f on $\mathbb{N}$ is θ-invariant if and only if it is constant. This also shows that the equality in the first part of Proposition 1 need not hold when the monoid of transformations $\mathcal{T}$ is not left-cancelable: in the present case, we have that $\mathcal{T}_\theta(m) = \{\theta^n(m) : n \geq 0\} = \{n \in \mathbb{N} : n \geq m\}$ is strictly included in the invariant atom $[m]_\theta = \mathbb{N}$ for all $m \geq 1$.

Another interesting case is that of $\mathcal{T}_{\mathcal{X}}$, the class of all transformations of $\mathcal{X}$. This a monoid, but it is not generally a group, nor Abelian. Moreover, it is not generally left-cancelable. We have, for any element x of $\mathcal{X}$ that $\{Tx : T \in \mathcal{T}_{\mathcal{X}}\} = \mathcal{X}$, and from Proposition 1 we deduce in a trivial manner that $[x]_{\mathcal{T}_{\mathcal{X}}} = \mathcal{X}$: the only invariant events under all transformations of $\mathcal{X}$ are $\emptyset$ and $\mathcal{X}$. This shows that the left-cancelability condition in the second part of Proposition 1 is not generally necessary.

4   Symmetry and invariance for belief models

We now have the necessary mathematical tools for studying the issue of symmetry in relation to the belief models discussed in Section 2. We shall see that for these coherent sets of almost-desirable gambles, there is an important distinction between the concepts 'symmetry of models' (which we shall call weak invariance) and 'models of symmetry' (which we shall call strong invariance). Let us first turn to the discussion of symmetrical belief models.

4.1 Weak invariance: symmetry of models

Consider a monoid $\mathcal{T}$ of transformations of $\mathcal{X}$. We want to express that a belief model about the value that the random variable X assumes in $\mathcal{X}$, exhibits a symmetry that is characterized by the transformations in $\mathcal{T}$. Thus, the notion of (weak) invariance of belief models that we are about to introduce is in a sense a purely mathematical one: it expresses that these belief models are left invariant under the transformations in $\mathcal{T}$.

DEFINITION 2 (Weak invariance). A coherent set of almost-desirable gambles $\mathcal{D}$ is called *weakly $\mathcal{T}$-invariant* if it is $\mathcal{T}$-invariant as a set of gambles, i.e., if $T^t \mathcal{D} \subseteq \mathcal{D}$ for all T in $\mathcal{T}$.

Why don't we require equality rather than the weaker requirement of set inclusion in this definition? In linear algebra, invariance of a subset of a linear space with respect to a linear transformation of that space is generally defined using only the inclusion. If we recall from Section 3 that lifting turns any transformation T of $\mathcal{X}$ into a linear transformation T^t of the linear space $\mathcal{L}(\mathcal{X})$, we see that our definition of invariance is just a special case of a notion that is quite common in the mathematical literature.

A few additional comments are in order. First of all, any coherent set of almost-desirable gambles is weakly $\mathrm{id}_{\mathcal{X}}$-invariant, so we may indeed always assume without loss of generality that $\mathcal{T}$ is at least a monoid (contains $\mathrm{id}_{\mathcal{X}}$).

Secondly, we have given an invariance definition for almost-desirability, but the definition for coherent sets of really desirable gambles $\mathcal{R}$ is completely analogous: $T^t \mathcal{R} \subseteq \mathcal{R}$ for all T in $\mathcal{T}$. Observe that if $\mathcal{R}$ is weakly $\mathcal{T}$-invariant then the associated set of almost-desirable gambles $\mathcal{D}_{\mathcal{R}}$, given by (2.2), is weakly $\mathcal{T}$-invariant as well.

Thirdly, if $\mathcal{T}$ is a group (or at least left-cancelable), then the weak invariance condition is actually equivalent to $T^t \mathcal{D} = \mathcal{D}$ for all T in $\mathcal{T}$: given a transformation T in $\mathcal{T}$ and its (left-)inverse $S \in \mathcal{T}$, consider $f \in \mathcal{D}$; then $T^t(S^t f) = (ST)^t f = f$, so there is a gamble $g = S^t f$, which belongs to $\mathcal{D}$ by weak invariance, such that $f = T^t g$; this means that $f \in T^t \mathcal{D}$, so $\mathcal{D} \subseteq T^t \mathcal{D}$ as well.

In summary, weak invariance is a mathematical notion that states that a subject's behavioral dispositions, as represented by a belief model $\mathcal{D}$, are invariant under certain transformations. If we posit that a subject's dispositions are in some way a 'faithful' reflection of the evidence available to her, we see that weak invariance is a way to model 'symmetry of evidence'. The following examples try to argue that if there is 'symmetry of evidence', then corresponding belief models should at least be weakly invariant.

EXAMPLE 3 (SHIFT TRANSFORMATIONS). Suppose our subject is completely ignorant about the value of a random variable X that assumes only non-negative integer values, so $\mathcal{X} = \mathbb{N}$. If her belief model is to be a reflection of the available evidence (none), we should like it to be weakly invariant with respect to the shift transformations in $\mathcal{T}_\theta$ (which is an Abelian monoid, but not a group). Indeed, if she is ignorant about X, she is also ignorant about $\theta(X) = X + 1$, apart from the fact that she knows that $\theta(X)$ cannot assume the value 0, whereas X can. Therefore, if our subject almost-accepts a gamble f, she should almost-accept $\theta^t f$: $\theta^t f(X) = f(X + 1)$ may assume the same values as $f(X)$, apart from the value $f(0)$, and because of her ignorance, our subject has no reason to treat the shifted gamble differently.

EXAMPLE 4 (ROLLING A DIE (CONT.)). Let us go back to the die example. Suppose that whatever evidence our subject has about the outcome X of rolling the die, is left invariant by permutations π of $\mathcal{X}_6 = \{1, \ldots, 6\}$. Assume that our subject almost-accepts a gamble f, meaning that she is willing to accept the uncertain reward $f(X) + \epsilon$ for any $\epsilon > 0$. But since the evidence gives our subject no reason to distinguish between the random variables X and $\pi(X)$, she should also be willing to accept the uncertain reward $f(\pi(X)) + \epsilon$ for any $\epsilon > 0$, or in other words, she should almost-accept the gamble $\pi^t f$.

We now investigate the corresponding notions for weak invariance for the equivalent belief models: coherent lower previsions and weak*-closed convex sets of coherent previsions. In order to do this, it is convenient to define the transformation of a (lower) prevision under a transformation T on $\mathcal{X}$, by lifting T to yet a higher level.

DEFINITION 3 (Transformation of a functional). [24] Let T be a transformation of $\mathcal{X}$ and let Λ be a real-valued functional defined on a T-invariant set of gambles $\mathcal{K} \subseteq \mathcal{L}(\mathcal{X})$. Then the transformation $T\Lambda$ of Λ is the real-valued functional defined on $\mathcal{K}$ by $T\Lambda := \Lambda \circ T^t$, or equivalently, by $T\Lambda(f) := \Lambda(T^t f) = \Lambda(f \circ T)$ for all gambles f in $\mathcal{K}$.

THEOREM 4. *Let $\underline{P}$ be a coherent lower prevision on $\mathcal{L}(\mathcal{X})$, $\mathcal{D}$ a coherent set of almost-desirable gambles, and $\mathcal{M}$ a weak*-closed convex set of coherent previsions on $\mathcal{L}(\mathcal{X})$. Assume that these belief models are equivalent, in the sense that they correspond to one another using the bijective relations in Table 1. Then the following statements are equivalent.*

1. $\mathcal{D}$ is weakly $\mathcal{T}$-invariant, in the sense that $T^t \mathcal{D} \subseteq \mathcal{D}$ for all T in $\mathcal{T}$.

---

[24]We use the same notation T for the transformation of $\mathcal{X}$ and for the corresponding transformation of a functional, first of all because we do not want to overload the mathematical notation, and also because, in contrast with lifting only once, lifting twice preserves the order of application of transformations.

2. $\underline{P}$ is weakly $\mathcal{T}$-invariant, *in the sense that* $T\underline{P} \geq \underline{P}$ *for all T in $\mathcal{T}$, or equivalently* $\underline{P}(T^t f) \geq \underline{P}(f)$ *for all T in $\mathcal{T}$ and f in $\mathcal{L}(\mathcal{X})$;*

3. $\mathcal{M}$ is weakly $\mathcal{T}$-invariant, *in the sense that* $T\mathcal{M} \subseteq \mathcal{M}$ *for all T in $\mathcal{T}$, or equivalently, $TP \in \mathcal{M}$ for all P in $\mathcal{M}$ and all T in $\mathcal{T}$.*[25]

Proof. We give a circular proof. Assume that $\mathcal{D}$ is weakly $\mathcal{T}$-invariant. Consider any T in $\mathcal{T}$ and f in $\mathcal{K}$, and observe that for the corresponding lower prevision $\underline{P}$

$$\underline{P}(T^t f) = \max\left\{\mu \colon T^t f - \mu \in \mathcal{D}\right\} \geq \max\left\{\mu \colon f - \mu \in \mathcal{D}\right\} = \underline{P}(f),$$

where the inequality follows from the invariance assumption on $\mathcal{D}$. This shows that the first statement implies the second.

Next, assume that $\underline{P}$ is weakly $\mathcal{T}$-invariant, and consider any T in $\mathcal{T}$ and P in the corresponding $\mathcal{M} = \mathcal{M}(\underline{P}) = \{P \colon (\forall f)P(f) \geq \underline{P}(f)\}$. Then for any gamble f on $\mathcal{X}$ we have that $TP(f) = P(T^t f) \geq \underline{P}(T^t f) \geq \underline{P}(f)$, where the second inequality follows for the invariance assumption on $\underline{P}$. This tells us that indeed $TP \in \mathcal{M}(\underline{P})$, so the second statement implies the third.

Finally, assume that $\mathcal{M}$ is weakly $\mathcal{T}$-invariant. Consider any T in $\mathcal{T}$ and any gamble f in the corresponding $\mathcal{D} = \mathcal{D}_{\mathcal{M}} = \{f \colon (\forall P \in \mathcal{M})P(f) \geq 0\}$. Then we have for any P in $\mathcal{M}$ that $P(T^t f) = TP(f) \geq 0$, since TP belongs to $\mathcal{M}(\underline{P})$ by the invariance assumption on $\mathcal{M}$. Consequently $T^t f \in \mathcal{D}$, which proves that the third statement implies the first. ∎

A coherent prevision P on $\mathcal{L}(\mathcal{X})$ is weakly $\mathcal{T}$-invariant if and only if $TP = P$ for all T in $\mathcal{T}$. This is easiest to prove by observing that $\mathcal{M}(P) = \{P\}$.[26] So for coherent previsions, we have an equality in the weak invariance condition. As we argued before, we generally won't have such an equality for arbitrary monoids $\mathcal{T}$, but the following corollary gives another sufficient condition on $\mathcal{T}$.

COROLLARY 5. *If the monoid $\mathcal{T}$ is left-cancelable, then the first weak invariance condition in Theorem 4 becomes $T^t \mathcal{D} = \mathcal{D}$ for all T in $\mathcal{T}$. If*

---

[25] This shows that our notion of a weakly invariant belief model corresponds to Pericchi and Walley's [1991] notion of a 'reasonable (or invariant) class of priors', rather than a 'class of reasonable (or invariant) priors', the latter being what our notion of strong invariance will correspond to. On the other hand, Walley [1991, Definition 3.5.1] defines a $\mathcal{T}$-invariant lower prevision $\underline{P}$ as one for which $\underline{P}(T^t f) = \underline{P}(f)$ for all $T \in \mathcal{T}$ and all gambles f, so he requires equality rather than inequality, as we do here.

[26] See Proposition 11 for a more direct proof.

$\mathcal{T}$ *is right-cancelable, then the second and third weak invariance conditions become* $T\underline{P} = \underline{P}$ *and* $T\mathcal{M} = \mathcal{M}$ *for all T in $\mathcal{T}$.*[27]

Proof. We have already proven the first statement near the beginning of Section 4.1. To prove the second statement, it suffices to show that when $\mathcal{T}$ is right-cancelable, $\mathcal{T}$-invariance implies that $\underline{P} \geq T\underline{P}$ and $\mathcal{M} \subseteq T\mathcal{M}$ for all T in $\mathcal{T}$. Consider any transformation T in the monoid $\mathcal{T}$, and let R be a right-inverse for T, i.e., $TR = \mathrm{id}_{\mathcal{X}}$. Consider a gamble h on $\mathcal{X}$, then $\underline{P}(h) = \underline{P}((TR)^t h) = \underline{P}(R^t(T^t h)) \geq \underline{P}(T^t h)$, where the inequality follows from the weak invariance of $\underline{P}$. So indeed, $\underline{P} \geq T\underline{P}$. Similarly, consider P in $\mathcal{M}$. Then $RP \in \mathcal{M}$ by weak invariance, and for any gamble f on $\mathcal{X}$, $T(RP)(f) = RP(T^t f) = P(R^t(T^t f)) = P(f)$ since $R^t(T^t f) = (TR)^t f = f$. So there is a $Q = RP$ in $\mathcal{M}$ such that $P = TQ$, meaning that $P \in T\mathcal{M}$. So indeed $\mathcal{M} \subseteq T\mathcal{M}$. ∎

We see from the definition that if a coherent set of almost-desirable gambles $\mathcal{D}$ (or a coherent lower prevision, or a weak*-closed convex set of coherent previsions) is weakly $\mathcal{T}$-invariant, it is also weakly $\mathcal{T}'$-invariant for any sub-monoid of transformations $\mathcal{T}' \subseteq \mathcal{T}$. Hence, as we add transformations, the collection of weakly invariant belief models will not increase. The limit case is when we consider the class $\mathcal{T}_{\mathcal{X}}$ of all transformations on $\mathcal{X}$. The following theorem shows that the vacuous belief models are the only ones that are *completely weakly invariant*, i.e., weakly $\mathcal{T}_{\mathcal{X}}$-invariant.

THEOREM 6. *Let $\mathcal{T}_{\mathcal{X}}$ be the monoid of all transformations of $\mathcal{X}$. Then the vacuous coherent set of almost-desirable gambles $\mathcal{C}_+$ (or equivalently, the vacuous lower prevision $\underline{P}_{\mathcal{X}}$, or equivalently, the weak*-closed convex set of all coherent previsions $\mathbb{P}(\mathcal{X})$) is the only coherent set of almost-desirable gambles (coherent lower prevision, weak*-closed convex set of coherent previsions) that is weakly $\mathcal{T}_{\mathcal{X}}$-invariant.*

Proof. We give the proof for coherent sets of almost-desirable gambles. It is obvious that $\mathcal{C}_+$ is $\mathcal{T}_{\mathcal{X}}$-invariant. So, consider any $\mathcal{T}_{\mathcal{X}}$-invariant coherent set of almost-desirable gambles $\mathcal{D}$. It follows from coherence [axiom (M2)] that $\mathcal{C}_+ \subseteq \mathcal{D}$. Assume *ex absurdo* that $\mathcal{C}_+ \subset \mathcal{D}$ and let f be any gamble in $\mathcal{D} \setminus \mathcal{C}_+$. This means that there is some x_0 in $\mathcal{X}$ such that $f(x_0) < 0$. Consider the transformation T_{x_0} of $\mathcal{X}$ that maps all elements of $\mathcal{X}$ to x_0, then $T_{x_0}^t f = f(x_0)$ and it follows from the T_{x_0}-invariance of $\mathcal{D}$ that the

---

[27]The reason for the difference in terms of left- versus right-cancelability lies of course in the fact that in the first condition, we work with transformations T^t of gambles, and in the second and third condition we work with transformations T of functionals, which are liftings of the former; simply recall that lifting reverses the order of application of transformations.

constant gamble $f(x_0) \in \mathcal{D}$, which violates coherence axiom (M1), so $\mathcal{D}$ cannot be coherent, a contradiction.[28] ∎

This result also tells us in particular that the vacuous belief model is always $\mathcal{T}$-invariant for any monoid of transformations $\mathcal{T}$. This implies that for any monoid of transformations $\mathcal{T}$, there always are $\mathcal{T}$-invariant belief models.

What are the behavioral consequences of weak invariance with respect to a monoid of transformations $\mathcal{T}$? It seems easiest to study this in terms of coherent lower previsions. First of all, we have that for any gamble f on $\mathcal{X}$ and any T in $\mathcal{T}$, our subject's supremum buying price $\underline{P}(T^t f)$ for the transformed gamble $T^t f$ should not be strictly smaller that her supremum price $\underline{P}(f)$ for buying f itself.

But there is also a more interesting consequence. Indeed, it follows from the coherence of $\underline{P}$ that

$$\underline{P}(f - T^t f) \le \underline{P}(f) - \underline{P}(T^t f) \le 0.$$

Walley [1991, Section 3.8.1] suggests that a subject *strictly prefers* a gamble f to a gamble g, which we denote as $f \succ g$, if $f > g$, or also if she accepts to pay some (strictly) positive price for exchanging g with f, so if $\underline{P}(f-g) > 0$. This means that weak $\mathcal{T}$-invariance implies that

$$f \not\succ T^t f \text{ for all } f \text{ in } \mathcal{L}(\mathcal{X}) \text{ and all } T \text{ in } \mathcal{T} \text{ such that } f \not> T^t f$$

which models that our subject *has no reason* (or disposition) *to strictly prefer any gamble f to any of its transformations $T^t f$ that it doesn't strictly dominate.*

4.2 Strong invariance: models of symmetry

Next, suppose that our subject believes that the (phenomenon underlying the) random variable X is subject to symmetry with respect to the transformations T in $\mathcal{T}$, so that she has *reason not to distinguish* between a gamble f and its transformation $T^t f$. Let us give an example to get a more intuitive understanding of what this means.

EXAMPLE 5 (ROLLING A DIE (CONT.)). Again, let us go back to the die example. Consider the gambles $I_{\{x\}}$, for $x \in \mathcal{X}_6 := \{1, \ldots, 6\}$. Since our subject believes the die (and the rolling mechanism behind it) to be symmetrical, she will be willing to exchange any gamble $I_{\{x\}}$ for any other gamble $I_{\{y\}}$ in return for any strictly positive amount of utility: $I_{\{x\}} - I_{\{y\}}$

---

[28] A similar argument tells us that the same result holds for complete weak invariance of coherent sets of really desirable gambles, where now the axiom (D1) will be violated.

should therefore be almost-desirable to her, or in other words, in terms of her lower prevision $\underline{P}$:

$$\underline{P}(I_{\{x\}} - I_{\{y\}}) \geq 0 \text{ for all } x \text{ and } y \text{ in } \mathcal{X}_6.$$

This is equivalent to stating that $I_{\{x\}} - \pi^t I_{\{x\}}$ should be almost-desirable, or that $\underline{P}(I_{\{x\}} - \pi^t I_{\{x\}}) \geq 0$ for all $x \in \mathcal{X}_6$ and all permutations π of $\mathcal{X}_6$. Now the only coherent lower prevision that satisfies these requirements is the uniform (precise) prevision, which assigns precise probability $\frac{1}{6}$ to each event $\{x\}$ [simply observe that for any coherent prevision P in $\mathcal{M}(\underline{P})$ it follows from these requirements that $P(I_{\{x\}}) = P(I_{\{y\}})$].

Let us now try and formalize the intuitive requirements in this example into a more formal definition. We stated above that if our subject believes that the (phenomenon underlying the) random variable X is subject to symmetry with respect to the transformations T in $\mathcal{T}$, then she has reason not to distinguish between a gamble f and its transformation $T^t f$. Suppose she has the gamble f in her possession, then she should be willing to exchange this for the gamble $T^t f$ in return for any strictly positive price, and *vice versa*. This means that she should almost-accept both $f - T^t f$ and $T^t f - f$, or in the language of Section 2.5, that she is *indifferent between f and $T^t f$*: $f \approx T^t f$. If $\mathcal{D}$ is her coherent set of almost-desirable gambles, this means that

$$f - T^t f \in \mathcal{D} \text{ and } T^t f - f \in \mathcal{D} \text{ for all } f \text{ in } \mathcal{L}(\mathcal{X}) \text{ and all } T \text{ in } \mathcal{T}.$$

If we define

$$\mathcal{D}_{\mathcal{T}} := \{f - T^t f \colon f \in \mathcal{L}(\mathcal{X}), T \in \mathcal{T}\} = \{T^t f - f \colon f \in \mathcal{L}(\mathcal{X}), T \in \mathcal{T}\},$$

this leads to the following definition.

DEFINITION 7.   A coherent set of almost-desirable gambles $\mathcal{D}$ is called *strongly $\mathcal{T}$-invariant* if $f - T^t f \in \mathcal{D}$ for all f in $\mathcal{L}(\mathcal{X})$ and all T in $\mathcal{T}$, or equivalently, if $\mathcal{D}_{\mathcal{T}} \subseteq \mathcal{D}$.

The following theorem gives equivalent characterizations of strong invariance in terms of the alternative types of belief models.

THEOREM 8.   *Let $\underline{P}$ be a coherent lower prevision on $\mathcal{L}(\mathcal{X})$, $\mathcal{D}$ a coherent set of almost-desirable gambles, and $\mathcal{M}$ a weak*-closed convex set of coherent previsions on $\mathcal{L}(\mathcal{X})$. Assume that these belief models are equivalent, in the sense that they correspond to one another using the bijective relations in Table 1. Then the following statements are equivalent:*

1. *$\mathcal{D}$ is strongly $\mathcal{T}$-invariant, in the sense that $\mathcal{D}_{\mathcal{T}} \subseteq \mathcal{D}$;*

2. $\underline{P}$ *is strongly* $\mathcal{T}$*-invariant, in the sense that* $\underline{P}(f-T^tf) \geq 0$ *and* $\underline{P}(T^tf - f) \geq 0$, *and therefore* $\underline{P}(f - T^tf) = \underline{P}(T^tf - f) = 0$ *for all* f *in* $\mathcal{L}(\mathcal{X})$ *and* T *in* $\mathcal{T}$;

3. $\mathcal{M}$ *is strongly* $\mathcal{T}$*-invariant, in the sense that* $TP = P$ *for all* P *in* $\mathcal{M}$ *and all* T *in* $\mathcal{T}$.[29]

Proof. We give a circular proof. Assume that $\mathcal{D}$ is strongly $\mathcal{T}$-invariant, and consider any gamble f on $\mathcal{X}$ and any T in $\mathcal{T}$. Then for the associated coherent lower prevision $\underline{P}$ we find $\underline{P}(f-T^tf) = \max\{s: f - T^tf - s \in \mathcal{D}\} \geq 0$, and similarly $\underline{P}(T^tf - f) \geq 0$. But since $\underline{P}$ is coherent, we find that also $\underline{P}(f - T^tf) = -\overline{P}(T^tf - f) \leq -\underline{P}(T^tf - f) \leq 0$ and similarly $\underline{P}(T^tf-f) = -\overline{P}(f-T^tf) \leq -\underline{P}(f-T^tf) \leq 0$, whence indeed $\underline{P}(f-T^tf) = \underline{P}(T^tf - f) = 0$, so the first statement implies the second.

Next, assume that $\underline{P}$ is strongly $\mathcal{T}$-invariant and consider any P in the associated set of dominating coherent previsions $\mathcal{M}(\underline{P})$ and any T in $\mathcal{T}$. Then for any gamble f on $\mathcal{X}$ we see that $P(f-T^tf) \geq 0$ and $P(T^tf-f) \geq 0$, and since P is a coherent prevision, this implies that $P(T^tf) = P(f)$, so indeed $TP = P$. Hence, the second statement implies the third.

Finally, assume that $\mathcal{M}$ is strongly $\mathcal{T}$-invariant, and consider any gamble f on $\mathcal{X}$ and any T in $\mathcal{T}$. Then for all P in $\mathcal{M}$ we have that $P(f - T^tf) = P(T^tf - f) = 0$, so both $f - T^tf$ and $T^tf - f$ belong to the associated set of almost-desirable gambles $\mathcal{D} = \{g: (\forall P \in \mathcal{M})P(g) \geq 0\}$. This tells us that the third statement implies the first. ∎

Let us now study in more detail the relationship between weak and strong invariance. First of all, strong invariance implies weak invariance, but generally not the other way around. It is easiest to see this using weak*-closed convex sets of coherent previsions $\mathcal{M}$. If $\mathcal{M}$ is strongly $\mathcal{T}$-invariant, we have that $TP = P$ and consequently $TP \in \mathcal{M}$ for all P in $\mathcal{M}$, so $\mathcal{M}$ is also weakly $\mathcal{T}$-invariant. To see that the converse doesn't generally hold, consider the set of all coherent previsions $\mathbb{P}(\mathcal{X})$ (the vacuous belief model), which is weakly invariant with respect to any monoid of transformations, but not necessarily strongly so, as, unless $\mathcal{X}$ contains only one element, we can easily find transformations T and coherent previsions P such that TP is different from P (also see Theorem 9 below).

But the theorem above, when interpreted well, also tells us a number of very interesting things on this issue. First of all, we see that a coherent prevision P on $\mathcal{L}(\mathcal{X})$ is strongly $\mathcal{T}$-invariant if and only if it is weakly $\mathcal{T}$-invariant,

---

[29]So strongly invariant belief models correspond to Pericchi and Walley's [1991] notion of a 'class of reasonable (or invariant) priors'.

so both notions of invariance coincide for coherent previsions. *So anyone who insists on modeling beliefs with Bayesian belief models (coherent previsions) only, cannot distinguish between the two types of invariance.* This confirms in general what we claimed in the Introduction about Bayesian belief models. From now on, we shall therefore no longer distinguish between strong and weak invariance for coherent previsions, and simply call them *invariant*.

Furthermore, we see that a coherent lower prevision $\underline{P}$ is strongly $\mathcal{T}$-invariant if and only if all its dominating coherent lower previsions are, or equivalently, if all its dominating coherent previsions, i.e., all the coherent previsions in $\mathcal{M}(\underline{P})$, are $\mathcal{T}$-invariant. Or even stronger, it is easy to see that a coherent lower prevision is strongly invariant if and only if it is a lower envelope of some (not necessarily weak*-closed nor convex) set of invariant coherent previsions.

The notions of weak and strong invariance, and the motivation for introducing them, are tailored to the direct behavioral interpretation of lower previsions, or the equivalent belief models. But what happens if we give a lower prevision $\underline{P}$ a Bayesian sensitivity analysis interpretation? We then hold that there is some actual precise coherent prevision P_a modeling the subject's uncertainty about the random variable X, that we have only imperfect information about in the sense that we only know that $P_a \geq \underline{P}$, or equivalently, that $P_a \in \mathcal{M}(\underline{P})$. Assume that we want the imperfect model $\underline{P}$ to capture that there is 'symmetry of evidence' with respect to a monoid of transformations $\mathcal{T}$. The actual model P_a then should be weakly $\mathcal{T}$-invariant, but since this is a (precise) coherent prevision, we can not distinguish between weak and strong invariance, and it should therefore simply be $\mathcal{T}$-invariant: $TP_a = P_a$ for all $T \in \mathcal{T}$. Since $\mathcal{M}(\underline{P})$ is interpreted as the set of candidate models for P_a, all of the coherent previsions P in $\mathcal{M}(\underline{P})$ must be $\mathcal{T}$-invariant too, or equivalently $\underline{P}$ must be *strongly* $\mathcal{T}$-invariant. A completely analogous course of reasoning shows that if we want $\underline{P}$ to capture 'evidence of symmetry', $\underline{P}$ must be strongly $\mathcal{T}$-invariant as well. So in contradistinction with the direct behavioral interpretation, *on a Bayesian sensitivity analysis interpretation of $\underline{P}$, we cannot distinguish between 'symmetry of evidence' and 'evidence of symmetry', and strong invariance is the proper symmetry property to use in both cases.*[30]

As is the case for weak invariance, a belief model that is strongly $\mathcal{T}$-invariant, is also strongly $\mathcal{T}'$-invariant for any sub-monoid $\mathcal{T}' \subseteq \mathcal{T}$. But in contrast with weak invariance, given any monoid $\mathcal{T}$, there do not always exist coherent belief models that are strongly invariant with respect

---

[30]See Walley [1991, Section 9.5] for related comments about the difference between permutability and exchangeability. These notions will be briefly discussed in Section 9.2.

to T. This is an immediate consequence of the following theorem, which makes an even stronger claim: it is totally *irrational* to require *complete* strong invariance, i.e., strong invariance with respect to the monoid T_X of all transformations of X.

THEOREM 9. *Assume that X contains more than one element. Then any belief model that is strongly T_X-invariant incurs a sure loss.*

Proof. We shall give a proof for lower previsions. Assume *ex absurdo* that $\underline{P}$ avoids sure loss, so $\mathcal{M}(\underline{P})$ is non-empty. Consider any P in $\mathcal{M}(\underline{P})$ and any non-constant gamble f on X [there is at least one such gamble because X contains more than one element]. This implies that there are (different) x_1 and x_2 in X such that $f(x_1) \neq f(x_2)$. For any y in X, consider the transformation T_y that maps all elements of X to y. Then we find that $T_y^t f = f(y)$, whence $P(f(y)-f) \geq \underline{P}(f(y)-f) \geq 0$ and $P(f-f(y)) \geq \underline{P}(f-f(y)) \geq 0$, since $\underline{P}$ is by assumption in particular strongly T_y-invariant. Consequently $P(f) = f(y)$. But this holds in particular for $y = x_1$ and for $y = x_2$, so we infer that $f(x_1) = P(f) = f(x_2)$, a contradiction. ∎

In fact, we easily see in this proof that given the transformation T_y that maps all elements of X to y, the only strongly T_y-invariant belief model that avoids sure loss is the constant prevision on y. Consequently, if we consider a monoid T that includes two different constant transformations, any belief model that is strongly T-invariant incurs a sure loss.

As a result, we see that there are monoids T for which there are no strongly invariant coherent (lower) previsions. Under which conditions, then, are there strongly T-invariant coherent (lower) previsions? It seems easiest, and yields most insight, if we look at this problem in terms of sets of almost-desirable gambles: indeed if we consider a coherent lower prevision $\underline{P}$ on $\mathcal{L}(X)$, then it is strongly T-invariant if and only if for its associated set of almost-desirable gambles $\mathcal{D}_{\underline{P}} = \{f \in \mathcal{L}(X): \underline{P}(f) \geq 0\}$ we have that $\mathcal{D}_T \subseteq \mathcal{D}_{\underline{P}}$. We can consider $\mathcal{D}_T$ itself as a set of almost-desirable gambles, but at this point, we do not know whether $\mathcal{D}_T$ is coherent, or whether it even avoids sure loss. Interestingly, the set of coherent previsions that is associated with $\mathcal{D}_T$ is given by

$$\mathcal{M}(\mathcal{D}_T) = \{P \in \mathbb{P}(X): (\forall g \in \mathcal{D}_T)(P(g) \geq 0)\}$$
$$= \{P \in \mathbb{P}(X): (\forall f \in \mathcal{L}(X))(\forall T \in T)(P(f) = P(T^t f))\}.$$

So $\mathcal{M}(\mathcal{D}_T)$ is precisely the convex and weak*-closed set of all T-invariant coherent previsions, and $\underline{P}$ is strongly T-invariant if and only if $\mathcal{M}(\underline{P}) \subseteq \mathcal{M}(\mathcal{D}_T)$, or in other words, if and only if all coherent previsions that dominate $\underline{P}$ are T-invariant. So there are strongly T-invariant coherent lower previsions if and only if $\mathcal{M}(\mathcal{D}_T) \neq \emptyset$, i.e., if there are T-invariant coherent previsions, and in this case the lower envelope of $\mathcal{M}(\mathcal{D}_T)$ is the point-wise smallest strongly T-invariant coherent lower prevision.

In summary, we see that there are T-invariant coherent previsions if and only if the set of almost-desirable gambles $\mathcal{D}_T$ avoids sure loss,[31] which, taking into account (2.6), is equivalent[32] to the condition[33]

$$\sup \sum_{k=1}^{n} \left[f_k - T_k^t f_k \right] \geq 0$$

$$\text{for all } n \geq 0, \ f_1, \ldots, f_n \text{ in } \mathcal{L}(\mathcal{X}) \text{ and } T_1, \ldots, T_n \text{ in } T. \quad (4.1)$$

In that case, the natural extension $\mathcal{E}_T := \mathcal{E}_{\mathcal{D}_T}^m$ of $\mathcal{D}_T$ to a coherent set of almost-desirable gambles is given by[34]

$$\mathcal{E}_T = \bigcap_{\epsilon > 0} \left\{ f \in \mathcal{L}(\mathcal{X}) : f - \epsilon \geq \sum_{k=1}^{n} \left[f_k - T_k^t f_k \right] \right.$$

$$\left. \text{for some } n \geq 0, \ f_k \in \mathcal{L}(\mathcal{X}), \ T_k \in T \right\} \quad (4.2)$$

This is the smallest coherent and strongly T-invariant set of almost-desirable gambles, or in other words, the belief model that represents evidence of symmetry involving the monoid T. The corresponding lower prevision, defined by [it is easy to see that $\mathcal{M}(\mathcal{D}_T) = \mathcal{M}(\mathcal{E}_T)$]

$$\underline{E}_T(f) = \min \left\{ P(f) : P \in \mathcal{M}(\mathcal{D}_T) \right\} \quad (4.3)$$

$$= \max \left\{ \mu \in \mathbb{R} : f - \mu \in \mathcal{E}_T \right\} \quad (4.4)$$

is then, by virtue of Eq. (4.3) [see also Theorem 15 further on], the point-wise smallest (most conservative) strongly T-invariant coherent lower prevision on $\mathcal{L}(\mathcal{X})$, and if we combine Eqs. (4.2) and (4.4), we find that[35]

---

[31] Also see Walley's [1991, Lemma 3.3.2] Separation Lemma.

[32] Observe that the set $\mathcal{D}_T$ is a *cone*, i.e., closed under scalar multiplication with non-negative real numbers.

[33] The same condition was derived by Walley [1991, Theorem 3.5.2 and Corollary 3.5.4] using an argument that works directly with coherent lower previsions. Although our argument strongly plays on the connection between the three equivalent types of belief models of Table 1, we believe that it produces more insight, once this connection is fully understood.

[34] Again, observe that $\mathcal{D}_T$ is a cone.

[35] Again, Walley [1991, Theorem 3.5.2 and Corollary 3.5.4] proves the same result in a different manner, see also footnote 33.

$$(4.5) \quad \underline{E}_{\mathcal{T}}(f) = \sup \left\{ \inf \left[f - \sum_{k=1}^{n} \left[f_k - T_k^t f_k \right] \right] : n \geq 0, f_k \in \mathcal{L}(\mathcal{X}), T_k \in \mathcal{T} \right\}.$$

Remember that this lower prevision is only well-defined (assumes finite real values) whenever the condition (4.1) is satisfied. Taking into account Theorem 15 further on, we deduce that a coherent (lower) prevision is (strongly) $\mathcal{T}$-invariant if and only if it dominates $\underline{E}_{\mathcal{T}}$. Also, $\underline{E}_{\mathcal{T}}$ is the belief model we should use if nothing else but the evidence of symmetry is given. Finally, this formula for the *lower* prevision is constructive, but usually the existence of invariant previsions (on infinite spaces) is proven in a non-constructive (Hahn–Banach) way; see Section 8, and also Agnew and Morse [1938] and Bhaskara Rao and Bhaskara Rao [1983, Section 2.1.3(8)]. So we cannot usually get to the coherent invariant previsions by construction, but we can always construct their lower envelope explicitly!

We shall have much more to say about the existence of strongly invariant belief models in Section 7, where we show that this existence is guaranteed in particular if the monoid $\mathcal{T}$ is Abelian, or if it is a finite group. The following counterexample tells us that there is no such guarantee for infinite groups.

EXAMPLE 6 (PERMUTATION INVARIANCE ON THE NATURAL NUMBERS). Consider the set $\mathcal{P}_{\mathbb{N}}$ of all permutations of the set of natural numbers $\mathbb{N}$. We show that there are no (strongly) $\mathcal{P}_{\mathbb{N}}$-invariant coherent (lower) previsions on $\mathcal{L}(\mathbb{N})$ by showing that the condition (4.1) doesn't hold. Indeed, consider the partition of $\mathbb{N}$ made up of the sets

$$R_3^r = \{3n + r : n \in \mathbb{N}\}, \quad r = 0, 1, 2,$$

and any permutations π_r for $r = 0, 1, 2$ such that for all $n \in \mathbb{N}$, $\pi_r(n) \in R_3^r$ if and only if $n \notin R_3^r$ [for instance, let π_r be involutive and such that it assigns the first element of R_3^r to the first of $(R_3^r)^c$, the second element of R_3^r to the second of $(R_3^r)^c$, etc.] Consider the gamble $G = \sum_{r=0}^{2} [I_{R_3^r} - \pi_r^t I_{R_3^r}]$ on $\mathbb{N}$, then we are done if we can show that $\sup G < 0$. Indeed, if $n \in R_3^r$ then $G(n) = 1 + 0 + 0 - (1 + 1 + 0) = -1$ for $r = 0, 1, 2$, so $\sup G = -1$.

These results expose another fundamental difference between weak and strong invariance: while strong invariance with respect to a greater number of transformations means that we must refine our beliefs (i.e, it make them more precise), this is not the case with weak invariance.

On the other hand, strong invariance is preserved by dominating lower previsions: if $\underline{P}_1$ is a coherent lower prevision that is strongly $\mathcal{T}$-invariant and $\underline{P}_2$ is a coherent lower prevision that dominates $\underline{P}_1$, then $\underline{P}_2$ is also strongly $\mathcal{T}$-invariant. It indeed seems reasonable that, if a subject has evidence of symmetry, and she has some additional information that allows her to make her judgments more precise, she can add assessments while still preserving strong invariance. But a similar result does not hold for weak invariance: since the vacuous lower prevision is weakly $\mathcal{T}_{\mathcal{X}}$-invariant, this would mean that any lower prevision should be weakly $\mathcal{T}_{\mathcal{X}}$-invariant, *quod non*.

In summary, there is an important conceptual difference between weak and strong invariance. Weakly invariant belief models capture in particular that a subject *has no reason to* strictly prefer a gamble f to its transformation $T^t f$ whenever $f \not> T^t f$. Strong invariance captures that a subject *has reason not to* distinguish between, i.e., to be indifferent between, the gambles f and $T^t f$. And it is only if you insist on using Bayesian belief models always that you must infer indifference from having no reason to (strictly) prefer. This is of particular relevance for models that try to represent a subject's complete ignorance, as we now proceed to show.

5   Modeling complete ignorance

Suppose our subject is completely ignorant about the value that X assumes in $\mathcal{X}$. Then she has no relevant information that would allow her to favor one possible value of X over another. This implies that the corresponding belief model should be symmetric in the possible values of X, or in other words it should be weakly invariant with respect to the group $\mathcal{P}_{\mathcal{X}}$ of all permutations of $\mathcal{X}$. This leads to a form of Walley's [1991, Section 5.5.1] Symmetry Principle.

PRINCIPLE 1 (SYMMETRY PRINCIPLE (SP)). If a subject is completely ignorant about the value of a random variable X in $\mathcal{X}$, then her corresponding belief model should be weakly invariant with respect to the group $\mathcal{P}_{\mathcal{X}}$ of all permutations of $\mathcal{X}$.

We have mentioned before that the appropriate belief model for complete ignorance about X seems to be the vacuous lower prevision $\underline{P}_{\mathcal{X}}$. But SP by itself is not sufficient to single out this lower prevision: if, for instance, $\mathcal{X}$ is finite, then the uniform precise prevision P^u, given by

$$P^u(f) = \frac{1}{|\mathcal{X}|} \sum_{x \in \mathcal{X}} f(x)$$

for each gamble f on $\mathcal{X}$, which assigns equal probability mass $1/|\mathcal{X}|$ to each element of $\mathcal{X}$, is also weakly permutation invariant. We shall also see in Examples 10 and 11 of Section 9 that there may be many more coherent lower previsions that share the same weak permutation invariance property. If, however, we strengthen the Symmetry Principle to require weak invariance with respect to *all transformations*, and not just all permutations, then

Theorem 6 tells us that the vacuous lower prevision $\underline{P}_{\mathcal{X}}$ is indeed the only coherent lower prevision that is compatible with the following

PRINCIPLE 2 (STRONG SYMMETRY PRINCIPLE (SSP)). *If a subject is completely ignorant about the value of a random variable X in $\mathcal{X}$, then her corresponding belief model should be weakly invariant with respect to the monoid $\mathcal{T}_{\mathcal{X}}$ of all transformations of $\mathcal{X}$.*

Walley [1991, Section 5.5.1 and note 7 on p. 526] has shown that for random variables X taking values in a finite set $\mathcal{X}$, the vacuous lower prevision $\underline{P}_{\mathcal{X}}$ is the only coherent lower prevision that is compatible with SP and the so-called[36]

PRINCIPLE 3 (EMBEDDING PRINCIPLE (EP)). *Consider a random variable X, and consider a set of possible values A for X. Then the (lower) probability assigned to the event A, i.e., the lower probability that $X \in A$, should not depend on the set $\mathcal{X}$ of all possible values for X in which A is embedded.*

So under coherence, SSP is equivalent to SP and EP taken together. Under coherence, it is also equivalent to the following rationality principle, as we shall shortly see.

PRINCIPLE 4 (REVISED PRINCIPLE OF INSUFFICIENT REASON (RPIR)). *If you have two different gambles f and g on a random variable X that you are completely ignorant about, then if $f \not\geq g$ you have no reason to prefer f to g.*

Indeed, the only coherent belief model that is compatible with this principle, is the vacuous one. We shall argue in terms of real desirability models[37] $\mathcal{R}$ (see Section 2.1). Say that a subject (really) *prefers* f to g whenever $f \neq g$ and $f - g \in \mathcal{R}$, i.e., she accepts to exchange g for f. Then RPIR implies that for all $f \neq 0$, $f \not\geq 0$ implies that $f \notin \mathcal{R}$, or equivalently, by contraposition, that $f \in \mathcal{R}$ implies $f \geq 0$. Hence $\mathcal{R} = \mathcal{C}_+$ is the vacuous belief model.

In summary, we have the following equivalences, under coherence, and the only belief model that is compatible with these three equivalent rationality requirements, is the vacuous one:

$$\text{SSP} \Leftrightarrow \text{SP+EP} \Leftrightarrow \text{RPIR}.$$

---

[36] For additional discussion of this principle, see also Walley [1996a] and Walley and Bernard [1999].

[37] A similar argument can be given for almost-desirability models $\mathcal{D}$ and lower previsions $\underline{P}$, using for preference Walley's [1991, Sections 3.7.7–3.7.9] corresponding notion of *strict preference*, which corresponds to the present argument by using $\mathcal{D}_{\underline{P}}^{+} \cup \{0\}$ as a coherent set of really desirable gambles.

RPIR is a revised version of the Principle of Insufficient Reason (PIR), which states that if you are completely ignorant about the value of a random variable X, then you have no reason to distinguish between the different possible values, and therefore should consider all these values to have equal probability. Indeed, from a historical point of view, the PIR was used extensively by Laplace (see for instance Howie [2002]) to justify using a uniform probability for modeling complete ignorance.

We are of course aware that our reformulation RPIR of Laplace's PIR is quite unusual and has little or no historical grounds, which is why we refer to it as a *revised*, or perhaps better, improved principle. It might have been preferable to call RPIR the 'Principle of Insufficient Reason to Prefer', but we decided against that for aesthetical reasons.

We think that RPIR is reasonable, but that PIR isn't. Indeed, one of the reasons for the critical attitudes of many researchers towards 'Bayesian methods' and inverse probability in the nineteenth and early twentieth century seem to lie in the indiscriminate use by many of Laplace's PIR in order to obtain uniform prior probabilities that can be plugged into Bayes's formula.[38] And by 'indiscriminate use' we mean precisely the confusion that exists between symmetry of evidence and evidence of symmetry: we have argued that it is only evidence of symmetry that justifies using strongly invariant belief models (and in many cases, such as permutation invariance for finite spaces, strong invariance singles out the uniform probability as the only compatible belief model, see also Section 9). If there is only symmetry of evidence, we should use weakly invariant belief models, and in the special case of complete ignorance, vacuous ones. Of course, as we said in the Introduction and proved in the previous section, for precise previsions (Bayesian belief models) there is no difference between weak and strong invariance, so if you insist on using a Bayesian belief model, symmetry of evidence leads you to a (strongly) invariant one! The problem with the PIR, therefore, is that the belief model is only allowed to be precise: there would be fewer or no difficulties if in its formulation we just replaced 'probability' with 'lower and upper probability', for instance.

6   Weakly invariant lower previsions

Let us now turn to a more involved mathematical study of the invariance of coherent lower previsions. So far, we have only looked at coherent lower previsions that were defined on all gambles. But of course, it will usually happen that our subject specifies a supremum acceptable buying price $\underline{P}(f)$ for only a limited number of gambles f, say those in a subset $\mathcal{K}$ of $\mathcal{L}(\mathcal{X})$.

---

[38] An interesting historical discussion of such attitudes can be found in Howie [2002] and Zabell [1989a].

And then we can ask ourselves whether such an assessment can be coherently extended to a weakly, or to a strongly, $\mathcal{T}$-invariant lower prevision on all gambles. We shall address these, and related, problems in this and the following section. Let us begin here with weak invariance. The following definition generalizes the already established notion of weak invariance to lower previsions defined on any $\mathcal{T}$-invariant domain, that are not necessarily coherent (they may even incur a sure loss).[39]

DEFINITION 10 (Weak invariance). A lower prevision $\underline{P}$ defined on a set of gambles $\mathcal{K} \subseteq \mathcal{L}(\mathcal{X})$ is called weakly $\mathcal{T}$-invariant if

(W1) $T^t f \in \mathcal{K}$ for all f in $\mathcal{K}$ and T in $\mathcal{T}$, i.e., $\mathcal{K}$ is $\mathcal{T}$-invariant;

(W2) $\underline{P}(T^t f) \geq \underline{P}(f)$ for all f in $\mathcal{K}$ and T in $\mathcal{T}$, i.e., all $T\underline{P}$ point-wise dominate $\underline{P}$.

As before, if $\mathcal{T}$ is right-cancelable (and in particular if it is a group), the inequality in the invariance definition is actually an equality: consider a gamble f in $\mathcal{K}$, a transformation T in $\mathcal{T}$ and its right-inverse R, we have $\underline{P}(f) = \underline{P}((TR)^t f) = \underline{P}(R^t(T^t f)) \geq \underline{P}(T^t f)$ in addition to $\underline{P}(T^t f) \geq \underline{P}(f)$.

Next, because taking convex combinations, lower envelopes, limits inferior and superior preserves inequalities, it is easy to see that convex combinations, lower envelopes and point-wise limits of weakly invariant lower previsions are also weakly invariant. Observe by the way that the same operations also preserve coherence.

The following proposition looks at weak invariance for (precise) previsions.

PROPOSITION 11. *Let P be a prevision, i.e., a self-conjugate lower prevision, defined on a negation-invariant domain $\mathcal{K} = -\mathcal{K}$. Assume that $\mathcal{K}$ is also $\mathcal{T}$-invariant. Then P is weakly $\mathcal{T}$-invariant if and only if $P(T^t f) = P(f)$ for all T in $\mathcal{T}$ and all f in $\mathcal{K}$.*

Proof. It is clear that the condition is sufficient. To show that it is also necessary, assume that P is $\mathcal{T}$-invariant, and consider any T in $\mathcal{T}$ and any gamble f in $\mathcal{K}$. Then it follows from the $\mathcal{T}$-invariance of P that on the one hand $P(T^t f) \geq P(f)$, and on the other hand, since $-f \in \mathcal{K}$ and $T^t(-f) = -T^t f \in \mathcal{K}$, that $P(-T^t f) = P(T^t(-f)) \geq P(-f)$, or equivalently, using the self-conjugacy of P, that $P(f) \geq P(T^t f)$. ∎

---

[39] Our notion of weak invariance for a lower prevision is weaker than Walley's [1991, Section 3.5.1] corresponding notion of invariance, which requires equality, and has the drawback that it is not preserved by natural extension.

We study next whether a weakly invariant lower prevision $\underline{P}$ with domain $\mathcal{K}$ can be extended to a coherent weakly invariant lower prevision on the set of all gambles, or more generally, whether there is a coherent weakly invariant lower prevision on all gambles that dominates $\underline{P}$. We already know from the material in Section 2.3 that a necessary condition for this is that $\underline{P}$ should avoid sure loss. Indeed, if $\underline{P}$ incurs sure loss then it has no dominating coherent lower prevision, let alone a weakly invariant one. The perhaps surprising result we prove next is that avoiding sure loss is also sufficient, and that all we have to do is consider the natural extension $\underline{E}_P$ of $\underline{P}$, as it preserves weak invariance. This natural extension is automatically guaranteed to be the point-wise smallest weakly $\mathcal{T}$-invariant coherent lower prevision that dominates $\underline{P}$.[40]

THEOREM 12 (Natural extension preserves weak invariance). *The natural extension $\underline{E}_P$ of a weakly $\mathcal{T}$-invariant lower prevision $\underline{P}$ on a set of gambles $\mathcal{K}$ that avoids sure loss is still weakly $\mathcal{T}$-invariant, i.e., for all gambles f on $\mathcal{X}$ and all T in $\mathcal{T}$,*

$$T\underline{E}_P(f) = \underline{E}_P(T^t f) \geq \underline{E}_P(f).$$

Consequently, $\underline{E}_P$ is the point-wise smallest weakly $\mathcal{T}$-invariant coherent lower prevision on $\mathcal{L}$ that dominates $\underline{P}$ on its domain $\mathcal{K}$.

Proof. Consider any gamble f on $\mathcal{X}$ and any T in $\mathcal{T}$. From the definition (2.8) of natural extension, and the fact that $T^t\mathcal{K} \subseteq \mathcal{K}$, we get

$$\underline{E}_P(T^t f) = \sup_{\substack{\lambda_k \geq 0, f_k \in \mathcal{K} \\ k=1\ldots,n, n \geq 0}} \left\{ \alpha : T^t f - \alpha \geq \sum_{k=1}^{n} \lambda_k \left[f_k - \underline{P}(f_k) \right] \right\}$$

$$\geq \sup_{\substack{\lambda_k \geq 0, g_k \in \mathcal{K} \\ k=1\ldots,n, n \geq 0}} \left\{ \alpha : T^t f - \alpha \geq \sum_{k=1}^{n} \lambda_k \left[T^t g_k - \underline{P}(T^t g_k) \right] \right\}. \quad (6.1)$$

Now it follows from the T-invariance of $\underline{P}$ that $\underline{P}(T^t g_k) \geq \underline{P}(g_k)$, whence

$$\sum_{k=1}^{n} \lambda_k \left[T^t g_k - \underline{P}(T^t g_k) \right] \leq T^t \sum_{k=1}^{n} \lambda_k \left[g_k - \underline{P}(g_k) \right],$$

and consequently $f - \alpha \geq \sum_{k=1}^{n} \lambda_k [g_k - \underline{P}(g_k)]$ implies that

$$T^t f - \alpha \geq T^t \sum_{k=1}^{n} \lambda_k \left[g_k - \underline{P}(g_k) \right] \geq \sum_{k=1}^{n} \lambda_k \left[T^t g_k - \underline{P}(T^t g_k) \right].$$

---

[40]This result is mentioned, with only a hint at the proof, by Walley [1991, Theorem 3.5.2].

So we may infer from the inequality (6.1) that

$$\underline{E}_{\underline{P}}(T^t f) \geq \sup_{\substack{\lambda_k \geq 0, g_k \in \mathcal{K} \\ k=1\dots,n, n \geq 0,}} \left\{ \alpha : f - \alpha \geq \sum_{k=1}^{n} \lambda_k \left[g_k - \underline{P}(g_k) \right] \right\} = \underline{E}_{\underline{P}}(f),$$

which completes the proof. ∎

Hence, if we start out with a lower prevision $\underline{P}$ on $\mathcal{K}$ that is weakly $\mathcal{T}$-invariant and already coherent, then its natural extension $\underline{E}_{\underline{P}}$ is the smallest coherent and weakly $\mathcal{T}$-invariant lower prevision on all gambles that agrees with $\underline{P}$ on $\mathcal{K}$. As we shall show further on, this result does not carry over to strong invariance.

7   Strongly invariant lower previsions

We now turn to the study of strong invariance for lower previsions on general domains.

7.1   Definition and immediate properties

The following definition generalizes the notion of strong invariance introduced in Section 4.2 to lower previsions that needn't be coherent, nor defined on all of $\mathcal{L}(\mathcal{X})$.

DEFINITION 13 (Strong invariance). A lower prevision $\underline{P}$ defined on a set of gambles $\mathcal{K} \subseteq \mathcal{L}(\mathcal{X})$ is called *strongly $\mathcal{T}$-invariant* if

(S1) $T^t f - f \in \mathcal{K}$ and $f - T^t f \in \mathcal{K}$ for all f in $\mathcal{K}$ and all $T \in \mathcal{T}$;

(S2) $\underline{P}(T^t f - f) \geq 0$ and $\underline{P}(f - T^t f) \geq 0$ for all f in $\mathcal{K}$ and all $T \in \mathcal{T}$.

As is the case for weak invariance, it is easy to see that strong $\mathcal{T}$-invariance is preserved under convex combinations, lower envelopes, and point-wise limits, simply because all these operations preserve inequalities.

PROPOSITION 14. *A strongly $\mathcal{T}$-invariant coherent lower prevision on a $\mathcal{T}$-invariant domain is also weakly $\mathcal{T}$-invariant.*

Proof. First of all, the coherence and strong invariance of $\underline{P}$ imply that $0 \leq \underline{P}(T^t f - f) \leq \underline{P}(T^t f) - \underline{P}(f)$, whence $\underline{P}(T^t f) \geq \underline{P}(f)$ and similarly, we derive from $\underline{P}(f - T^t f) \geq 0$ that $\underline{P}(f) \geq \underline{P}(T^t f)$. So we see that $\underline{P}$ is also weakly $\mathcal{T}$-invariant (with equality). ∎

To see that a converse result does not generally hold, so weak invariance is actually weaker than strong invariance, consider the vacuous lower prevision $\underline{P}_{\mathcal{X}}$ on $\mathcal{L}(\mathcal{X})$ and the transformation T_{x_0} that maps all elements x of $\mathcal{X}$ to x_0.

Then, for any gamble f such that $\inf f < f(x_0)$ we have $\underline{P}_{\mathcal{X}}(f - T_{x_0}^t f) < 0$. Hence, $\underline{P}_{\mathcal{X}}$ is not strongly T_{x_0}-invariant but Theorem 6 implies that it is weakly T_{x_0}-invariant. If we consider a finite space $\mathcal{X}$ and the vacuous lower prevision $\underline{P}_{\mathcal{X}}$ on $\mathcal{L}(\mathcal{X})$ and the class $\mathcal{P}_{\mathcal{X}}$ of all permutations of $\mathcal{X}$, we can see that weak invariance (with equality) does not imply strong invariance.

So weak invariance is indeed a weaker notion than strong invariance. The following theorem expresses the main difference between these two concepts: while the former means that the set of coherent previsions $\mathcal{M}(\underline{P})$ is invariant, the latter means that every element of this set is invariant.

THEOREM 15. *Let $\mathcal{K}$ be a negation invariant and $\mathcal{T}$-invariant set of gambles such that $T^t f - f$ is in $\mathcal{K}$ for all f in $\mathcal{K}$ and T in $\mathcal{T}$.*

1. *A coherent prevision P on $\mathcal{K}$ is weakly $\mathcal{T}$-invariant if and only if it is strongly $\mathcal{T}$-invariant. In either case we simply call it $\mathcal{T}$-invariant.*

2. *A coherent lower prevision $\underline{P}$ on $\mathcal{K}$ is strongly $\mathcal{T}$-invariant if and only if all its dominating coherent previsions are (strongly) $\mathcal{T}$-invariant on $\mathcal{K}$.*

Proof. We start with the first statement. We only need to prove the direct implication, so assume that P is weakly $\mathcal{T}$-invariant, and consider any f in $\mathcal{K}$. Then from the assumption and Proposition 11 we get $P(T^t f) = P(f)$, and it follows from the additivity of P that indeed $P(T^t f - f) = P(f - T^t f) = 0$.

We now turn to the second statement. Since any coherent lower prevision is the lower envelope of its dominating coherent previsions, the converse implications follow at once, since taking a lower envelope preserves strong invariance. To prove the direct implication, assume that $\underline{P}$ is strongly $\mathcal{T}$-invariant, and consider any coherent prevision P in $\mathcal{M}(\underline{P})$. For any T in $\mathcal{T}$ and any f in $\mathcal{K}$ we then find that

$$0 \leq \underline{P}(f - T^t f) \leq P(f - T^t f) = -P(T^t f - f) \leq -\underline{P}(T^t f - f) \leq 0,$$

whence indeed $P(f) = P(T^t f)$. ∎

7.2 Strongly invariant natural extension

We have shown when studying weak invariance that for any weakly $\mathcal{T}$-invariant lower prevision $\underline{P}$ on some domain $\mathcal{K}$ that avoids sure loss, there is a point-wise smallest weakly invariant coherent lower prevision defined on all gambles that dominates it: its natural extension $\underline{E}_P$. Let us now investigate whether something similar can be done for the notion of strong invariance. The question then is: Consider a monoid $\mathcal{T}$ of transformations of $\mathcal{X}$ and a lower prevision $\underline{P}$ on $\mathcal{K}$ that avoids sure loss, are there strongly

T-invariant coherent lower previsions on all $\mathcal{L}(\mathcal{X})$ that dominate $\underline{P}$, and if so, what is the point-wise smallest such lower prevision? Let us denote, as before, by

$$\mathcal{D}_{\underline{P}} = \left\{ f \in \mathcal{L}(\mathcal{X}) \colon \underline{E}_{\underline{P}}(f) \geq 0 \right\}$$

the set of almost-desirable gambles associated with $\underline{P}$, and by

$$\mathcal{M}(\underline{P}) = \{ P \in \mathbb{P}(\mathcal{X}) \colon (\forall f \in \mathcal{K})(P(f) \geq \underline{P}(f)) \}$$

its set of dominating coherent previsions, then clearly a coherent lower prevision $\underline{Q}$ on $\mathcal{L}(\mathcal{X})$ is strongly T-invariant and dominates $\underline{P}$ if and only if $\mathcal{M}(\underline{Q}) \subseteq \mathcal{M}(\underline{P}) \cap \mathcal{M}(\mathcal{D}_T)$, or equivalently, $\mathcal{D}_{\underline{P}} \cup \mathcal{D}_T \subseteq \mathcal{D}_{\underline{Q}}$. So there are strongly T-invariant coherent (lower) previsions that dominate $\underline{P}$ if and only if $\mathcal{M}(\underline{P}) \cap \mathcal{M}(\mathcal{D}_T) \neq \emptyset$, or equivalently, if the set of almost-desirable gambles $\mathcal{D}_{\underline{P}} \cup \mathcal{D}_T$ avoids sure loss, and in this case the lower envelope of $\mathcal{M}(\underline{P}) \cap \mathcal{M}(\mathcal{D}_T)$, or equivalently, the lower prevision associated with the natural extension of the set of almost-desirable gambles $\mathcal{D}_{\underline{P}} \cup \mathcal{D}_T$, is the smallest such lower prevision. In the language of coherent lower previsions, this leads to the following theorem.[41]

THEOREM 16 (Strongly invariant natural extension). *Consider a lower prevision $\underline{P}$ on $\mathcal{K}$ that avoids sure loss, and a monoid T of transformations of $\mathcal{X}$. Then there are strongly T-invariant coherent (lower) previsions on $\mathcal{L}(\mathcal{X})$ that dominate $\underline{P}$ on $\mathcal{K}$ if and only if*

(7.1)
$$\overline{E}_{\underline{P}} \left(\textstyle\sum_{k=1}^{n} [f_k - T_k^t f_k] \right) \geq 0$$
$$\text{for all } n \geq 0,\ f_1,\ \ldots,\ f_n \text{ in } \mathcal{L}(\mathcal{X}) \text{ and } T_1,\ \ldots,\ T_n \text{ in } T,$$

or equivalently, if

(7.2) $\overline{E}_T \left(\displaystyle\sum_{k=1}^{n} [f_k - \underline{P}(f_k)] \right) \geq 0 \quad$ *for all $n \geq 0$, and $f_1,\ \ldots,\ f_n$ in $\mathcal{K}$.*

In that case the smallest coherent and strongly T-invariant lower prevision on $\mathcal{L}(\mathcal{X})$ that dominates $\underline{P}$ on its domain $\mathcal{K}$ is given by

(7.3)
$$\underline{E}_{\underline{P},T}(f) = \sup \left\{ \underline{E}_{\underline{P}} (f - \textstyle\sum_{k=1}^{n} [f_k - T_k^t f_k]) \right.$$
$$\left. n \geq 0, f_k \in \mathcal{L}(\mathcal{X}), T_k \in T \right\}$$

---

[41]Walley [1991, Theorems 3.5.2 and 3.5.3] proves similar results involving Eqs. (7.1) and (7.3) for what we call weakly T-invariant $\underline{P}$ that avoid sure loss, in a different manner. See also footnotes 33 and 35.

(7.4)
$$= \sup \{ \underline{E}_T \, (f - \textstyle\sum_{k=1}^{n} \lambda_k \, [f_k - \underline{P}(f_k)]) \, n \geq 0,$$
$$f_k \in \mathcal{K}, \lambda_k \geq 0 \}$$

for all gambles f on $\mathcal{X}$; and $\mathcal{M}(\underline{E}_{\underline{P},T})$ is the set of all T-invariant coherent previsions that dominate $\underline{P}$ on $\mathcal{K}$.

Proof. We already know that there is a dominating coherent (lower) prevision if and only if $\mathcal{M}(\underline{P}) \cap \mathcal{M}(\mathcal{D}_T)$ is non-empty. Let us show that this is equivalent to the conditions (7.1) and (7.2). To see the equivalence between these two conditions, it suffices to notice [use Eq. (2.8) and the fact that $\overline{E}_{\underline{P}}(h) = -\underline{E}_{\underline{P}}(-h)$] that condition (7.1) is equivalent to

(7.5)
$$\sup \left[\textstyle\sum_{k=1}^{n} [f_k - T_k^t f_k] + \textstyle\sum_{j=1}^{m} [g_j - \underline{P}(g_j)] \right] \geq 0$$
$$\text{for all } n, m \geq 0, \ f_k \in \mathcal{L}(\mathcal{X}), \ T_k \in \mathcal{T}, \ g_j \in \mathcal{K},$$

and that this is in turn [use Eq. (4.5) and the fact that $\overline{E}_T(h) = -\underline{E}_T(-h)$] equivalent to condition (7.2). But, considering condition (2.6), we see that condition (7.5) holds if and only if the set of almost-desirable gambles $\mathcal{D}_{\underline{P}} \cup \mathcal{D}_T$ avoids sure loss, or equivalently, if the corresponding set of coherent previsions $\mathcal{M}(\underline{P}) \cap \mathcal{M}(\mathcal{D}_T)$ is non-empty.

We now prove the validity of the expression (7.4) for the lower envelope $\underline{E}_{\underline{P},T}$ of the set of coherent previsions $\mathcal{M}(\underline{P}) \cap \mathcal{M}(\mathcal{D}_T)$. The proof for the expression (7.3) is analogous. We know from the material in Section 2 that this lower envelope is also the coherent lower prevision associated with the natural extension of the set of almost-desirable gambles $\mathcal{D}_{\underline{P}} \cup \mathcal{D}_T$, so we get by applying Eq. (2.8) with $\mathcal{D} = \mathcal{D}_{\underline{P}} \cup \mathcal{D}_T$ that

$$\underline{E}_{\underline{P},T}(f) = \sup_{\substack{\lambda_k \geq 0, g_k \in \mathcal{D}_{\underline{P}} \\ k=1,\ldots,n, n \geq 0}} \ \sup_{\substack{\mu_\ell \geq 0, h_\ell \in \mathcal{D}_T \\ \ell=1,\ldots,m, m \geq 0}} \ \inf \left[f - \sum_{k=1}^{n} \lambda_k g_k - \sum_{\ell=1}^{m} \mu_\ell h_\ell \right]$$

$$= \sup_{\substack{\lambda_k \geq 0, g_k \in \mathcal{D}_{\underline{P}} \\ k=1,\ldots,n, n \geq 0}} \ \sup_{\substack{\mu_\ell \geq 0, h_\ell \in \mathcal{D}_T \\ \ell=1,\ldots,m, m \geq 0}} \ \inf \left[\left(f - \sum_{k=1}^{n} \lambda_k g_k \right) - \sum_{\ell=1}^{m} \mu_\ell h_\ell \right]$$

$$= \sup_{\substack{\lambda_k \geq 0, g_k \in \mathcal{D}_{\underline{P}} \\ k=1,\ldots,n, n \geq 0}} \underline{E}_T \left(f - \sum_{k=1}^{n} \lambda_k g_k \right)$$

$$= \sup_{\substack{\lambda_k \geq 0, f_k \in \mathcal{K} \\ k=1,\ldots,n, n \geq 0}} \underline{E}_T \left(f - \sum_{k=1}^{n} \lambda_k [f_k - \underline{P}(f_k)] \right),$$

for every gamble f on $\mathcal{X}$, also taking into account the definition (4.5) of $\underline{E}_T$. ∎

In conclusion, whenever the equivalent conditions (7.1) and (7.2) are satisfied for a lower prevision $\underline{P}$ that avoids sure loss, then (and only then) the functional $\underline{E}_{P,T}$, defined by Eqs. (7.3) and (7.4), is the point-wise smallest coherent and strongly T-invariant lower prevision that dominates $\underline{P}$. We shall call $\underline{E}_{P,T}$ the *strongly T-invariant natural extension* of $\underline{P}$, as it is the belief model that the assessments captured in $\underline{P}$ lead to if in addition a (so-called structural)[42] assessment of symmetry involving the monoid T is made.

7.3   The existence of strongly invariant coherent (lower) previsions

There is a beautiful and surprisingly simple argument to show that for some types of monoids T, there always are strongly T-invariant lower previsions that dominate a given lower prevision that is weakly T-invariant and avoids sure loss. It is based on the combination of a number of ideas in the literature: (i) Agnew and Morse [1938, Section 2] constructed some specific type of Minkowski functional and used this together with a Hahn–Banach extension result to prove the existence of linear functionals that are invariant with respect to certain groups of permutations; (ii) Day [1942, Theorem 3] showed, in a discussion of ergodic theorems, that a similar construction always works for Abelian semigroups of transformations; (iii) with crucially important insight, Walley [1991, Theorems 3.5.2 and 3.5.3] recognized that the Minkowski functional in the existence proofs of Agnew and Morse, and Day, is actually what we have called a strongly invariant lower prevision, and he used the ideas behind this construction to introduce what we shall call *mixture lower previsions* in Section 7.4; (iv) in another seminal discussion of mean ergodic theorems, Alaoglu and Birkhoff [1940] show that (Moore–Smith-like) convergence of convex mixtures of linear transformations is instrumental in characterizing ergodicity; and (v) Bhaskara Rao and Bhaskara Rao [1983, Section 2.1.3] use so-called Banach limits to generate shift-invariant probability charges. In this and the next section, we combine and extend these ideas to prove more general existence results for (strongly) invariant coherent (lower) previsions, and to investigate their relation to (generalized) Banach limits (Section 8). As we shall see in Section 7.4, Walley's [1991, Section 3.5] results can then be derived from our more general treatment.

Consider a monoid T of transformations of $\mathcal{X}$. We can, as before, consider the set of lifted transformations T^t as a monoid of linear transformations of the linear space $\mathcal{L}(\mathcal{X})$. A *convex combination* T^* of elements of T^t is a

---

[42]Structural assessments are discussed in general in Walley [1991, Chapter 9].

linear transformation of $\mathcal{L}(\mathcal{X})$ of the form

$$T^* = \sum_{k=1}^{n} \lambda_k T_k^t,$$

where $n \geq 1$, λ_1, ..., λ_n are non-negative real numbers that sum to one, and of course $T^* f = \sum_{k=1}^{n} \lambda_k T_k^t f$. We denote by $\mathcal{T}^*$ the set of all convex combinations of elements of $\mathcal{T}^t$. We have of course for any two elements $T_1^* = \sum_{k=1}^{m} \lambda_k U_k^t$ and $T_2^* = \sum_{k=1}^{n} \mu_k V_k^t$ of $\mathcal{T}^*$ that their composition

$$T_2^* T_1^* = \sum_{k=1}^{n} \mu_k V_k^t \left(\sum_{\ell=1}^{m} \lambda_\ell U_\ell^t \right) = \sum_{k=1}^{n} \sum_{\ell=1}^{m} \mu_k \lambda_\ell V_k^t U_\ell^t = \sum_{k=1}^{n} \sum_{\ell=1}^{m} \lambda_\ell \mu_k (U_\ell V_k)^t$$

again belongs to $\mathcal{T}^*$. This implies that $\mathcal{T}^*$ is a monoid of linear transformations of $\mathcal{L}(\mathcal{X})$ as well. We can now introduce invariance definitions involving transformations in $\mathcal{T}^*$ in precisely the same way as we defined them for $\mathcal{T}$ (or actually $\mathcal{T}^t$). We can also define, for any real functional Λ and $T^* \in \mathcal{T}^*$, the transformed functional $T^* \Lambda$ as $\Lambda \circ T^*$. We then have the following result.

PROPOSITION 17. *The following statements hold, where f is a gamble on $\mathcal{X}$, $\mathcal{K}$ is a convex set of gambles on $\mathcal{X}$, and $\underline{P}$ is a coherent lower prevision on $\mathcal{K}$:*

1. *f is $\mathcal{T}$-invariant if and only if f is $\mathcal{T}^*$-invariant;*

2. *$\mathcal{K}$ is $\mathcal{T}$-invariant if and only if $\mathcal{K}$ is $\mathcal{T}^*$-invariant;*

3. *$\underline{P}$ is weakly $\mathcal{T}$-invariant if and only if $\underline{P}$ is weakly $\mathcal{T}^*$-invariant;*

4. *$\underline{P}$ is strongly $\mathcal{T}$-invariant if and only if $\underline{P}$ is strongly $\mathcal{T}^*$-invariant.*

Proof. It suffices of course to prove the direct implications. Consider an arbitrary $T^* = \sum_k \lambda_k T_k \in \mathcal{T}^*$. For the first statement, let f be $\mathcal{T}$-invariant, then $T^* f = \sum_k \lambda_k T_k^t f = \sum_k \lambda_k f = f$, where the second equality follows from the $\mathcal{T}$-invariance of f. So f is $\mathcal{T}^*$-invariant. For the second statement, let $\mathcal{K}$ be $\mathcal{T}$-invariant and let $f \in \mathcal{K}$, then $T^* f = \sum_k \lambda_k T_k^t f \in \mathcal{K}$, because $T_k^t f \in \mathcal{K}$ for all k by the $\mathcal{T}$-invariance of $\mathcal{K}$ and because $\mathcal{K}$ is convex. So $\mathcal{K}$ is $\mathcal{T}^*$-invariant. For the third statement, assume that $\underline{P}$ is weakly $\mathcal{T}$-invariant. For any $f \in \mathcal{K}$,

$$\underline{P}(T^* f) = \underline{P} \left(\sum_k \lambda_k T_k^t f \right) \geq \sum_k \lambda_k \underline{P}(T_k^t f) \geq \sum_k \lambda_k \underline{P}(f) = \underline{P}(f),$$

where the first inequality follows from the coherence of $\underline{P}$, and the second from the weak $\mathcal{T}$-invariance of $\underline{P}$. Hence $\underline{P}$ is weakly $\mathcal{T}^*$-invariant. For the last statement, assume that $\underline{P}$ is strongly $\mathcal{T}$-invariant. For any $f \in \mathcal{K}$,

$$\underline{P}\left(\sum_k \lambda_k T_k^t f - f\right) = \underline{P}\left(\sum_k \lambda_k(T_k^t f - f)\right) \geq \sum_k \lambda_k \underline{P}(T_k^t f - f) \geq 0,$$

where the first inequality follows from the coherence of $\underline{P}$, and the second from the strong $\mathcal{T}$-invariance of $\underline{P}$. Similarly $\underline{P}(f - \sum_k \lambda_k T_k^t f) \geq 0$. Hence $\underline{P}$ is strongly $\mathcal{T}^*$-invariant. ∎

We now define the following binary relation $\geqslant$ on $\mathcal{T}^*$: for T_1^* and T_2^* in $\mathcal{T}^*$ we say that T_2^* *is a successor* of T_1^*, and we write $T_2^* \geqslant T_1^*$, if and only if there is some T^* in $\mathcal{T}^*$ such that $T_2^* = T^* T_1^*$. Clearly $\geqslant$ is a reflexive and transitive relation, because $\mathcal{T}^*$ is a monoid. We say that $\mathcal{T}^*$ has the *Moore–Smith property*, or is *directed by* $\geqslant$, if any two elements of $\mathcal{T}^*$ have a common successor, i.e., for any T_1^* and T_2^* in $\mathcal{T}^*$ there is some T^* in $\mathcal{T}^*$ such that $T^* \geqslant T_1^*$ and $T^* \geqslant T_2^*$. It is not difficult to see that if $\mathcal{T}$ is Abelian, or a finite group, then $\mathcal{T}^*$ is directed by the successor relation. This need not hold if $\mathcal{T}$ is an infinite group or a finite monoid, however.

Now, given a *net* α on $\mathcal{T}^*$, i.e., a mapping $\alpha \colon \mathcal{T}^* \to \mathbb{R}$, we can take the *Moore–Smith limit* of α with respect to the directed set $(\mathcal{T}^*, \geqslant)$ [Moore and Smith, 1922, Section I, p. 103], which, if it exists, is uniquely defined as the real number a such that, for every $\epsilon > 0$, there is a T_ϵ^* in $\mathcal{T}^*$, such that $|\alpha(T^*) - a| < \epsilon$ for all $T^* \geqslant T_\epsilon^*$. The Moore–Smith limit a of α is denoted by $\lim_{T^* \in \mathcal{T}^*} \alpha(T^*)$. This limit always exists if α is non-decreasing and bounded from above, or if α is non-increasing and bounded from below.

THEOREM 18. *Let $\underline{P}$ be a coherent and weakly $\mathcal{T}$-invariant lower prevision on $\mathcal{L}(\mathcal{X})$, and assume that $\mathcal{T}^*$ has the Moore–Smith property. Then for any gamble f on $\mathcal{X}$ the Moore–Smith limit $\lim_{T^* \in \mathcal{T}^*} \underline{P}(T^* f)$ converges to a real number $\underline{Q}_{\underline{P},\mathcal{T}}(f)$. Moreover, $\underline{Q}_{\underline{P},\mathcal{T}}$ is the point-wise smallest strongly $\mathcal{T}$-invariant coherent lower prevision on $\mathcal{L}(\mathcal{X})$ that dominates $\underline{P}$ on $\mathcal{L}(\mathcal{X})$, and*

$$(7.6) \quad \begin{aligned} \underline{Q}_{\underline{P},\mathcal{T}}(f) &= \sup\{\underline{P}(T^* f) : T^* \in \mathcal{T}^*\} \\ &= \sup\left\{\underline{P}\left(\tfrac{1}{n} \sum_{k=1}^n T_k^t f\right) : n \geq 1, T_1, \ldots, T_n \in \mathcal{T}\right\}. \end{aligned}$$

Proof. First, fix f in $\mathcal{L}(\mathcal{X})$. Consider T_1^* and T_2^* in $\mathcal{T}^*$, and assume that $T_2^* \geqslant T_1^*$. This means that there is some T^* in $\mathcal{T}^*$ such that $T_2^* = T^* T_1^*$, and consequently we find that $\underline{P}(T_2^* f) = \underline{P}(T^*(T_1^* f)) \geq \underline{P}(T_1^* f)$, where the inequality follows from the fact that $\underline{P}$ is in particular weakly

T^*-invariant [observe that $\mathcal{L}(\mathcal{X})$ is convex and that $\underline{P}$ is weakly $\mathcal{T}$-invariant, and apply Proposition 17]. This means that the net $\underline{P}(T^*f)$, $T^* \in \mathcal{T}^*$ is non-decreasing. Since this net is moreover bounded from above [by $\sup f$, since $\underline{P}$ is coherent], it converges to a real number $\underline{Q}_{\underline{P},\mathcal{T}}(f)$, and clearly

$$(7.7) \quad \underline{Q}_{\underline{P},\mathcal{T}}(f) = \lim_{T^* \in \mathcal{T}^*} \underline{P}(T^*f) = \sup\{\underline{P}(T^*f) \colon T^* \in \mathcal{T}^*\}.$$

This tells us that the net of coherent lower previsions $T^*\underline{P}$, $T^* \in \mathcal{T}^*$ converges point-wise to the lower prevision $\underline{Q}_{\underline{P},\mathcal{T}}$, so $\underline{Q}_{\underline{P},\mathcal{T}}$ is a coherent lower prevision as well [taking a point-wise limit preserves coherence]. Since $\mathrm{id}_\mathcal{X}^t \in \mathcal{T}^*$, it follows from Eq. (7.7) that $\underline{Q}_{\underline{P},\mathcal{T}}(f) \geq \underline{P}(\mathrm{id}_\mathcal{X}^t f) = \underline{P}(f)$, so $\underline{Q}_{\underline{P},\mathcal{T}}$ dominates $\underline{P}$ on $\mathcal{L}(\mathcal{X})$. We now show that $\underline{Q}_{\underline{P},\mathcal{T}}$ is strongly $\mathcal{T}$-invariant.[43] Consider any f in $\mathcal{L}(\mathcal{X})$ and T in $\mathcal{T}$. Then for any $n \geq 1$, $T_n^* := \frac{1}{n}\sum_{k=1}^n (T^k)^t$ belongs to $\mathcal{T}^*$, and it follows from the coherence of $\underline{P}$ that

$$\underline{P}(T_n^*(f - T^t f)) = \frac{1}{n}\underline{P}(T^t f - (T^{n+1})^t f) \geq \frac{1}{n}\inf\left[T^t f - (T^{n+1})^t f\right]$$

$$= -\frac{1}{n}\sup\left[(T^{n+1})^t f - T^t f\right] \geq -\frac{2}{n}\sup|f|,$$

and consequently $\underline{Q}_{\underline{P},\mathcal{T}}(f - T^t f) \geq \sup\{-\frac{2}{n}\sup|f| \colon n \geq 1\} = 0$. A similar argument can be given for $\underline{Q}_{\underline{P},\mathcal{T}}(T^t f - f) \geq 0$, so $\underline{Q}_{\underline{P},\mathcal{T}}$ is indeed strongly $\mathcal{T}$-invariant.

Next, consider any strongly $\mathcal{T}$-invariant and coherent lower prevision $\underline{Q}$ on $\mathcal{L}(\mathcal{X})$, and assume that it dominates $\underline{P}$. Then we get for any gamble f on $\mathcal{X}$ and any T^* in $\mathcal{T}^*$:

$$\underline{Q}(f) = \underline{Q}(f - T^*f + T^*f) \geq \underline{Q}(f - T^*f) + \underline{Q}(T^*f) \geq \underline{Q}(T^*f) \geq \underline{P}(T^*f),$$

where the first inequality follows from the coherence of $\underline{Q}$, the second inequality from its strong $\mathcal{T}$-invariance [use Proposition 17], and the last inequality from the fact that $\underline{Q}$ dominates $\underline{P}$. We then deduce from Eq. (7.7) that $\underline{Q}$ dominates $\underline{Q}_{\underline{P},\mathcal{T}}$. So $\underline{Q}_{\underline{P},\mathcal{T}}$ is indeed the point-wise smallest strongly $\mathcal{T}$-invariant coherent lower prevision on $\mathcal{L}(\mathcal{X})$ that dominates $\underline{P}$ on $\mathcal{L}(\mathcal{X})$.

Finally, let us prove the second equality in Eq. (7.6). Consider a gamble f and any $\epsilon > 0$. Then, by Eq. (7.7), there is some T^* in $\mathcal{T}^*$ such that $\underline{Q}_{\underline{P},\mathcal{T}}(f) \leq \underline{P}(T^*f) + \frac{\epsilon}{2}$. For this T^*, there are $n \geq 1$, $T_1, \ldots, T_n$ in $\mathcal{T}$ and $\lambda_1, \ldots, \lambda_n \geq 0$ that sum to one, such that $T^* = \sum_{k=1}^n \lambda_k T_k^t$. Let $\rho_1, \ldots, \rho_n$

---

[43] The idea for this part of the proof is due to Walley [1991, Point (iv) of the proof of Theorem 3.5.3].

be non-negative rational numbers satisfying $|\rho_i - \lambda_i| \leq \frac{\epsilon}{2n \sup|f|}$ such that moreover $\sum_{i=1}^n \rho_i = 1$.[44] Now it follows from the coherence of $\underline{P}$ that

$$\underline{P}(T^* f) = \underline{P}\left(\sum_{i=1}^n \lambda_i T_i^t f\right) \leq \underline{P}\left(\sum_{i=1}^n \rho_i T_i^t f\right) - \underline{P}\left(\sum_{i=1}^n (\rho_i - \lambda_i) T_i^t f\right),$$

and also

$$\underline{P}\left(\sum_{i=1}^n (\rho_i - \lambda_i) T_i^t f\right) \geq \sum_{i=1}^n \underline{P}((\rho_i - \lambda_i) T_i^t f) \geq \sum_{i=1}^n \inf (\rho_i - \lambda_i) T_i^t f$$

$$\geq \sum_{i=1}^n - \frac{\epsilon}{2n \sup|f|} \sup|f| = -\frac{\epsilon}{2},$$

whence

$$\underline{Q}_{\underline{P},\mathcal{T}}(f) \leq \underline{P}(T^* f) + \frac{\epsilon}{2} \leq \underline{P}\left(\sum_{i=1}^n \rho_i T_i^t f\right) + \epsilon,$$

and consequently

$$\underline{Q}_{\underline{P},\mathcal{T}}(f) = \sup\left\{\underline{P}\left(\sum_{i=1}^n \rho_i T_i^t f\right) : n \geq 1, T_k \in \mathcal{T}, \rho_k \in \mathbb{Q}^+, \sum_{i=1}^n \rho_i = 1\right\},$$

where $\mathbb{Q}^+$ denotes the set of non-negative rational numbers. It is easy to see [just consider the least common multiple of the denominators of $\rho_1, \ldots, \rho_n$] that this supremum coincides with the right-hand side of Eq. (7.6). ∎

This result allows us to establish the following corollary. It gives a sufficient condition for the existence of strongly $\mathcal{T}$-invariant lower previsions dominating a given coherent lower prevision $\underline{P}$. The smallest such lower prevision reflects how initial behavioral dispositions, reflected in $\underline{P}$, are modified (strengthened) to $\underline{E}_{\underline{P},\mathcal{T}}$ when we add the extra assessment of strong invariance with respect to a monoid $\mathcal{T}$ of transformations.

COROLLARY 19. *Consider a monoid $\mathcal{T}$ of transformations of $\mathcal{X}$ and let $\underline{P}$ be a weakly $\mathcal{T}$-invariant lower prevision on some set of gambles $\mathcal{K}$, that avoids sure loss. Assume that $\mathcal{T}^*$ has the Moore–Smith property. Then there*

---

[44]To see that such rational numbers exist, it suffices to consider non-negative rational numbers $\rho_1, \ldots, \rho_{n-1}$ such that $0 \leq \rho_i \leq \lambda_i \leq 1$ and $|\rho_i - \lambda_i| \leq \frac{\epsilon}{2n^2 \sup|f|}$ for $i = 1, \ldots, n-1$, and to let $\rho_n := 1 - \sum_{i=1}^{n-1} \rho_i \geq 1 - \sum_{i=1}^{n-1} \lambda_i = \lambda_n \geq 0$. Then $\rho_n \in [0,1]$, and for n big enough, and unless we are in the trivial case where $\lambda_i = 1$ for some i, we get $|\rho_n - \lambda_n| \leq \frac{\epsilon}{2n \sup|f|}$.

are strongly T-invariant coherent lower previsions on $\mathcal{L}(\mathcal{X})$ that dominate $\underline{P}$ on $\mathcal{L}(\mathcal{X})$, and the smallest such lower prevision, which is called the strongly T-invariant natural extension of $\underline{P}$, is given by $\underline{E}_{\underline{P},T} = \underline{Q}_{\underline{E}_{\underline{P}},T}$. Moreover, for every T-invariant gamble f we have that $\underline{E}_{\underline{P},T}(f) = \underline{E}_{\underline{P}}(f)$.

Proof. The first part of the proof follows at once from the observation that a coherent lower prevision $\underline{Q}$ on $\mathcal{L}(\mathcal{X})$ dominates $\underline{P}$ on $\mathcal{K}$ if and only if it dominates $\underline{E}_{\underline{P}}$ on all gambles. For the second part of the proof, simply observe that if f is a T-invariant gamble, then $T^* f = f$ and therefore $\underline{E}_{\underline{P}}(T^* f) = \underline{E}_{\underline{P}}(f)$ for all T^* in $\mathcal{T}^*$. ∎

Let us show in particular how this result applies when we consider the monoid $\mathcal{T}_T$ generated by a single transformation T:

COROLLARY 20. *Let T be a transformation of $\mathcal{X}$ and consider the Abelian monoid $\mathcal{T}_T = \{T^n : n \geq 0\}$. Then for any weakly T-invariant lower prevision $\underline{P}$ on some set of gambles $\mathcal{K}$ that avoids sure loss, there are strongly T-invariant coherent (lower) previsions on $\mathcal{L}(\mathcal{X})$ that dominate $\underline{P}$, and the point-wise smallest such lower prevision $\underline{E}_{\underline{P},T}$ is given by*

$$\underline{E}_{\underline{P},T}(f) = \lim_{n\to\infty} \underline{E}_{\underline{P}}\left(\frac{1}{n}\sum_{k=0}^{n-1}(T^k)^t f\right) = \sup_{n\geq 1}\underline{E}_{\underline{P}}\left(\frac{1}{n}\sum_{k=0}^{n-1}(T^k)^t f\right).$$

Proof. The existence of strongly T-invariant coherent (lower) previsions on $\mathcal{L}(\mathcal{X})$ that dominate $\underline{P}$ follows from Corollary 19, and the fact that for any Abelian monoid $\mathcal{T}$, $\mathcal{T}^*$ has the Moore–Smith property. It also follows from this corollary that for any gamble f on $\mathcal{X}$,

$$\underline{E}_{\underline{P},T}(f) = \sup\left\{\underline{E}_{\underline{P}}(T^* f) : T^* \in \mathcal{T}_T^*\right\} \geq \sup_{n\geq 1}\underline{E}_{\underline{P}}\left(\frac{1}{n}\sum_{k=0}^{n-1}(T^k)^t f\right).$$

To prove the converse inequality, fix any T^* in $\mathcal{T}_T^*$ and any gamble f on $\mathcal{X}$. Then there is some $N \geq 1$ and non-negative $\lambda_0, \ldots, \lambda_{N-1}$ that sum to one, such that $T^* = \sum_{k=0}^{N-1}\lambda_k(T^k)^t$. Consider the element $S_M^* = \frac{1}{M}\sum_{\ell=0}^{M-1}(T^\ell)^t$ of $\mathcal{T}^*$, where M is any natural number such that $M \geq N$. Observe that

$$S_M^* T^* = \sum_{\ell=0}^{M-1}\sum_{k=0}^{N-1}\frac{\lambda_k}{M}(T^{k+\ell})^t = \sum_{m=0}^{M+N-2}\mu_m(T^m)^t,$$

where we let, for $0 \leq m \leq M + N - 2$,

$$\mu_m := \sum_{k=0}^{N-1}\sum_{\ell=0}^{M-1}\frac{\lambda_k}{M}\delta_{m,k+\ell} = \begin{cases} \sum_{k=0}^{m}\frac{\lambda_k}{M} & \text{if } 0 \leq m \leq N-2 \\ \frac{1}{M} & \text{if } N-1 \leq m \leq M-1 \\ \sum_{k=m-M+1}^{N-1}\frac{\lambda_k}{M} & \text{if } M \leq m \leq M+N-2. \end{cases}$$

This tells us that $\mu_m = \frac{1}{M}$ for $N - 1 \leq m \leq M - 1$, and $0 \leq \mu_m \leq \frac{1}{M}$ for all other m. If we let $\delta_m := \mu_m - \frac{1}{N+M-1}$, it follows at once that

$$|\delta_m| \leq \begin{cases} \dfrac{N-1}{M(M+N-1)} & \text{if } N - 1 \leq m \leq M - 1 \\[2mm] \dfrac{1}{M+N-1} & \text{if } 0 \leq m \leq N - 2 \text{ or } M \leq m \leq M + N - 2 \end{cases}$$

Consequently, it follows from the weak $\mathcal{T}$-invariance and the coherence of $\underline{E}_P$ that

$$\underline{E}_P(T^* f)$$

$$\leq \underline{E}_P(S_M^* T^* f) = \underline{E}_P\left(S_{M+N-1}^* f + \sum_{m=0}^{M+N-2} \delta_m (T^m)^t f \right)$$

$$\leq \underline{E}_P(S_{M+N-1}^* f) + \sum_{m=0}^{M+N-2} |\delta_m| \sup|f|$$

$$\leq \underline{E}_P(S_{M+N-1}^* f) + \sup|f| \left[\frac{(N-1)(M-N+1)}{M(M+N-1)} + \frac{2N-2}{M+N-1} \right]$$

$$= \underline{E}_P(S_{M+N-1}^* f) + \sup|f| \frac{(N-1)(3M-N+1)}{M(M+N-1)}.$$

Recall that f and T^*, and therefore also N are fixed. Consider any $\epsilon > 0$, then there is some $M_\epsilon \geq N$ such that $\sup|f| \frac{(N-1)(3M-N+1)}{M(M+N-1)} < \epsilon$ for all $M \geq M_\epsilon$, whence

$$\underline{E}_P(T^* f) \leq \underline{E}_P(S_{M_\epsilon+N-1}^* f) + \epsilon \leq \sup_{n \geq 1} \underline{E}_P(S_n^* f) + \epsilon.$$

Since this holds for all $\epsilon > 0$, we get $\underline{E}_P(T^* f) \leq \sup_{n \geq 1} \underline{E}_P(S_n^* f)$. Taking the supremum over all T^* in $\mathcal{T}^*$ leads to the desired inequality. ∎

7.4 Mixture lower previsions

The condition established in Theorem 18 is fairly general, and guarantees for instance the existence of $\mathcal{T}$-invariant coherent previsions whenever the monoid $\mathcal{T}$ is Abelian, or a finite group. In case $\mathcal{T}^*$ is not directed, however, as may happen for instance for groups $\mathcal{T}$ that are not finite nor Abelian, there may still be $\mathcal{T}$-invariant coherent previsions, as we shall see in Example 7 below. So we see that the directedness of $\mathcal{T}^*$ is not a necessary condition for the existence of $\mathcal{T}$-invariant coherent previsions.

But consider a weakly $\mathcal{T}$-invariant lower prevision $\underline{P}$ defined on some domain $\mathcal{K}$, that avoids sure loss. Even if $\mathcal{T}^*$ is not directed,[45] we may still associate with $\underline{P}$ a lower prevision $\underline{Q}_{\underline{P},\mathcal{T}}$ on $\mathcal{L}(\mathcal{X})$ through Eq. (7.6):

$$\underline{Q}_{\underline{P},\mathcal{T}}(f) = \sup_{T^* \in \mathcal{T}^*} \underline{E}_{\underline{P}}(T^* f) = \sup \left\{ \underline{E}_{\underline{P}} \left(\frac{1}{n} \sum_{k=1}^{n} T_k^t f \right) : n \geq 1, T_k \in \mathcal{T} \right\},$$

where we have replaced the Moore–Smith limit by a supremum (with which it would coincide in case $\mathcal{T}^*$ were directed), and where $\underline{E}_{\underline{P}}$ is the natural extension of $\underline{P}$ to all gambles. We shall call this lower prevision the *mixture lower prevision* associated with the weakly invariant $\underline{P}$. The supremum in this expression is finite, since it is dominated by $\sup f$. This mixture lower prevision is not necessarily coherent, but it is still strongly $\mathcal{T}^*$-invariant.[46] Moreover, this mixture lower prevision dominates $\underline{E}_{\underline{P}}$, and therefore also $\underline{P}$ [observe that $\underline{E}_{\underline{P}}$ is weakly invariant because $\underline{P}$ is]; and if there are $\mathcal{T}$-invariant coherent previsions, it is dominated by the strongly $\mathcal{T}$-invariant natural extension $\underline{E}_{\underline{P},\mathcal{T}}$ of $\underline{P}$.[47] This shows that $\mathcal{M}(\underline{Q}_{\underline{P},\mathcal{T}}) = \mathcal{M}(\underline{E}_{\underline{P},\mathcal{T}})$, since all coherent previsions that dominate the strongly $\mathcal{T}$-invariant $\underline{Q}_{\underline{P},\mathcal{T}}$ are necessarily $\mathcal{T}$-invariant. And clearly then, if this mixture lower prevision is coherent, it coincides with the strongly invariant natural extension. So we see that the mixture lower prevision, even if it is not coherent, still allows us to characterize all $\mathcal{T}$-invariant coherent previsions. In particular, there are such invariant coherent previsions if and only if it avoids sure loss.

EXAMPLE 7 (DIRECTEDNESS IS NOT NECESSARY). Let us consider the space $\mathcal{X}_3 := \{1, 2, 3\}$, and let T_1 and T_2 be the transformations of $\mathcal{X}$ given by $T_1(1) = 1$, $T_1(2) = 2$, $T_1(3) = 2$ and $T_2(1) = 1$, $T_2(2) = 3$, $T_2(3) = 3$, respectively. Since $T_1 T_1 = T_1$, $T_2 T_2 = T_2$, $T_2 T_1 = T_2$ and $T_1 T_2 = T_1$, we deduce that the set of transformations $\mathcal{T} = \{\mathrm{id}_{\mathcal{X}}, T_1, T_2\}$ is a monoid. Let $P_{\{1\}}$ be the coherent prevision on $\mathcal{L}(\mathcal{X})$ given by $P_{\{1\}}(f) = f(1)$ for any gamble f, i.e., all of whose probability mass lies in 1. Then we have $P_{\{1\}}(f) = P_{\{1\}}(T_1^t f) = P_{\{1\}}(T_2^t f)$ for any gamble f, so $P_{\{1\}}$ is $\mathcal{T}$-invariant. Let us show that $\mathcal{T}^*$ does not have the Moore–Smith property.

---

[45]This is the general situation that Walley [1991, Section 3.5] considers, and he doesn't discuss the directedness of $\mathcal{T}^*$. He does consider the special case that $\mathcal{T}$ is Abelian for which he proves that the existence of invariant coherent previsions is guaranteed. The results in this section were first proven by him.

[46]Simply observe that the relevant part (near the end) of the proof of Theorem 18 is not based on the directedness of $\mathcal{T}^*$.

[47]To prove that the mixture lower prevision dominates $\underline{P}$, consider $T^* = \mathrm{id}_{\mathcal{X}}$ in its definition. To prove that it is dominated by the strongly invariant natural extension, take $f_k = f/n$ in the expression (7.3) for this natural extension.

Consider T_1^* and T_2^* in $\mathcal{T}^*$ given by $T_1^* = \lambda T_1^t + (1 - \lambda)T_2^t$ and $T_2^* = \mu T_1^t + (1 - \mu)T_2^t$, with $\lambda \neq \mu$. Let T^* be another element of $\mathcal{T}^*$, so there are non-negative α_1, α_2 and α_3 such that $\alpha_1 + \alpha_2 + \alpha_3 = 1$ and $T^* = \alpha_1 \operatorname{id}_\mathcal{X}^t + \alpha_2 T_1^t + \alpha_3 T_2^t$. Now

$$
\begin{aligned}
T^* T_1^* &= \alpha_1 \lambda \operatorname{id}_\mathcal{X}^t T_1^t + \alpha_1 (1 - \lambda) \operatorname{id}_\mathcal{X}^t T_2^t \\
&\quad + \alpha_2 \lambda T_1^t T_1^t + \alpha_2 (1 - \lambda) T_1^t T_2^t + \alpha_3 \lambda T_2^t T_1^t + \alpha_3 (1 - \lambda) T_2^t T_2^t \\
&= \alpha_1 \lambda T_1^t + \alpha_1 (1 - \lambda) T_2^t + \alpha_2 \lambda T_1^t \\
&\quad + \alpha_2 (1 - \lambda) T_2^t + \alpha_3 \lambda T_1^t + \alpha_3 (1 - \lambda) T_2^t \\
&= \lambda T_1^t + (1 - \lambda) T_2^t = T_1^*.
\end{aligned}
$$

Similarly, $T^* T_2^* = T_2^*$ for any $T^* \in \mathcal{T}^*$. This means that T_1^* is the only possible successor of T_1^*, and T_2^* is the only possible successor of T_2^*. Hence, $\mathcal{T}^*$ cannot have the Moore–Smith property. Nevertheless, there is a $\mathcal{T}$-invariant coherent prevision $P_{\{1\}}$.

Let us consider the vacuous, and therefore weakly $\mathcal{T}$-invariant and coherent, lower prevision $\underline{P}_{\mathcal{X}_3}$ on $\mathcal{L}(\mathcal{X}_3)$, and the mixture lower prevision $\underline{Q}_{\underline{P}_{\mathcal{X}_3}, \mathcal{T}}$ that corresponds with it. It is easy to show that for any gamble f, $\underline{Q}_{\underline{P}_{\mathcal{X}_3}, \mathcal{T}}(f) = \min\{f(1), \max\{f(2), f(3)\}\}$ and this lower prevision avoids sure loss, and is therefore strongly $\mathcal{T}$-invariant, but it is not coherent [it is not super-additive]. It is easy to see that $P_{\{1\}}$ is the only coherent prevision that dominates $\underline{Q}_{\underline{P}_{\mathcal{X}_3}, \mathcal{T}}$, and is therefore the only $\mathcal{T}$-invariant coherent prevision.

7.5 Invariance and Choquet integration

Until now, we have explored the relation between coherence and (weak or strong) invariance. To complete this section, we intend to explore this relation for the particular case of the n-monotone lower previsions and probabilities introduced near the end of Section 2.4.

Consider an n-monotone lower probability $\underline{P}$ defined on a lattice of events $\mathcal{K}$ containing $\emptyset$ and $\mathcal{X}$. Then its natural extension to all events coincides with its inner set function $\underline{P}_*$, given by $\underline{P}_*(A) = \sup\{\underline{P}(B) : B \in \mathcal{K}, B \subseteq A\}$. Furthermore, the natural extension to all gambles is given by the Choquet integral with respect to $\underline{P}_*$:

$$
\underline{E}_{\underline{P}}(f) = (C) \int_\mathcal{X} f \, \mathrm{d}\underline{P}_* := \inf f + (R) \int_{\inf f}^{\sup f} \underline{P}_*(\{x \in \mathcal{X} : f(x) \geq \alpha\}) \, \mathrm{d}\alpha
$$

for all gambles f on $\mathcal{X}$, where the integral on the right-hand side is a Riemann integral. This natural extension (and therefore also the inner set function) is still n-monotone (see De Cooman *et al.* [2005a; 2005b]). Since we have proven in Theorem 12 that natural extension preserves weak invariance, we can deduce that the inner set function of a n-monotone weakly invariant coherent lower probability, and the associated Choquet functional, are still weakly invariant, n-monotone and coherent. We now show that weak invariance of the inner set function and the associated Choquet integral is still guaranteed if the lower probability $\underline{P}$ is not coherent or 2-monotone, but only monotone. In what follows, it is important to remember that for a transformation T of $\mathcal{X}$ and a subset A of $\mathcal{X}$, $T^t I_A = I_{T^{-1}(A)}$.

PROPOSITION 21. *Let $\underline{P}$ be a weakly T-invariant monotone lower probability, defined on a T-invariant lattice of events $\mathcal{K}$ that contains $\emptyset$ and $\mathcal{X}$, and such that $\underline{P}(\emptyset) = 0$ and $\underline{P}(\mathcal{X}) = 1$. Then*

1. *the inner set function $\underline{P}_*$ of $\underline{P}$ is weakly T-invariant; and*

2. *the Choquet integral with respect to $\underline{P}_*$ is weakly T-invariant.*

Proof. To prove the first statement, consider any $A \subseteq \mathcal{X}$, and let $B \in \mathcal{K}$ be a any subset of A. Then for any T in $\mathcal{T}$, $T^{-1}(B) \in \mathcal{K}$ and $T^{-1}(B) = \{x \colon Tx \in B\} \subseteq \{x \colon Tx \in A\} = T^{-1}(A)$, whence $\underline{P}(B) \leq \underline{P}(T^{-1}(B)) \leq \underline{P}_*(T^{-1}(A))$, where the first inequality follows from the weak invariance of $\underline{P}$, and the second from the fact that $\underline{P}_*$ is monotone and coincides with $\underline{P}$ on its domain, because $\underline{P}$ is assumed to be monotone. Consequently $\underline{P}_*(A) = \sup_{B \in \mathcal{K}, B \subseteq A} \underline{P}(B) \leq \underline{P}_*(T^{-1}(A))$. Hence, $\underline{P}_*$ is also weakly T-invariant.

To prove the second statement, let f be any gamble on $\mathcal{X}$. Define, for any α in $\mathbb{R}$, the level set $f_\alpha := \{x \colon f(x) \geq \alpha\}$. Then by the first statement,

$$\underline{P}_*(f_\alpha) \leq \underline{P}_*(T^{-1}(f_\alpha)) = \underline{P}_*(\{x \colon Tx \in f_\alpha\})$$
$$= \underline{P}_*(\{x \colon f(Tx) \geq \alpha\}) = \underline{P}_*((T^t f)_\alpha).$$

Hence

$$(C) \int f \, d\underline{P}_* = \inf f + (R) \int_{\inf f}^{\sup f} \underline{P}_*(f_\alpha) \, d\alpha$$

$$\leq \inf f + (R) \int_{\inf f}^{\sup f} \underline{P}_*((T^t f)_\alpha) \, d\alpha = (C) \int T^t f \, d\underline{P}_*,$$

also taking into account for the last equality that $\underline{P}_*((T^t f)_\alpha) = 1$ for all α in $[\inf f, \inf T^t f)$, and that $\underline{P}_*(f_\alpha) = 0$ for all α in $(\sup T^t f, \sup f]$. ∎

As we said before, natural extension does not preserve strong invariance in general, and a simple example shows that this continues to hold in particular for n-monotone lower previsions: the unique coherent lower prevision defined on $\{\emptyset, \mathcal{X}\}$ is trivially completely monotone and strongly invariant with respect to any monoid of transformations $\mathcal{T}$, but its natural extension, the vacuous lower prevision $\underline{P}_{\mathcal{X}}$ (which is completely monotone), is not strongly $\mathcal{T}$-invariant unless in the trivial case that $\mathcal{T} = \{\mathrm{id}_{\mathcal{X}}\}$.

It is nonetheless interesting that if we restrict ourselves to coherent previsions (which constitute a particular instance of completely monotone lower previsions), natural extension from events to gambles does preserve strong invariance. This is a consequence of the following theorem.

THEOREM 22. *Let $\underline{P}$ be a coherent lower prevision on $\mathcal{L}(\mathcal{X})$ and let $\mathcal{T}$ be a monoid of transformations on $\mathcal{X}$. Then $\underline{P}$ is strongly $\mathcal{T}$-invariant if and only if any P in $\mathcal{M}(\underline{P})$, its restriction to events is (weakly) $\mathcal{T}$-invariant, in the sense that $P(T^{-1}(A)) = P(A)$ for all $A \subseteq \mathcal{X}$ and all $T \in \mathcal{T}$.*

Proof. We start with the direct implication. If $\underline{P}$ is strongly $\mathcal{T}$-invariant, then any P in $\mathcal{M}(\underline{P})$ is $\mathcal{T}$-invariant by Theorem 15. Hence, given $A \subseteq \mathcal{X}$ and $T \in \mathcal{T}$, we get $P(A) = P(T^{-1}(A))$.

Conversely, consider P in $\mathcal{M}(\underline{P})$. Recall that a coherent prevision on all events has only one coherent extension from all events to all gambles, namely its natural extension, or Choquet functional; see [De Cooman *et al.*, 2005b]. So for any gamble f on $\mathcal{X}$ and any T in $\mathcal{T}$, taking into account that P is assumed to be invariant on events, and that $T^{-1}(f_\alpha) = (T^t f)_\alpha$ [see the proof of Proposition 21], we get

$$P(f) = (C) \int_{\mathcal{X}} f \, dP = \inf f + (R) \int_{\inf f}^{\sup f} P(f_\alpha) \, d\alpha$$

$$= \inf f + (R) \int_{\inf f}^{\sup f} P((T^t f)_\alpha) \, d\alpha = (C) \int_{\mathcal{X}} T^t f \, dP = P(T^t f).$$

Hence, P is strongly $\mathcal{T}$-invariant and, applying Theorem 15, so is the lower envelope $\underline{P}$ of $\mathcal{M}(\underline{P})$. ∎

We see that, although the condition of strong invariance cannot be considered for lower probabilities, in the sense that $I_A - T^t I_A$ will not be in general the indicator of an event, it is still to some extent characterized by behavior on events. Moreover, we may deduce the following result.

COROLLARY 23. *Let $\underline{P}$ be a strongly $\mathcal{T}$-invariant lower prevision on a $\mathcal{T}$-invariant set of gambles $\mathcal{K}$ that includes all indicators of events. Assume that $\underline{P}$ avoids sure loss. Then its natural extension to all gambles* *is strongly $\mathcal{T}$-invariant, and coincides therefore with the strongly invariant natural extension of $\underline{P}$.*

Proof. Since $\underline{P}$ avoids sure loss, $\mathcal{M}(\underline{P})$ is non-empty. Since $\underline{P}$ is strongly invariant on a domain that includes all events, any element P of $\mathcal{M}(\underline{P})$ is (strongly) invariant on all events. Hence, by the previous theorem, P is also (strongly) invariant on all gambles, since a coherent prevision on all events has only one coherent extension from all events to all gambles (namely its natural extension, or Choquet functional). Therefore, the natural extension of $\underline{P}$ is a lower envelope of invariant coherent previsions, and is therefore strongly invariant. ∎

This result provides further insight into the existence problem for strongly invariant coherent lower previsions. The existence of strongly invariant coherent lower previsions on all gambles is equivalent to the existence of invariant coherent previsions on all gambles, which in turn is equivalent to the existence of invariant coherent previsions on all events (or in other words, invariant finitely additive probabilities). And it is the impossibility of satisfying invariance with finitely additive probabilities in some cases (for instance for the class $\mathcal{T}_{\mathcal{X}}$ of all transformations) that prevents the existence of coherent strongly invariant belief models.

We also infer that if the restriction Q of a coherent lower prevision $\underline{P}$ on $\mathcal{L}(\mathcal{X})$ to gambles of the type $I_A - T^t I_A$ and $T^t I_A - I_A$, involving only indicators of events, is strongly invariant, then $\underline{P}$ is strongly invariant on all of $\mathcal{L}(\mathcal{X})$: it will dominate the natural extension $\underline{E}_Q$ of Q, which is strongly invariant by Corollary 23, and consequently it will also be strongly invariant.

We can also deduce the following result. Recall that a linear lattice of gambles $\mathcal{K}$ is a set of gambles that is at once a lattice of gambles and a linear subspace of $\mathcal{L}(\mathcal{X})$. If in addition $\mathcal{K}$ contains all constant gambles, then for any coherent prevision P defined on $\mathcal{K}$, its natural extension to all gambles [Walley, 1991, Theorem 3.1.4] is given by the *inner extension* $P_*(f) := \sup\{P(g): g \in \mathcal{K}, g \leq f\}$. Let us denote by P^* the conjugate upper prevision of P_*.

COROLLARY 24. *Let $\mathcal{T}$ be a monoid of transformations of $\mathcal{X}$, and let $\underline{P}$ be a strongly $\mathcal{T}$-invariant lower prevision on a linear lattice of gambles $\mathcal{K}$ that contains all constant gambles. The natural extension $\underline{E}_P$ of $\underline{P}$ to all gambles is strongly $\mathcal{T}$-invariant if and only if for any coherent prevision P on $\mathcal{K}$ that dominates $\underline{P}$, we have $P_*(A \setminus T^{-1}(A)) = P^*(A \setminus T^{-1}(A)) = P_*(T^{-1}(A) \setminus A) = P^*(T^{-1}(A) \setminus A)$ for all $A \subseteq \mathcal{X}$ and all $T \in \mathcal{T}$.*

Proof. It follows from Walley [1991, Theorem 3.4.2] that $\underline{E}_P$ is the lower envelope of the coherent lower previsions P_*, where P is any coherent prevision on $\mathcal{K}$ that dominates $\underline{P}$ on $\mathcal{K}$. But then, clearly, $\underline{E}_P$ will be strongly $\mathcal{T}$-invariant if and only if all the P_* are. Consider any such P_*. By Theorem 22, P_* is strongly invariant if and only if for all $A \subseteq \mathcal{X}$ and $T \in \mathcal{T}$:

$$Q(A) = Q(T^{-1}(A)) \quad \text{for all } Q \text{ in } \mathcal{M}(P_*)$$

which is obviously equivalent to $P_*(I_A - T^t I_A) = P_*(T^t I_A - I_A) = 0$. Now observe that $I_A - T^t I_A = I_A - I_{T^{-1}(A)} = I_{A \setminus T^{-1}(A)} - I_{T^{-1}(A) \setminus A}$, and that the functions $I_{A \setminus T^{-1}(A)}$ and $-I_{T^{-1}(A) \setminus A}$ are comonotone. Since P is a coherent prevision on $\mathcal{K}$, it is completely monotone. Hence, its inner extension P_* is coherent and completely monotone on all gambles, and therefore comonotone additive [De Cooman et al., 2005b]. This means that $P_*(I_A - T^t I_A)$ is equal to

$$
\begin{aligned}
P_*(I_{A \setminus T^{-1}(A)} - I_{T^{-1}(A) \setminus A}) &= P_*(I_{A \setminus T^{-1}(A)}) + P_*(-I_{T^{-1}(A) \setminus A}) \\
&= P_*(I_{A \setminus T^{-1}(A)}) - P^*(I_{T^{-1}(A) \setminus A}) \\
&= P_*(A \setminus T^{-1}(A)) - P^*(T^{-1}(A) \setminus A)
\end{aligned}
$$

and similarly $P_*(T^t I_A - I_A) = P_*(T^{-1}(A) \setminus A) - P^*(A \setminus T^{-1}(A))$. The rest of the proof is now immediate. ∎

8   Shift-invariance and its generalizations

8.1   Strongly shift-invariant coherent lower previsions on $\mathcal{L}(\mathbb{N})$

Let us consider, as an example, the case of the shift-invariant, i.e., $\mathcal{T}_\theta$-invariant, coherent previsions on $\mathcal{L}(\mathbb{N})$. These are usually called *Banach limits* in the literature, see for instance, Bhaskara Rao and Bhaskara Rao [1983, Section 2.1.3] or Walley [1991, Sections 2.9.5 and 3.5.7]. We know from Corollary 19 that there are always Banach limits that dominate a given weakly shift-invariant lower prevision—so we know that there actually are Banach limits. Let us denote by $\mathbb{P}_\theta(\mathbb{N})$ the set of all Banach limits. We also know that a coherent lower prevision on $\mathcal{L}(\mathbb{N})$ is strongly shift-invariant if and only if it is a lower envelope of such Banach limits. The smallest strongly shift-invariant coherent lower prevision $\underline{E}_\theta$ on $\mathcal{L}(\mathbb{N})$ is the lower envelope of all Banach limits, and it is given by:[48]

$$(8.1) \quad \underline{E}_\theta(f) = \sup_{\substack{m_1,\dots m_n \geq 0 \\ n \geq 0}} \inf_{k \geq 0} \frac{1}{n} \sum_{\ell=1}^{n} f(k + m_\ell) = \lim_{n \to \infty} \inf_{k \geq 0} \frac{1}{n} \sum_{\ell=k}^{k+n-1} f(\ell),$$

---

[48]See also Walley [1991, Section 3.5.7]. The expression on the right hand side is not a limit inferior!

for any gamble f on $\mathbb{N}$ (or in other words, for any bounded sequence $f(n)_{n \in \mathbb{N}}$ of real numbers). The first equality follows from Corollary 19, and the second from Corollary 20. $\underline{E}_\theta(f)$ is obtained by taking the infimum sample mean of f over 'moving windows' of length n, and then letting the window length n go to infinity. Since this is the lower prevision on $\mathcal{L}(\mathbb{N})$ that can be derived *solely* using considerations of coherence and the evidence of shift-invariance, we believe that this $\underline{E}_\theta$ is a natural candidate for a '*uniform distribution*' on $\mathbb{N}$. It is the belief model to use if we only have evidence of shift-invariance, as all other strongly shift-invariant coherent lower previsions will point-wise dominate $\underline{E}_\theta$, and will therefore represent stronger behavioral dispositions than warranted by the mere evidence of shift-invariance.[49]

We could also sample f over the set $\{1, \dots, n\}$ leading to a coherent 'sampling' prevision

$$S_n(f) = \frac{1}{n} \sum_{\ell=0}^{n-1} f(\ell),$$

but the problem here is that for any given f the sequence of sampling averages $S_n(f)$ is not guaranteed to converge. Taking the limits inferior of such sequences (one for each gamble f), however, yields a coherent lower prevision[50] $\underline{S}_\theta$ given by

$$\underline{S}_\theta(f) = \liminf_{n \to \infty} S_n(f) = \liminf_{n \to \infty} \frac{1}{n} \sum_{\ell=0}^{n-1} f(\ell)$$

for any gamble f on $\mathbb{N}$. For any event $A \subseteq N$, or equivalently, any zero-one-valued sequence, we have that $S_n(A) = \frac{1}{n}|A \cap \{0, \dots, n-1\}|$ is the 'relative frequency' of ones in the sequence $I_A(n)$ and

$$\underline{S}_\theta(A) = \liminf_{n \to \infty} S_n(A) = \liminf_{n \to \infty} \frac{1}{n}|A \cap \{0, \dots, n-1\}|.$$

Let $\overline{S}_\theta$ denote the conjugate of $\underline{S}_\theta$, given by $\overline{S}_\theta(f) = \limsup_n S_n(f)$. Those events A for which $\underline{S}_\theta(A) = \overline{S}_\theta(A)$ have a 'limiting relative frequency' equal

---

[49] But this belief model has the important defect that, like the lower prevision $\underline{S}_\theta$ defined further on, it is not fully conglomerable; see Walley [1991, Section 6.6.7] and observe that the counterexample that Walley gives for $\underline{S}_\theta$, also applies to $\underline{E}_\theta$. Walley's remark there that his example shows that there are no (what we call) fully conglomerable (strongly) shift-invariant (lower) previsions that dominate $\underline{S}_\theta$, can be extended in a straightforward manner to $\underline{E}_\theta$ to show that *there are no fully conglomerable (strongly) shift-invariant (lower) previsions*.

[50] A limit inferior of a sequence of coherent lower previsions is always coherent, see Walley [1991, Corollary 2.6.7].

to this common value. It is not difficult to show that the coherent 'limiting relative frequency' lower prevision $\underline{S}_\theta$ is actually also strongly shift-invariant.[51] This implies that all the coherent previsions that dominate $\underline{S}_\theta$ are strongly shift-invariant. But it is easy to see (see Example 8 below) that $\underline{E}_\theta$ is strictly dominated by $\underline{S}_\theta$, so there are Banach limits that do not dominate $\underline{S}_\theta$.

PROPOSITION 25. *Let L be any Banach limit on $\mathcal{L}(\mathbb{N})$, let f be any gamble on $\mathbb{N}$. Then the following statements hold.*

1. $\liminf_{n\to\infty} f(n) \le \underline{E}_\theta(f) \le \underline{S}_\theta(f) \le \overline{S}_\theta(f) \le \overline{E}_\theta(f) \le \limsup_{n\to\infty} f(n)$

2. *If $\lim_{n\to\infty} f(n)$ exists, then*

$$\underline{E}_\theta(f) = \underline{S}_\theta(f) = \overline{E}_\theta(f) = \overline{S}_\theta(f) = L(f) = \lim_{n\to\infty} f(n).$$

3. *If f is θ^m-invariant (has period $m \ge 1$), then*

$$\underline{E}_\theta(f) = \underline{S}_\theta(f) = \overline{E}_\theta(f) = \overline{S}_\theta(f) = L(f) = \frac{1}{m}\sum_{r=1}^{m-1} f(r).$$

4. *If f is zero except in a finite number of elements of $\mathbb{N}$, then $\underline{E}_\theta(f) = \underline{S}_\theta(f) = \overline{E}_\theta(f) = \overline{S}_\theta(f) = L(f) = 0$. In particular, this holds for the indicator of any finite subset A of $\mathbb{N}$.*

Proof. We begin with the first statement. By conjugacy, we can concentrate on the lower previsions. We have already argued that $\underline{S}_\theta$ is a strongly shift-invariant coherent lower prevision, so $\underline{S}_\theta$ will dominate the smallest strongly shift-invariant coherent lower prevision $\underline{E}_\theta$. So it remains to prove that $\underline{E}_\theta$ dominates the limit inferior. Consider the first equality in Eq. (8.1). Fix the natural numbers $n \ge 1$, $m_1, \ldots m_n$. We can assume without loss of generality that the m_1 is the smallest of all the m_ℓ. Observe that

$$\inf_{k\ge0} \frac{1}{n}\sum_{\ell=1}^{n} f(k+m_\ell) \ge \inf_{k\ge0} \min_{\ell=1}^{n} f(k+m_\ell) = \min_{\ell=1}^{n} \inf_{k\ge m_\ell} f(k) = \inf_{k\ge m_1} f(k),$$

and therefore $\underline{E}_\theta(f) \ge \sup_{m_1\ge0} \inf_{k\ge m_1} f(k) = \liminf_{n\to\infty} f(n)$.

The second statement is an immediate consequence of the first, and the third follows easily from the definition of $\underline{E}_\theta$ and $\overline{E}_\theta$. Finally, the fourth statement follows at once from the second. ∎

---

[51]The following simple proof is due to Walley [1991, Section 3.5.7]. Observe that $S_n(\theta^t f - f) = [f(n) - f(0)]/n \to 0$ as $n \to \infty$, so $\underline{S}_\theta(\theta^t f - f) = \overline{S}_\theta(\theta^t f - f) = 0$.

EXAMPLE 8 (NOT ALL BANACH LIMITS DOMINATE $\underline{S}_\theta$). Consider the event

$$A = \left\{ n^2 + k \colon n \geq 1, k = 0, \ldots, n - 1 \right\}.$$

Then A has 'limiting relative frequency' $\underline{S}_\theta(A) = \overline{S}_\theta(A) = 1/2$, whereas $\underline{E}_\theta(A) = 0$ and $\overline{E}_\theta(A) = 1$. This shows that $\underline{S}_\theta$ strictly dominates $\underline{E}_\theta$, so not all Banach limits dominate $\underline{S}_\theta$.

Indeed, for the limiting relative frequency, consider the subsequence $S_{m^2-1}(A)$, $m \geq 2$ of $S_n(A)$. Then $S_{m^2-1}(A)$ is equal to

$$\frac{|A \cap \{0, \ldots, m^2 - 2\}|}{m^2 - 1} = \frac{1 + 2 + \cdots + m - 1}{m^2 - 1} = \frac{\frac{1}{2}m(m-1)}{m^2 - 1} = \frac{1}{2}\frac{m}{m+1},$$

so this subsequence converges to $\frac{1}{2}$. Now the 'integer intervals' $[m^2 - 1, (m+1)^2 - 1]$, $m \geq 1$ cover the set of all natural numbers, and as n varies over such an interval, $S_n(A)$ starts at $S_{m^2-1}(A) = \frac{1}{2}\frac{m}{m+1} < \frac{1}{2}$, increases to $S_{m^2+m}(A) = \frac{1}{2}\frac{m^2+m}{m^2+m} = \frac{1}{2}$, and then again decreases to $S_{(m+1)^2-1}(A) = \frac{1}{2}\frac{m+1}{m+2} < \frac{1}{2}$. Both the lower and upper bounds converge to $\frac{1}{2}$ as $m \to \infty$, and therefore the sequence $S_n(A)$ converges to $\frac{1}{2}$ as well.

To calculate $\underline{E}_\theta(A)$, we consider the second equality in Eq. (8.1). Fix $n \geq 1$ and let $k = n^2 + n$, then $k + n - 1 = (n+1)^2 - 2$, so

$$\frac{1}{n} \sum_{\ell=k}^{k+n-1} I_A(\ell) = \frac{1}{n} \sum_{\ell=n^2+n}^{(n+1)^2-2} I_A(\ell) = 0,$$

whence $\inf_{k \geq 0} \frac{1}{n} \sum_{\ell=k}^{k+n-1} I_A(\ell) = 0$ for all $n \geq 1$, and therefore $\underline{E}_\theta(A) = 0$. To calculate $\overline{E}_\theta(A)$, fix $n \geq 1$ and let $k = n^2$ then

$$\frac{1}{n} \sum_{\ell=k}^{k+n-1} I_A(\ell) = \frac{1}{n} \sum_{\ell=n^2}^{n^2+n-1} I_A(\ell) = 1,$$

whence $\sup_{k \geq 0} \frac{1}{n} \sum_{\ell=k}^{k+n-1} I_A(\ell) = 1$ for all $n \geq 1$, and therefore $\overline{E}_\theta(A) = 1$.

In an interesting paper, Kadane and O'Hagan [1995] study candidates for the 'uniform distribution' on $\mathbb{N}$. They consider, among others, all the finitely additive probabilities (or equivalently, all coherent previsions) that coincide with the limiting relative frequency on all events for which this limit exists. One could also consider as such candidates the coherent previsions that dominate the sampling lower prevision $\underline{S}_\theta$, which have the benefit of being strongly shift-invariant. But, we actually believe that *all* Banach limits (or actually, their lower envelope) are good candidates for being called 'uniform distributions on N' and not just the ones that dominate $\underline{S}_\theta$. Kadane and O'Hagan also propose to consider other coherent previsions, and their idea is to consider the 'residue sets', which are the subsets

$$R_m^r = \{km + r : k \geq 0\} = \{\ell \in \mathbb{N} : \ell = r \mod m\}$$

of $\mathbb{N}$, where $m \geq 1$ and $r = 1, \ldots, m - 1$. These sets are θ^m-invariant, so we already know from Proposition 25 that $\underline{E}_\theta(R_m^r) = \underline{S}_\theta(R_m^r) = \overline{S}_\theta(R_m^r) = \overline{E}_\theta(R_m^r) = \frac{1}{m}$ for all $m \geq 1$ and $r = 1, \ldots, m - 1$. Now what Kadane and O'Hagan do, is consider the set of all coherent previsions (finitely additive probabilities in their paper, but that is equivalent) that extend the probability assessments $P(R_m^r) = 1/m$ for all events R_m^r. In other words, they consider the natural extension $\underline{E}_{\text{res}}$ of all such assessments, i.e., the lower envelope of all such coherent previsions. It is not difficult to prove that this natural extension is given by[52]

$$\underline{E}_{\text{res}}(f) = \lim_{m \to \infty} \frac{1}{m} \sum_{r=0}^{m-1} \inf_{k \in \mathbb{N}} f(km + r).$$

This coherent lower prevision is completely monotone [as a point-wise limit of completely monotone lower previsions, even (natural extensions to gambles of so-called) belief functions [Shafer, 1976]], and weakly shift-invariant [since the natural extension of any weakly shift-invariant lower prevision is]. Since the assessments $P(R_m^r) = \frac{1}{m}$ coincide with the values given by $\underline{E}_\theta$, we see that $\underline{E}_\theta$ will point-wise dominate the natural extension $\underline{E}_{\text{res}}$ of these assessments to all gambles. But as we shall shortly prove in Example 9, $\underline{E}_{\text{res}}$ is not strongly shift-invariant, meaning that among the coherent previsions that extend these assessments, there also are coherent previsions that are not Banach limits (not shift-invariant).

EXAMPLE 9. Here we show by means of a counterexample that $\underline{E}_{\text{res}}$ is not strongly shift-invariant. Let $B_m := \{0, \ldots, m - 1\}$ and $A := \bigcup_{m \geq 1} \{m\} \times B_m$, and consider the map

$$\phi \colon A \to \mathbb{N} \colon (m, r) \mapsto \phi(m, r) := \frac{m(m - 1)}{2} + r + 1.$$

It is easy to see that ϕ is a bijection (one-to-one and onto). Also define the map

$$\kappa \colon A \to \mathbb{N} \colon (m, r) \mapsto \kappa(m, r) := Nm\phi(m, r) + r.$$

for some fixed $N \geq 2$. We consider the strict order $<$ on A induced by the bijection ϕ, i.e., $(m, r) < (m', r')$ if and only if $\phi(m, r) < \phi(m', r')$ [if and

---

[52] See De Cooman *et al.* [2006] for a proof.

only if $m < m$, or $m = m'$ and $r < r'$, so $<$ is the lexicographic order]. Then κ is an increasing map with respect to this order. To see this, assume that $(m, r) < (m', r')$. If $m < m'$, then

$$\kappa(m, r) = Nm\phi(m, r) + r < Nm\phi(m', 0) + r$$
$$< Nm'\phi(m', 0) + 0 \leq Nm'\phi(m', r') + r' = \kappa(m, r').$$

If on the other hand $m = m'$ and $r < r'$, then $\kappa(m, r) = Nm\phi(m, r) + r < Nm\phi(m, r') + r' = \kappa(m, r')$.

Moreover, given $(m, r) < (m', r')$, we see that $\kappa(m', r') - \kappa(m, r) \geq N$. Indeed, since κ is increasing, it suffices to prove this for consecutive pairs in the order $<$ we have defined on A. There are only two possible expressions of consecutive pairs (m, r) and (m', r'): either we have $(m', r') = (m, r+1)$, and then we get

$$\kappa(m, r+1) - \kappa(m, r) = Nm[\phi(m, r+1) - \phi(m, r)] + 1 = Nm + 1 \geq N;$$

or we have $r = m - 1, (m', r') = (m + 1, 0)$, and then we get

$$\kappa(m + 1, 0) - \kappa(m, m - 1)$$
$$= Nm[\phi(m + 1, 0) - \phi(m, m - 1)] + N\phi(m + 1, 0) - (m - 1)$$
$$= Nm + N\phi(m + 1, 0) - (m - 1) \geq Nm \geq N,$$

taking into account that $\phi(m + 1, 0) \geq m - 1$ by definition of ϕ.

Consider the set $C = \kappa(A)^c$. Then

$$\underline{E}_{\text{res}}(C) = \lim_{m \to \infty} \frac{1}{m} \sum_{r=0}^{m-1} \inf_{k \in \mathbb{N}} I_C(km + r).$$

Since for every $m \in \mathbb{N}$ and $r \in B_m$ the value $\kappa(m, r) = Nm\phi(m, r) + r$ does not belong to C, we deduce that $\frac{1}{m} \sum_{r=0}^{m-1} \inf_{k \in \mathbb{N}} I_C(km + r) = 0$ for all m, and consequently $\underline{E}_{\text{res}}(C) = 0$.

On the other hand, $\underline{E}_\theta(C) = \lim_{n \to \infty} \inf_{k \geq 0} \frac{1}{n} \sum_{\ell=k}^{k+n-1} I_C(\ell)$. Since by construction any two elements in $\kappa(A)$ differ in at least N elements, we deduce that $\inf_{k \geq 0} \frac{1}{n} \sum_{\ell=k}^{k+n-1} I_C(\ell) \geq 1 - \frac{2}{N+1}$, and this for all $n \in \mathbb{N}$. This implies that $\underline{E}_\theta(C) \geq 1 - \frac{2}{N+1} > 0$. Hence, $\underline{E}_{\text{res}}$ is strictly smaller than the smallest strongly shift-invariant natural extension $\underline{E}_\theta$, and therefore not strongly shift-invariant.

8.2   Strong T-invariance

Now consider an arbitrary non-empty set $\mathcal{X}$. Also consider a transformation T of $\mathcal{X}$ and the Abelian monoid $\mathcal{T}_T = \{T^n : n \geq 0\}$ generated by T. We shall characterize the strongly T-invariant coherent lower previsions on $\mathcal{L}(\mathcal{X})$ using the Banach limits on $\mathcal{L}(\mathbb{N})$.

First of all, consider any coherent lower prevision $\underline{P}$ on $\mathcal{L}(\mathcal{X})$, and any gamble f on $\mathcal{X}$. Define the gamble $f_{\underline{P}}$ on $\mathbb{N}$ as

$$(8.2) \quad f_{\underline{P}}(n) := \underline{P}((T^t)^n f) = \underline{P}(f \circ T^n).$$

[This is indeed a gamble, as for all n we deduce from the coherence of $\underline{P}$ that $f_{\underline{P}}(n) = \underline{P}(f \circ T^n) \leq \sup[f \circ T^n] \leq \sup f$ and similarly $f_{\underline{P}}(n) \geq \inf f$.] On the one hand $(T^t f)_{\underline{P}}(n) = \underline{P}(T^t f \circ T^n)) = \underline{P}(T^t(f \circ T^n)) = f_{T\underline{P}}(n)$ and on the other hand $(T^t f)_{\underline{P}}(n) = \underline{P}(f \circ T^{n+1}) = f_{\underline{P}}(n+1) = f_{\underline{P}}(\theta n)$, so

$$(8.3) \quad (T^t f)_{\underline{P}} = f_{T\underline{P}} = \theta^t f_{\underline{P}},$$

and this observation allows us to establish a link between the transformation T on $\mathcal{X}$ and the shift transformation θ on $\mathbb{N}$. This makes us think of the following trick, inspired by what Bhaskara Rao and Bhaskara Rao [1983, Section 2.1.3(9)] do for probability charges, rather than coherent lower previsions. Let L be any shift-invariant coherent prevision on $\mathcal{L}(\mathbb{N})$, or in other words, a Banach limit on $\mathcal{L}(\mathbb{N})$. Define the real-valued functional $\underline{P}_L$ on $\mathcal{L}(\mathcal{X})$ by $\underline{P}_L(f) := L(f_{\underline{P}})$. We show that this functional has very special properties.

PROPOSITION 26. *Let L be a shift-invariant coherent prevision on $\mathcal{L}(\mathbb{N})$, let $\underline{P}$ be a coherent lower prevision on $\mathcal{L}(\mathcal{X})$, and let T be a transformation of $\mathcal{X}$. Then the following statements hold.*

1. *$\underline{P}_L$ is a weakly T-invariant coherent lower prevision on $\mathcal{L}(\mathcal{X})$ (with equality).*

2. *If $\underline{P}$ dominates a weakly T-invariant coherent lower prevision $\underline{Q}$ on $\mathcal{L}(\mathcal{X})$, then $\underline{P}_L$ dominates $\underline{Q}$.*

3. *If $\underline{P} = P$ is a coherent prevision, then P_L is a (strongly) T-invariant coherent prevision on $\mathcal{L}(\mathcal{X})$.*

4. *If $\underline{Q}$ is a weakly T-invariant coherent lower prevision on $\mathcal{L}(\mathcal{X})$, then the (strongly) T-invariant coherent prevision P_L dominates $\underline{Q}$ for any P in $\mathcal{M}(\underline{Q})$.*

5. *If $\underline{P} = P$ is a T-invariant coherent prevision, then $P_L = P$.*

Proof. We first prove the first statement. Consider gambles f and g on $\mathcal{X}$. Since $\inf f \leq f_{\underline{P}}$, it follows from the coherence of L that $\inf f \leq L(f_{\underline{P}}) =$

$\underline{P}_L(f)$. Moreover, we have for any n in $\mathbb{N}$ that

$$(f+g)_{\underline{P}}(n) = \underline{P}((f+g) \circ T^n) = \underline{P}(f \circ T^n + g \circ T^n)$$
$$\geq \underline{P}(f \circ T^n) + \underline{P}(g \circ T^n) = f_{\underline{P}}(n) + g_{\underline{P}}(n),$$

where the inequality follows from the coherence [super-additivity] of $\underline{P}$. Since L is coherent, we see that $\underline{P}_L(f+g) \geq L(f_{\underline{P}}) + L(g_{\underline{P}}) = \underline{P}_L(f) + \underline{P}_L(g)$. Finally, for any $\lambda \geq 0$, we have that $(\lambda f)_{\underline{P}}(n) = \underline{P}((\lambda f) \circ T^n) = \underline{P}(\lambda(f \circ T^n)) = \lambda \underline{P}(f \circ T^n) = \lambda f_{\underline{P}}(n)$, since $\underline{P}$ is coherent. Consequently $\underline{P}_L(\lambda f) = L(\lambda f_{\underline{P}}) = \lambda L(f_{\underline{P}}) = \lambda \underline{P}_L(f)$, since L is coherent. This proves that $\underline{P}_L$ is a coherent lower prevision on $\mathcal{L}(\mathcal{X})$ [because (P1)–(P3) are satisfied]. To show that it is weakly T-invariant, recall that $(T^t f)_{\underline{P}} = \theta^t f_{\underline{P}}$, whence

$$\underline{P}_L(T^t f) = L((T^t f)_{\underline{P}}) = L(\theta^t f_{\underline{P}}) = L(f_{\underline{P}}) = \underline{P}_L(f),$$

since L is shift-invariant.

To prove the second statement, assume that $\underline{P}$ dominates the weakly T-invariant coherent lower prevision $\underline{Q}$ on $\mathcal{L}(\mathcal{X})$. Then for any gamble f on $\mathcal{X}$, we see that

$$f_{\underline{P}}(n) = \underline{P}(f \circ T^n) \geq \underline{Q}(f \circ T^n) \geq \underline{Q}(f),$$

where the last inequality follows from the weak T-invariance of $\underline{Q}$. Consequently, since L is coherent, we get $\underline{P}_L(f) = L(f_{\underline{P}}) \geq \underline{Q}(f)$.

The third statement follows immediately from the first and the fact that P_L is a self-conjugate coherent lower prevision (and therefore a coherent prevision) because P and L are.

The fourth statement follows at once from the second and the third. The fifth is an immediate consequence of the definition of P_L. ∎

We can use the results in this proposition to characterize all strongly T-invariant coherent lower previsions using Banach limits on $\mathcal{L}(\mathbb{N})$.

THEOREM 27. *Let $\underline{P}$ be a weakly T-invariant coherent lower prevision defined on some T-invariant domain $\mathcal{K}$, that avoids sure loss. Then the set of all T-invariant coherent previsions on $\mathcal{L}(\mathcal{X})$ that dominate $\underline{P}$ on $\mathcal{K}$ is given by*

$$\{P_L \colon P \in \mathcal{M}(\underline{P}) \text{ and } L \in \mathbb{P}_\theta(\mathbb{N})\},$$

so the smallest strongly T-invariant coherent lower prevision $\underline{E}_{\underline{P},T}$ on $\mathcal{L}(\mathcal{X})$ that dominates $\underline{P}$, i.e., the strongly T-invariant natural extension of $\underline{P}$, is the lower envelope of this set, and also given by

$$\underline{E}_{\underline{P},T}(f) = \inf_{P \in \mathcal{M}(\underline{P})} \underline{E}_\theta(f_P) = \inf_{P \in \mathcal{M}(\underline{P})} \sup_{n \geq 1} \inf_{k \geq 0} \left[\frac{1}{n} \sum_{\ell=k}^{k+n-1} P((T^\ell)^t f) \right]$$

for any gamble f on $\mathcal{X}$. As a consequence, the set $\mathbb{P}_T(\mathcal{X})$ of all T-invariant coherent previsions on $\mathcal{L}(\mathcal{X})$ is given by

$$\mathbb{P}_T(\mathcal{X}) = \{P_L \colon P \in \mathbb{P} \text{ and } L \in \mathbb{P}_\theta(\mathbb{N})\}.$$

This tells us that all T-invariant coherent previsions can be constructed using Banach limits on $\mathcal{L}(\mathbb{N})$. The smallest strongly T-invariant coherent lower prevision $\underline{E}_T$ on $\mathcal{L}(\mathcal{X})$ is the lower envelope of this set, and also given by

$$\underline{E}_T(f) = \inf_{P \in \mathbb{P}(\mathcal{X})} \underline{E}_\theta(f_P) = \inf_{P \in \mathbb{P}(\mathcal{X})} \sup_{n \geq 1} \inf_{k \geq 0} \left[\frac{1}{n} \sum_{\ell=k}^{k+n-1} P((T^\ell)^t f) \right]$$

for any gamble f on $\mathcal{X}$.

Proof. First of all, a coherent prevision P on $\mathcal{L}(\mathcal{X})$ belongs to $\mathcal{M}(\underline{P})$, i.e., dominates $\underline{P}$ on its domain $\mathcal{K}$, if and only if P dominates the natural extension $\underline{E}_P$ on all gambles. Moreover, $\underline{E}_P$ is weakly T-invariant by Theorem 12. Now consider any $P \in \mathcal{M}(\underline{P})$. Use the above observations together with Proposition 26 [statements 3 and 4] to show that for any Banach limit L on $\mathcal{L}(\mathbb{N})$, P_L is a T-invariant coherent prevision that dominates $\underline{P}$. Conversely, if P is a T-invariant coherent prevision on $\mathcal{L}(\mathcal{X})$ that dominates $\underline{P}$ on $\mathcal{K}$, then by Proposition 26 [statement 5], $P = P_L$ for any Banach limit L on $\mathcal{L}(\mathbb{N})$. This shows that $\{P_L \colon P \in \mathcal{M}(\underline{P}), L \in \mathbb{P}_\theta(\mathbb{N})\}$ is indeed the set of T-invariant coherent previsions on $\mathcal{L}(\mathcal{X})$ that dominate $\underline{P}$ on $\mathcal{K}$. Consequently, $\underline{E}_{\underline{P},T}$ is the lower envelope of this set, whence for any gamble f on $\mathcal{X}$

$$\underline{E}_{\underline{P},T}(f) = \inf_{P \in \mathcal{M}(\underline{P})} \inf_{L \in \mathbb{P}_\theta(\mathbb{N})} P_L(f) = \inf_{P \in \mathcal{M}(\underline{P})} \inf_{L \in \mathbb{P}_\theta(\mathbb{N})} L(f_P)$$

$$= \inf_{P \in \mathcal{M}(\underline{P})} \underline{E}_\theta(f_P) = \inf_{P \in \mathcal{M}(\underline{P})} \sup_{n \geq 1} \inf_{k \geq 0} \left[\frac{1}{n} \sum_{\ell=k}^{k+n-1} P((T^\ell)^t f) \right].$$

where the third equality follows since $\underline{E}_\theta$ is the lower envelope of $\mathbb{P}_\theta(\mathbb{N})$, and the last equality follows from Eqs. (8.1) and (8.2). The rest of the proof is now immediate. ∎

8.3 Generalized Banach limits

The above results on monoids $\mathcal{T}_T$ generated by a single transformation T can be generalized towards more general monoids $\mathcal{T}$ of transformations of $\mathcal{X}$, such that the set $\mathcal{T}^*$ of convex mixtures of the lifted linear transformations in $\mathcal{T}^t$ is directed by the successor relation $\geqslant$ on $\mathcal{T}^*$. The following discussion establishes an interesting connection between strong invariance and the notion of a generalized Banach limit.

We can consider T^* as a monoid of transformations of itself, as follows: with any element T^* we associate a transformation of T^*, also denoted by T^*, such that $T^*(S^*) := S^*T^* \in T^*$, for any S^* in T^*.[53] We can, in the usual fashion, lift T^* to a transformation $(T^*)^t$ on $\mathcal{L}(T^*)$ by letting $(T^*)^t g = g \circ T^*$, or in other words

(8.4) $(T^*)^t g(S^*) = g(T^*(S^*)) = g(S^*T^*),$

for any S^* in T^* and any gamble g on T^*, i.e., $g \in \mathcal{L}(T^*)$.

Now a *generalized Banach limit* [Schechter, 1997, Sections 12.33–12.38] on $\mathcal{L}(T^*)$ is defined as any linear functional on $\mathcal{L}(T^*)$ that dominates the limit inferior operator with respect to the directed set T^*. Let us take a closer look at this limit inferior operator. It is defined by

$$\liminf_{T^*} g = \liminf_{T^* \in T^*} g(T^*) := \sup_{S^* \in T^*} \inf_{T^* \geqslant S^*} g(T^*),$$

for any gamble g on T^*. Now recall that $T^* \geqslant S^*$ if and only if there is some R^* in T^* such that $T^* = R^*S^*$, so we get, using Eq. (8.4), that

$$\liminf_{T^* \in T^*} g(T^*) = \sup_{S^* \in T^*} \inf_{R^* \in T^*} g(R^*S^*)$$

$$= \sup_{S^* \in T^*} \inf_{R^* \in T^*} (S^*)^t g(R^*) = \lim_{S^* \in T^*} \underline{P}_{T^*}((S^*)^t g),$$

where $\underline{P}_{T^*}$ is the vacuous lower prevision on $\mathcal{L}(T^*)$. If we look at Corollary 19 for the special case $\mathcal{X} = T^*$ and the monoid of transformations T^*, recall that we need to lift transformations in T^* before we can apply them to gambles, and that the lifted transformations of T^* already constitute a convex set[54], we easily get to the following conclusion.

PROPOSITION 28. *The limit inferior operator on $\mathcal{L}(T^*)$ is the point-wise smallest strongly T^*-invariant coherent lower prevision, and the generalized Banach limits on $\mathcal{L}(T^*)$ are the T^*-invariant coherent previsions on $\mathcal{L}(T^*)$.*

We can now apply arguments similar to the ones in the previous section, for general monoids T of transformations of $\mathcal{X}$ such that T^* is directed.

---

[53]Usually, $T^*(S^*)$ is defined as T^*S^*, see for instance Walley [1991, Note 1 of Section 3.5.1]. But we have to take a different route here because the elements of T^* are convex mixtures of *lifted* transformations, and as we have seen, lifting reverses the order of application of transformations.

[54]In general, even if T^t is directed by the successor relation $\geqslant$, the limit inferior operator on $\mathcal{L}(T^t)$ will not be strongly invariant. But convexification, or going from T^t to T^*, makes the limit inferior strongly invariant. Observe in this respect that the limit inferior operator on $\mathcal{L}(\mathbb{N})$ is not strongly shift-invariant, but its 'convexified' counterpart $\underline{E}_\theta$ is.

Consider any coherent lower prevision $\underline{P}$ on $\mathcal{L}(\mathcal{X})$ and any gamble f, and define the following gamble $f_{\underline{P}}$ on $\mathcal{T}^*$:

$$f_{\underline{P}}(S^*) := \underline{P}(S^* f)$$

for any S^* in $\mathcal{T}^*$, which generalizes Eq. (8.2). Observe that, using Eq. (8.4),

$$(T^* f)_{\underline{P}}(S^*) = \underline{P}(S^* T^* f) = f_{\underline{P}}(S^* T^*) = (T^*)^t f_{\underline{P}}(S^*),$$

so

$$(T^* f)_{\underline{P}} = (T^*)^t f_{\underline{P}},$$

which generalizes Eq. (8.3). If we consider any $\mathcal{T}^*$-invariant coherent prevision L on $\mathcal{L}(\mathcal{T}^*)$, or in other words a generalized Banach limit on $\mathcal{L}(\mathcal{T}^*)$, we can now define a new lower prevision $\underline{P}_L$ on $\mathcal{L}(\mathcal{X})$ by $\underline{P}_L(f) := L(f_{\underline{P}})$, and Proposition 26, as well as Theorem 27, can now easily be generalized from monoids of transformations with a single generator to arbitrary directed monoids. In particular, we find that

$$\underline{E}_{\underline{P},\mathcal{T}}(f) = \inf_{P \in \mathcal{M}(\underline{P})} \liminf_{T^* \in \mathcal{T}^*} P(T^* f) \text{ and } \underline{E}_{\mathcal{T}}(f) = \inf_{P \in \mathbb{P}(\mathcal{X})} \liminf_{T^* \in \mathcal{T}^*} P(T^* f)$$

for any gamble f on $\mathcal{X}$, where $\underline{P}$ is any weakly $\mathcal{T}$-invariant lower prevision that avoids sure loss.

9   Permutation invariance on finite spaces

Assume now that $\mathcal{T}$ is a finite group $\mathcal{P}$ of permutations of $\mathcal{X}$. Then we have the following characterization result for the weakly $\mathcal{P}$-invariant coherent lower previsions.

THEOREM 29. *Let $\mathcal{P}$ be a finite group of permutations of $\mathcal{X}$. All weakly $\mathcal{P}$-invariant coherent lower previsions $\underline{Q}$ on $\mathcal{L}(\mathcal{X})$ have the form*

$$(9.1) \quad \underline{Q} = \frac{1}{|\mathcal{P}|} \sum_{\pi \in \mathcal{P}} \pi \underline{P},$$

where $|\mathcal{P}|$ is the number of permutations in $\mathcal{P}$, and $\underline{P}$ is any coherent lower prevision on $\mathcal{L}(\mathcal{X})$.

Proof. Consider a coherent lower prevision $\underline{P}$ on $\mathcal{L}(\mathcal{X})$, and let $\underline{Q}$ be the corresponding lower prevision, given by Eq. (9.1). Then $\underline{Q}$ is coherent, as a convex mixture of coherent lower previsions $\pi \underline{P}$. Moreover, let ϖ be any element of $\mathcal{P}$, then

$$\varpi \underline{Q} = \frac{1}{|\mathcal{P}|} \sum_{\pi \in \mathcal{P}} (\varpi \pi) \underline{P}, = \frac{1}{|\mathcal{P}|} \sum_{\pi \in \varpi \mathcal{P}} \pi \underline{P},$$

where $\varpi\mathcal{P} = \{\varpi\pi \colon \pi \in \mathcal{P}\} = \mathcal{P}$, because $\mathcal{P}$ is a group of permutations. Consequently $\varpi\underline{Q} = \underline{Q}$, so $\underline{Q}$ is weakly $\mathcal{P}$-invariant.

Conversely, let $\underline{Q}$ be any weakly $\mathcal{P}$-invariant coherent lower prevision, then we recover $\underline{Q}$ on the left-hand side if we insert $\underline{Q}$ in the right-hand side of Eq. (9.1). So any weakly $\mathcal{P}$-invariant coherent lower prevision is indeed of the form (9.1). ∎

Next, we give an interesting representation result for the strongly $\mathcal{P}$-invariant coherent lower previsions, when in addition, $\mathcal{X}$ is a finite set.[55] As we shall see further on, this essentially simple result has many interesting consequences, amongst which a generalization to coherent lower previsions of de Finetti's [1937] representation result for finite sequences of exchangeable random variables (see Section 9.2). Recall that $\mathcal{A}_\mathcal{P}$ is the set of all $\mathcal{P}$-invariant atoms of $\mathcal{X}$. For each A in $\mathcal{A}_\mathcal{P}$, define $P^u(\cdot|A)$ as the coherent prevision on $\mathcal{L}(\mathcal{X})$ all of whose probability mass is uniformly distributed over A, i.e., for all gambles f on $\mathcal{X}$:

$$P^u(f|A) = \frac{1}{|A|} \sum_{x \in A} f(x).$$

Finally, let $P^u(f|\mathcal{A}_\mathcal{P})$ denote the gamble on $\mathcal{A}_\mathcal{P}$ that assumes the value $P^u(f|\mathcal{A}_\mathcal{P})(A) := P^u(f|A)$ in any element A of $\mathcal{A}_\mathcal{P}$.

THEOREM 30. *Let $\mathcal{P}$ be a group of permutations of the finite set $\mathcal{X}$. A coherent lower prevision on $\mathcal{L}(\mathcal{X})$ is strongly $\mathcal{P}$-invariant if and only if $\underline{P}(f) = \underline{P}_0(P^u(f|\mathcal{A}_\mathcal{P}))$ for all f in $\mathcal{L}(\mathcal{X})$, where $\underline{P}_0$ is an arbitrary coherent lower prevision on $\mathcal{L}(\mathcal{A}_\mathcal{P})$.*

Proof. We begin with the 'if' part. Let $\underline{P}_0$ be an arbitrary coherent lower prevision on $\mathcal{L}(\mathcal{A}_\mathcal{P})$, and suppose that $\underline{P} = \underline{P}_0(P^u(\cdot|\mathcal{A}_\mathcal{P}))$. Then it is easy to see that $\underline{P}$ is coherent. We show that $\underline{P}$ is strongly $\mathcal{P}$-invariant. Consider any gamble f on $\mathcal{X}$ and any $\pi \in \mathcal{P}$. Then for any A in $\mathcal{A}_\mathcal{P}$ and any gamble f on $\mathcal{X}$,

$$P^u(f - \pi^t f|A) = \frac{1}{|A|} \sum_{x \in A} [f(x) - f(\pi x)] = 0,$$

because $x \in A$ is equivalent to $\pi x \in A$. So we see that $\underline{P}(f - \pi^t f) = \underline{P}_0(0) = 0$, since $\underline{P}_0$ is coherent. In a similar way, we can prove that $\underline{P}(\pi^t f - f) = 0$, so $\underline{P}$ is indeed strongly $\mathcal{P}$-invariant.

---

[55]We find the 'permutation symmetry' between Theorems 29 and 30 quite surprising: the former states that a weakly $\mathcal{P}$-invariant coherent lower prevision is a uniform prevision (or mixture) of coherent lower previsions, and the latter that a strongly $\mathcal{P}$-invariant coherent lower prevision is a coherent lower prevision of uniform previsions.

To prove the 'only if' part, we first concentrate on the case of a $\mathcal{P}$-invariant coherent prevision P on $\mathcal{L}(\mathcal{X})$. Fix any gamble f on $\mathcal{X}$. Since P is a coherent prevision, we find that

$$f = \sum_{A \in \mathcal{A}_{\mathcal{P}}} f I_A \quad \text{and} \quad P(f) = \sum_{A \in \mathcal{A}_{\mathcal{P}}} P(f I_A) = \sum_{A \in \mathcal{A}_{\mathcal{P}}} P(f|A)P(A),$$

where we have used Bayes's rule to define $P(f|A) := P(f I_A)/P(A)$ if $P(A) > 0$ and $P(f|A)$ is arbitrary otherwise.

Now assume that P is $\mathcal{P}$-invariant. Fix any $\mathcal{P}$-invariant atom A in $\mathcal{A}_{\mathcal{P}}$ such that $P(A) > 0$ and let $\pi \in \mathcal{P}$. For any gamble f on $\mathcal{X}$, we see that $\pi^t(f I_A) = (\pi^t f)I_A$, since A is in particular π-invariant. Consequently

$$P(\pi^t f|A) = P((\pi^t f)I_A)/P(A)$$
$$= P(\pi^t(f I_A))/P(A) = P(f I_A)/P(A) = P(f|A),$$

so $P(\cdot|A)$ is $\mathcal{P}$-invariant as well.[56] Now let for any y in the finite set A, $p(y|A) := P(\{y\}|A) \geq 0$, then on the one hand $\sum_{x \in A} p(x|A) = P(A|A) = 1$. On the other hand, it follows from the π-invariance of $P(\cdot|A)$ that $p(x|A) = p(\pi x|A)$ for any x in A. Since we know from Proposition 1 that $A = \{\pi x : \pi \in \mathcal{P}\}$, we see that $p(\cdot|A)$ is constant on A, so $p(x|A) = 1/|A|$ for all x in A, and consequently $P(f|A) = P^u(f|A)$, whence $P(f) = \sum_{A \in \mathcal{A}_{\mathcal{P}}} P^u(f|A)P(A)$. So indeed there is a coherent prevision P_0 on $\mathcal{L}(\mathcal{A}_{\mathcal{P}})$, defined by $P_0(\{A\}) = P(A)$ for all $A \in \mathcal{A}_{\mathcal{P}}$, such that $P = P_0(P^u(\cdot|\mathcal{A}_{\mathcal{P}}))$.

Finally, let $\underline{P}$ be any strongly $\mathcal{P}$-invariant coherent lower prevision. Then any P in $\mathcal{M}(\underline{P})$ is $\mathcal{P}$-invariant and can be written as $P = P_0(P^u(\cdot|\mathcal{A}_{\mathcal{P}}))$. If we let $\underline{P}_0$ be the (coherent) lower envelope of the set $\{P_0 : P \in \mathcal{M}(\underline{P})\}$, then since $\underline{P}$ is the lower envelope of $\mathcal{M}(\underline{P})$, we find that $\underline{P} = \underline{P}_0(P^u(\cdot|\mathcal{A}_{\mathcal{P}}))$. ∎

As an immediate corollary, we see that that the uniform coherent prevision P^u on $\mathcal{L}(\mathcal{X})$ is the only strongly $\mathcal{P}$-invariant coherent lower prevision on $\mathcal{L}(\mathcal{X})$ if and only if $\mathcal{X}$ is the only $\mathcal{P}$-invariant atom, i.e., if $\mathcal{A}_{\mathcal{P}} = \{\mathcal{X}\}$. This is for instance the case if $\mathcal{P}$ is the group of all permutations of $\mathcal{X}$, or more generally if $\mathcal{P}$ includes the cyclic group of permutations of $\mathcal{X}$. It should therefore come as no surprise that, since symmetry of beliefs is so often confused with beliefs of symmetry, the uniform distribution is so often (but wrongly so) considered to be a good model for complete ignorance.

Another immediate corollary is that the smallest strongly $\mathcal{P}$-invariant coherent lower prevision on $\mathcal{L}(\mathcal{X})$ is given by $\underline{P}(f) = \inf_{A \in \mathcal{A}_{\mathcal{P}}} \frac{1}{|A|} \sum_{x \in A} f(x)$,

---

[56]This is an instance of a more general result, namely that coherent conditioning of a coherent lower prevision on an invariant event preserves both weak and strong invariance. A proof of this statement is not difficult, but outside the scope of this paper.

which of course agrees with the uniform distribution when we let $\mathcal{P}$ be the group of all permutations.

These results do not extend to the case where we have transformations of $\mathcal{X}$ that are not permutations; as we have said before, as soon as we have two different constant transformations in the monoid $\mathcal{T}$, there are no strongly invariant belief models.

9.1   A few simple examples

We now apply the theorems above in a number of interesting and simple examples.

EXAMPLE 10. Let $\mathcal{X} = \mathcal{X}_2 := \{1, 2\}$, then all coherent lower previsions on $\mathcal{L}(\mathcal{X}_2)$ are so-called *linear-vacuous mixtures*, i.e., convex combinations of a coherent (linear) prevision and the vacuous lower prevision, and therefore given by

$$\underline{P}(f) = \epsilon\left[\alpha f(1) + (1 - \alpha)f(2)\right] + (1 - \epsilon)\min\{f(1), f(2)\},$$

where $0 \leq \alpha \leq 1$ and $0 \leq \epsilon \leq 1$. Let $\mathcal{P}_2$ be the set of all permutations of $\mathcal{X}_2$. Then the only strongly $\mathcal{P}_2$-invariant coherent lower prevision is the uniform coherent prevision

$$P_{\frac{1}{2}}(f) = \frac{1}{2}[f(1) + f(2)],$$

corresponding to $\alpha = \frac{1}{2}$ and $\epsilon = 1$. The weakly $\mathcal{P}_2$-invariant coherent lower previsions are given by

$$\underline{P}(f) = \epsilon P_{\frac{1}{2}}(f) + (1 - \epsilon)\min\{f(1), f(2)\},$$

where $0 \leq \epsilon \leq 1$, so they are all the convex mixtures of the uniform coherent prevision and the vacuous lower prevision.

EXAMPLE 11. Let $\mathcal{X} = \mathcal{X}_3 := \{1, 2, 3\}$, then all 2-monotone coherent lower previsions on $\mathcal{L}(\mathcal{X}_3)$ are given by[57]

$$
\begin{aligned}
\underline{P}(f) = {} & m_1 f(1) + m_2 f(2) + m_3 f(3) \\
& + m_4 \min\{f(1), f(2)\} + m_5 \min\{f(2), f(3)\} + m_6 \min\{f(3), f(1)\} \\
& + m_7 \min\left\{ \frac{f(1)+f(2)}{2}, \frac{f(2)+f(3)}{2}, \frac{f(3)+f(1)}{2} \right\} \\
& + m_8 \min\{f(1), f(2), f(3)\}.
\end{aligned}
$$

where $0 \le m_k \le 1$ and $\sum_{k=1}^{8} m_k = 1$. Let $\mathcal{P}_3$ be the set of all permutations of $\mathcal{X}_3$. Then the only strongly $\mathcal{P}_3$-invariant coherent lower prevision is the uniform coherent prevision

$$
P(f) = \frac{1}{3}[f(1) + f(2) + f(3)],
$$

corresponding to $m_1 = m_2 = m_3 = \frac{1}{3}$ and $m_4 = m_5 = m_6 = m_7 = m_8 = 0$ [Observe that a coherent prevision is always 2-monotone.]. Weak $\mathcal{P}_3$- invariance, on the other hand, requires only that $m_1 = m_2 = m_3$ and $m_4 = m_5 = m_6$, so all the weakly $\mathcal{P}_3$-invariant and 2-monotone coherent lower previsions are given by

$$
\begin{aligned}
\underline{P}(f) = {} & \frac{M_1}{3}[f(1) + f(2) + f(3)] \\
& + \frac{M_2}{3}\left[\min\{f(1), f(2)\} + \min\{f(2), f(3)\} + \min\{f(3), f(1)\}\right] \\
& + M_3 \min\left\{ \frac{f(1)+f(2)}{2}, \frac{f(2)+f(3)}{2}, \frac{f(3)+f(1)}{2} \right\} \\
& + M_4 \min\{f(1), f(2), f(3)\}.
\end{aligned}
$$

where $0 \le M_k \le 1$ and $M_1 + M_2 + M_3 + M_4 = 1$. The weakly $\mathcal{P}_3$-invariant and completely monotone coherent lower previsions (natural extensions of belief functions) correspond to the choice $M_3 = 0$.

---

[57] An explicit proof of this statement is beyond the scope of this paper, but it runs along the following lines: (i) any coherent lower probability on the set of all events of a three-element space is 2-monotone [Walley, 1981, p. 58]; (ii) all 2-monotone coherent lower probabilities make up a convex set, and are convex mixtures of the extreme points of this set [Maaß, 2003, Chapter 2] (By the way, an argument similar to that given by Maaß shows that all strongly $\mathcal{T}$-invariant coherent lower previsions are (infinite) convex mixtures of the extreme strongly $\mathcal{T}$-invariant coherent lower previsions.); (iii) the 2-monotone coherent lower previsions on all gambles are natural extensions of the 2-monotone coherent lower previsions on all events [Walley, 1981; De Cooman et al., 2006; De Cooman et al., 2005a; De Cooman et al., 2005b]; and (iv) natural extension to gambles of 2-monotone coherent lower probabilities preserves convex mixtures.

EXAMPLE 12. Consider rolling a die for which there is evidence of symmetry between all even numbers, on the one hand, and between all odd numbers on the other. Let $\mathcal{X} = \mathcal{X}_6 := \{1, \ldots, 6\}$ and let $\mathcal{P}_{eo}$ be the set of all permutations of $\mathcal{X}_6$ that map even numbers to even numbers and odd numbers to odd numbers. The $\mathcal{P}_{eo}$-invariant atoms are $\{1, 3, 5\}$ and $\{2, 4, 6\}$. By Theorem 30, the strongly $\mathcal{P}_{eo}$-invariant coherent previsions on $\mathcal{L}(\mathcal{X}_6)$, which are the precise models that are compatible with the subject's beliefs of symmetry, are given by

$$P(f) = \frac{\alpha}{3}[f(1) + f(3) + f(5)] + \frac{1-\alpha}{3}[f(2) + f(4) + f(6)],$$

where $0 \leq \alpha \leq 1$, and more generally, the strongly $\mathcal{P}_{eo}$-invariant coherent lower previsions on $\mathcal{L}(\mathcal{X}_6)$ are [apply Theorem 30 and use the results in Example 10]

$$\underline{P}(f) = \epsilon \left[\frac{\alpha}{3}[f(1) + f(3) + f(5)] + \frac{1-\alpha}{3}[f(2) + f(4) + f(6)] \right]$$
$$+ (1 - \epsilon) \min \left\{ \frac{f(1) + f(3) + f(5)}{3}, \frac{f(2) + f(4) + f(6)}{3} \right\}$$

for $0 \leq \epsilon \leq 1$ and $0 \leq \alpha \leq 1$.

EXAMPLE 13. Let us show that the point-wise smallest strongly invariant coherent lower prevision extension is not necessarily 2-monotone. Consider $\mathcal{X}_4 := \{1, 2, 3, 4\}$, and let π be the permutation of $\mathcal{X}_4$ defined by $\pi(1) = 2$, $\pi(2) = 1$, $\pi(3) = 4$ and $\pi(4) = 3$. Observe that π is its own inverse, so $\mathcal{T}_\pi = \{\mathrm{id}_{\mathcal{X}_4}, \pi\}$ is a group. From Theorem 30 we infer that the point-wise smallest strongly π-invariant coherent lower prevision on all gambles is given by

$$\underline{E}_\pi(f) = \min \left\{ \frac{f(1) + f(2)}{2}, \frac{f(3) + f(4)}{2} \right\}.$$

Let us now consider the gambles f_1 and f_2 on $\mathcal{X}_4$, given by $f_1(1) = 0$, $f_1(2) = -1$, $f_1(3) = 1$, $f_1(4) = -1$ and $f_2(1) = -1$, $f_2(2) = -0.25$, $f_2(3) = -1.5$, $f_2(4) = 0$. Check that

$$\underline{E}_\pi(f_1 \wedge f_2) + \underline{E}_\pi(f_1 \vee f_2) = -1.25 - 0.125 = -1.375$$
$$< -0.5 - 0.75 = \underline{E}_\pi(f_1) + \underline{E}_\pi(f_2).$$

Hence, $\underline{E}_\pi$ is not 2-monotone.

The following example shows that possibility measures are not very useful for modeling permutation invariance.

EXAMPLE 14. Consider a possibility measure Π defined on all events of a finite space $\mathcal{X}$. Then there is a map $\lambda \colon \mathcal{X} \to \mathbb{R}^+$, called the *possibility distribution* of Π, such that $\lambda(x) := \Pi(\{x\})$ and moreover $\Pi(A) = \max_{x \in A} \lambda(x)$ for all non-empty events $A \subseteq \mathcal{X}$. We have mentioned before that Π is a coherent upper probability if and only if $\Pi(\mathcal{X}) = \max_{x \in \mathcal{X}} \lambda(x) = 1$. We shall assume this is the case. Now consider any group $\mathcal{P}$ of permutations of $\mathcal{X}$. Then clearly Π is weakly $\mathcal{P}$-invariant if and only if λ is constant on the $\mathcal{P}$-invariant atoms of $\mathcal{X}$. In particular, Π is weakly invariant with respect to all permutations if and only is λ is everywhere equal to one, so Π is the vacuous upper probability.

For strong $\mathcal{P}$-invariance, let $\overline{P}$ be any strongly $\mathcal{P}$-invariant coherent lower prevision whose domain contains at least all events. Let x be any element of $\mathcal{X}$, and let $[x]_{\mathcal{P}}$ be the $\mathcal{P}$-invariant atom that contains x. Then it follows from Theorem 30 that $\overline{P}(\{x\}) \leq 1/|[x]_{\mathcal{P}}|$. So for $\overline{P}$ to extend a possibility measure, it is necessary (but not sufficient) that there is at least one element z of $\mathcal{X}$ such that $\overline{P}(\{z\}) = 1$, implying that z should be left invariant by all the permutations in $\mathcal{P}$, or equivalently, $[z]_{\mathcal{P}} = \{z\}$.

9.2   Exchangeable lower previsions

As another example, we now discuss the case of so-called exchangeable coherent lower previsions. Consider a non-empty finite set $\mathcal{X}_\kappa := \{1, \dots, \kappa\}$ of categories, and N random variables $X_1, \dots, X_N$ taking values in the same set $\mathcal{X}_\kappa$, where κ and N are natural numbers with $\kappa \geq 2$ and $N \geq 1$. The joint random variable $\mathbf{X} := (X_1, \dots, X_N)$ assumes values in the set $\mathcal{X} := \mathcal{X}_\kappa^N$.[58] We want to model a subject's beliefs about the value that $\mathbf{X}$ assumes in $\mathcal{X}_\kappa^N$, and generally, we use a coherent lower prevision $\underline{P}$ on $\mathcal{L}(\mathcal{X}_\kappa^N)$ to represent such beliefs.

Now assume that our subject believes that all random variables X_k are generated by the same process at different times k, and that the properties of this process do not depend on the time k. So, the subject assesses that there is permutation symmetry between the different times k. How can such *beliefs of symmetry* be modeled?

With a permutation π of $\{1, \dots, N\}$, we can associate (by the usual procedure of lifting) a permutation of $\mathcal{X} = \mathcal{X}_\kappa^N$, also denoted by π, that maps any $\mathbf{x} = (x_1, \dots, x_N)$ in $\mathcal{X}_\kappa^N$ to $\pi\mathbf{x} := (x_{\pi(1)}, \dots, x_{\pi(N)})$. The belief models that are compatible with the subject's beliefs of symmetry, are therefore the coherent lower previsions on (subsets of) $\mathcal{L}(\mathcal{X}_\kappa^N)$ that are *strongly* $\mathcal{P}_\kappa^N$-invariant, where $\mathcal{P}_\kappa^N$ is the group of liftings to $\mathcal{X}_\kappa^N$ of all permutations of $\{1, \dots, N\}$. Walley [1991, Chapter 9] calls such lower previsions *exchangeable*, as they generalize de Finetti's [1937] notion of exchangeable coherent

---

[58]This means that we assume these N random variables to be *logically independent*.

previsions. We intend to characterize the exchangeable lower previsions using Theorem 30. This will lead us to a generalization (Eq. (9.2)) of de Finetti's [1937] representation result for finite numbers of exchangeable random variables.

It should be mentioned here that we should, as always, clearly distinguish between 'beliefs of symmetry' and 'symmetry of beliefs'. The latter imposes much weaker requirements on coherent lower previsions, namely those of weak $\mathcal{P}_\kappa^N$-invariance, which is called *permutability* by Walley [1991, Chapter 9].[59] In particular, the permutation symmetry that goes along with ignorance can only be invoked to justify permutability, but not, of course, exchangeability. Observe in this respect that the vacuous lower prevision on $\mathcal{L}(\mathcal{X}_\kappa^N)$ is permutable, but not exchangeable. It is well-known (see for instance Zabell [1989b; 1992]), that Laplace's Rule of Succession can be obtained by updating a particular exchangeable coherent prevision, but it should be clear from the discussion in this paper that ignorance alone (the Principle of Insufficient Reason) cannot be invoked to justify using such an exchangeable prevision, as (with considerable hindsight) Laplace implicitly seems to have done (see for instance Howie [2002] and Zabell [1989b; 1992]).

For any $\mathbf{x} = (x_1, \ldots, x_N)$ in $\mathcal{X}_\kappa^N$, the $\mathcal{P}_\kappa^N$-invariant atom $[\mathbf{x}]_{\mathcal{P}_\kappa^N}$ is the set of all permutations of (the components of) $\mathbf{x}$. If we define the set of possible *count vectors*

$$\mathcal{N}_\kappa^N = \left\{ (m_1, \ldots, m_\kappa) \colon m_k \in \mathbb{N}^+ \text{ and } \sum_{k=1}^{\kappa} m_k = N \right\}$$

and the *counting map* $\mathbf{T} \colon \mathcal{X}_\kappa^N \to \mathcal{N}_\kappa^N$ such that $\mathbf{T}(x_1, \ldots, x_N)$ is the κ-tuple, whose k-th component is given by

$$T_k(x_1, \ldots, x_N) = |\{\ell \in \mathcal{X}_\kappa \colon x_\ell = k\}|,$$

i.e., the number of components of x whose value is k, then the number of elements of the invariant atom $[\mathbf{x}]_{\mathcal{P}_\kappa^N}$ is precisely

$$\nu(\mathbf{T}(\mathbf{x})) := \binom{N}{T_1(\mathbf{x}) \ldots T_\kappa(\mathbf{x})} = \frac{N!}{T_1(\mathbf{x})! \ldots T_\kappa(\mathbf{x})!}$$

and $\mathbf{T}$ is a bijection (one-to-one and onto) between $\mathcal{A}_{\mathcal{P}_\kappa^N}$ and $\mathcal{N}_\kappa^N$. An invariant atom is therefore completely identified by the count vector $\mathbf{T}(\mathbf{x})$ of any of its elements $\mathbf{x}$, and we shall henceforth denote the invariant atoms

---

[59] See Walley [1991, Chapter 9] for a much more detailed discussion of the difference between permutability and exchangeability.

of $\mathcal{X}_\kappa^N$ by $[\mathbf{m}]$, where $\mathbf{m} = (m_1, \ldots, m_\kappa) \in \mathcal{N}_\kappa^N$, and $\mathbf{x} \in [\mathbf{m}]$ if and only if $\mathbf{T}(\mathbf{x}) = \mathbf{m}$.

The coherent prevision $P^u(\cdot|\mathbf{m})$ on $\mathcal{L}(\mathcal{X}_\kappa^N)$ whose probability mass is uniformly distributed over the invariant atom $[\mathbf{m}]$ is given by

$$P^u(f|\mathbf{m}) = \frac{1}{\nu(\mathbf{m})} \sum_{\mathbf{x} \in [\mathbf{m}]} f(\mathbf{x}).$$

Interestingly, this is the precise prevision that is associated with taking N a-select drawings without replacement from an urn with N balls, m_1 of which are of type 1, ..., and m_κ of which are of type κ. Theorem 30 now tells us that any exchangeable coherent lower prevision $\underline{P}$ on $\mathcal{L}(\mathcal{X}_\kappa^N)$ can be written as

(9.2) $\underline{P}(f) = \underline{P}_\kappa^N(P^u(f|\mathcal{N}_\kappa^N)),$

where $\underline{P}_\kappa^N$ is some coherent lower prevision on $\mathcal{L}(\mathcal{N}_\kappa^N)$. This means that such *an exchangeable lower prevision can be associated with N a-select drawings from an urn with N balls of types 1, ..., κ, whose composition $\mathbf{m}$ is unknown, but for which the available information about the unknown composition is modeled by a coherent lower prevision $\underline{P}_\kappa^N$.*

That exchangeable coherent previsions can be interpreted in terms of sampling without replacement from an urn with unknown composition, is actually well-known, and essentially goes back to de Finetti [1937]. Heath and Sudderth [1976] give a simple proof for random variables that may assume two values. But we believe our proof[60] for the more general case of exchangeable coherent *lower* previsions and random variables that may assume *more than two values*, is conceptually even simpler than Heath and Sudderth's proof, even though it is a special case of a much more general representation result (Theorem 30). The essence of the present proof in the special case of coherent previsions P is captured wonderfully well by Zabell's [1992, Section 3.1] succinct statement: "Thus P is exchangeable if and only if two sequences having the same frequency vector have the same probability."

Our subject's beliefs could, in addition, be symmetrical in the categories in $\mathcal{X}_\kappa = \{1, \ldots, \kappa\}$, for instance as a result of her ignorance about the process that generates the outcomes X_k at each time k. As we have seen, this will be typically represented by using a type of *weakly* invariant belief model, in this case with respect to permutations of the categories, rather

---

[60]Walley [1991, Chapter 9] also mentions this result for exchangeable coherent lower previsions. The essence of his argument is similar to what we do in the last paragraph of the proof of Theorem 30.

than the times. Any permutation ϖ of $\mathcal{X}_\kappa$ induces a permutation of $\mathcal{X}_\kappa^N$, also denoted by ϖ, through

$$\varpi \mathbf{x} = \varpi(x_1, \ldots, x_N) := (\varpi(x_1), \ldots, \varpi(x_N)).$$

What happens if we require that $\underline{P}$, in addition to being exchangeable, should also be *weakly* invariant under all such permutations? It is not difficult to prove that

$$\underline{P}^u(\varpi^{-1} f | \mathbf{m}) = \underline{P}^u(f | \varpi \mathbf{m}),$$

where we let $\varpi \mathbf{m} = \varpi(m_1, \ldots, m_\kappa) := (m_{\varpi(1)}, \ldots, m_{\varpi(\kappa)})$ in the usual fashion. This implies that there is such weak invariance if and only if the coherent lower prevision $\underline{P}_\kappa^N$ on $\mathcal{L}(\mathcal{N}_\kappa^N)$ is *weakly* invariant with respect to all category permutations! In particular, this weak invariance is satisfied for the vacuous lower prevision on $\mathcal{L}(\mathcal{N}_\kappa^N)$. Another type of lower coherent prevision that exhibits such a combination of strong invariance for time permutations and weak invariance for category permutations, and which also has other very special and interesting properties, is constructed by taking lower envelopes of specific sets of Dirichlet-Multinomial distributions, leading to the so-called Imprecise Dirichlet-Multinomial Model (IDMM, see Walley and Bernard [1999]).

In the literature, however, it is sometimes required that a coherent precise prevision should be invariant with respect to the combined action of the permutations of times and categories. These are the so-called *partition exchangeable* previsions (see Zabell [1992] for an interesting discussion and historical overview). Of course, the generalization of this notion to coherent lower previsions should be strongly invariant with respect to such combined permutations, and therefore be a lower envelope of partition exchangeable previsions. For such *partition exchangeable* lower previsions, Theorem 30 can be invoked to prove a representation result that is similar to that for coherent lower previsions that are only exchangeable. It should be clear that they correspond to exchangeable lower previsions for which the corresponding coherent lower prevision $\underline{P}_\kappa^N$ on $\mathcal{L}(\mathcal{N}_\kappa^N)$ is *strongly* rather than just weakly invariant with respect to all category permutations. Of course, any justification for such models should be based on beliefs that there is permutation symmetry in the categories behind the process that generates the outcomes X_k at different times k, and *cannot be justified by mere ignorance about this process*.

9.3 Updating exchangeable lower previsions: predictive inference

Finally, let us discuss possible applications of the discussion in this paper to predictive inference. Assume that we have n^* random variables X_1,

$\ldots X_{n^*}$, that may assume values in the set $\mathcal{X}_\kappa = \{1, \ldots, \kappa\}$. We assume that these random variables are assessed to be exchangeable, in the sense that any coherent lower prevision that describes the available information about the values that the joint random variable $\mathbf{X}^* = (X_1, \ldots, X_{n^*})$ assumes in $\mathcal{X}_\kappa^{n^*}$ should be exchangeable, i.e., strongly $\mathcal{P}_\kappa^{n^*}$-invariant. This requirement could be called *pre-data exchangeability*. So we know from the previous section that such a coherent lower prevision must be of the form $\underline{P} = \underline{P}_\kappa^{n^*}(P^u(\cdot|\mathcal{N}_\kappa^{n^*}))$, where $\underline{P}_\kappa^{n^*}$ is some coherent lower prevision on $\mathcal{L}(\mathcal{N}_\kappa^{n^*})$. We shall assume that $\underline{P}_\kappa^{n^*}$ is a lower envelope of a set of coherent previsions $\mathcal{M}_\kappa^{n^*}$ on $\mathcal{L}(\mathcal{N}_\kappa^{n^*})$.

Suppose we now observe the values $\mathbf{x} = (x_1, \ldots, x_n)$ of the first n random variables $\mathbf{X} = (X_1, \ldots, X_n)$, where $1 \leq n < n^*$. We ask ourselves how we should coherently update the belief model $\underline{P}$ to a new belief model $\underline{P}(\cdot|\mathbf{x})$ which describes our beliefs about the values of the remaining random variables $\mathbf{X}' = (X_{n+1}, \ldots, X_{n^*})$. This is, generally speaking, the problem of *predictive inference*. In order to make things as easy as possible, we shall assume that $\underline{P}(\{\mathbf{x}\}) > 0$, so our subject has some reason, prior to observing $\mathbf{x}$, to believe that this observation will actually occur, because she is willing to bet on its occurrence at non-trivial odds.

Let us denote by $n' = n^* - n$ the number of remaining random variables, then we know that $\mathbf{X}'$ assumes values in $\mathcal{X}_\kappa^{n'}$, and $\underline{P}(\cdot|\mathbf{x})$ will be a lower prevision on $\mathcal{L}(\mathcal{X}_\kappa^{n'})$.

We shall first look at the problem of updating the coherent prevision $P = Q(P^u(\cdot|\mathcal{N}_\kappa^{n^*}))$ for any coherent prevision Q in $\mathcal{M}_\kappa^{n^*}$. So consider any gamble g on $\mathcal{X}_\kappa^{n'}$. It follows from coherence requirements (Bayes's rule) that the updated coherent prevision $P(\cdot|\mathbf{x})$ is given by

$$(9.3) \quad P(g|\mathbf{x}) = \frac{P(gI_\mathbf{x})}{P(I_\mathbf{x})} = \frac{Q(P^u(gI_\mathbf{x}|\mathcal{N}_\kappa^{n^*}))}{Q(P^u(I_\mathbf{x}|\mathcal{N}_\kappa^{n^*}))},$$

where $I_\mathbf{x}(\mathbf{x}^*) = 1$ if the first n components of the vector $\mathbf{x}^* \in \mathcal{X}_\kappa^{n^*}$ are given by the vector $\mathbf{x}$, and zero otherwise. Observe, by the way, that by assumption, $P(I_\mathbf{x}) \geq \underline{P}(I_\mathbf{x}) = \underline{P}(\{\mathbf{x}\}) > 0$.

Now for any $\mathbf{m}^*$ in $\mathcal{N}_\kappa^{n^*}$ we find that, with obvious notations,

$$(9.4) \quad P^u(gI_\mathbf{x}|\mathbf{m}^*) = \frac{1}{\nu(\mathbf{m}^*)} \sum_{\mathbf{T}'(\mathbf{x}')+\mathbf{m}=\mathbf{m}^*} g(\mathbf{x}') = \frac{\nu(\mathbf{m}^* - \mathbf{m})}{\nu(\mathbf{m}^*)} P^u(g|\mathbf{m}^* - \mathbf{m})$$

where we let $\mathbf{m} = \mathbf{T}(\mathbf{x})$, and where $\mathbf{T}'$ maps samples $\mathbf{x}'$ in $\mathcal{X}_\kappa^{n'}$ to their corresponding count vectors $\mathbf{T}'(\mathbf{x}')$ in $\mathcal{N}_\kappa^{n'}$. Of course $\nu(\mathbf{m}^* - \mathbf{m})$ is non-zero only if $\mathbf{m}^* \geq \mathbf{m}$, or equivalently if $\mathbf{m}^* - \mathbf{m} \in \mathcal{N}_\kappa^{n'}$, or in other words if it is possible to select n balls of composition $\mathbf{m}$ without replacement from an urn with composition $\mathbf{m}^*$. In this expression, $P^u(\cdot|\mathbf{m}')$ stands for the coherent prevision on $\mathcal{L}(\mathcal{X}_\kappa^{n'})$ whose probability mass is uniformly distributed over the $\mathcal{P}_\kappa^{n'}$-invariant atom $[\mathbf{m}']$, for any $\mathbf{m}'$ in $\mathcal{N}_\kappa^{n'}$. Now for $g = 1$ we find that

$$(9.5)\quad P^u(I_\mathbf{x}|\mathbf{m}^*) = \frac{\nu(\mathbf{m}^* - \mathbf{m})}{\nu(\mathbf{m}^*)} = p(\mathbf{m}|\mathbf{m}^*) =: L_\mathbf{m}(\mathbf{m}^*)$$

is the probability of observing a sample of size n with composition $\mathbf{m}$ by sampling without replacement from an urn with composition $\mathbf{m}^*$. $L_\mathbf{m}$ is the corresponding likelihood function on $\mathcal{N}_\kappa^{n^*}$. We may as well consider $L_\mathbf{m}$ as a likelihood function on $\mathcal{N}_\kappa^{n'}$, and for any $\mathbf{m}'$ in $\mathcal{N}_\kappa^{n'}$ we let

$$L_\mathbf{m}(\mathbf{m}') := L_\mathbf{m}(\mathbf{m} + \mathbf{m}') = \frac{\nu(\mathbf{m}')}{\nu(\mathbf{m} + \mathbf{m}')}$$

be the probability that there remain n' balls of composition $\mathbf{m}'$ after drawing (without replacement) n balls of composition $\mathbf{m}$ from an urn with n^* balls. We may then rewrite Eq. (9.3), using Eqs. (9.4) and (9.5), as

$$(9.6)\quad P(g|\mathbf{x}) = \frac{Q(L_\mathbf{m}P^u(g|\mathcal{N}_\kappa^{n'}))}{Q(L_\mathbf{m})} = Q(P^u(g|\mathcal{N}_\kappa^{n'})|\mathbf{m}),$$

where $Q(L_\mathbf{m}) = P(I_\mathbf{x}) > 0$ by assumption, and $Q(\cdot|\mathbf{m})$ is the coherent prevision on $\mathcal{L}(\mathcal{N}_\kappa^{n'})$ defined by

$$(9.7)\quad Q(h|\mathbf{m}) := \frac{Q(L_\mathbf{m}h)}{Q(L_\mathbf{m})},$$

for any gamble h on $\mathcal{N}_\kappa^{n'}$, i.e., $Q(\cdot|\mathbf{m})$ is the coherent prevision obtained after using Bayes's rule to update Q with the likelihood function $L_\mathbf{m}$. This means that *if Q is a belief model for the unknown composition of an urn with n^* balls, then $Q(\cdot|\mathbf{m})$ is the corresponding belief model for the unknown composition of the remaining n' balls in the urn, after n balls with composition $\mathbf{m}$ have been taken from it.*

Now if we have a coherent lower prevision $\underline{P}_\kappa^{n^*}$ on $\mathcal{L}(\mathcal{N}_\kappa^{n^*})$ that is a lower envelope of a set $\mathcal{M}_\kappa^{n^*}$ of coherent previsions Q, then coherence[61] tells us that the updated lower prevision $\underline{P}(\cdot|\mathbf{x})$ is precisely the lower envelope of the corresponding updated coherent previsions $P(\cdot|\mathbf{x})$, and consequently, using Eqs. (9.6) and (9.7), we find that

$$(9.8)\quad \underline{P}(g|\mathbf{x}) = \underline{P}_\kappa^{n^*}(P^u(g|\mathcal{N}_\kappa^{n'})|\mathbf{m}),$$

where $\underline{P}_\kappa^{n^*}(\cdot|\mathbf{m})$ is the coherent lower prevision on $\mathcal{L}(\mathcal{N}_\kappa^{n'})$ given by

---

[61]This follows from Walley's [1991, Section 6.5] Generalized Bayes Rule.

$$(9.9) \quad \underline{P}_\kappa^{n^*}(h|\mathbf{m}) := \inf\left\{\frac{Q(L_\mathbf{m} h)}{Q(L_\mathbf{m})} : Q \in \mathcal{M}_\kappa^{n^*}\right\} = \inf\left\{Q(h|\mathbf{m}) : Q \in \mathcal{M}_\kappa^{n^*}\right\},$$

for any gamble h on $\mathcal{N}_\kappa^{n'}$. In other words, $\underline{P}_\kappa^{n^*}(\cdot|\mathbf{m})$ is the coherent lower prevision obtained after using coherence (the so-called Generalized Bayes Rule) to update $\underline{P}_\kappa^{n^*}$ with the likelihood function $L_\mathbf{m}$. This means again that *if $\underline{P}_\kappa^{n^*}$ is a belief model for the unknown composition of an urn with n^* balls, then $\underline{P}_\kappa^{n^*}(\cdot|\mathbf{m})$ is the corresponding belief model for the unknown composition of the remaining n' balls in the urn, after n balls with composition $\mathbf{m}$ have been taken from it.*

If we compare Eq. (9.8) with Eq. (9.2), we see that the updated belief model $\underline{P}(\cdot|\mathbf{x})$ is still strongly $\mathcal{P}_\kappa^{n'}$-invariant,[62] so there still is *post-data exchangeability* for the remaining random variables $\mathbf{X}' = (X_{n+1}, \ldots, X_{n^*})$. Moreover, by looking at Eq. (9.3) and Eqs. (9.8) and (9.9), we see that the updated (lower) previsions $P(\cdot|\mathbf{x})$ and $\underline{P}(\cdot|\mathbf{x})$ only depend on the observed sample $\mathbf{x}$ through the *likelihood function* $L_{\mathbf{T}(\mathbf{x})}$. This tells us that this type of predictive inference satisfies the so-called *likelihood principle*, and moreover that the count vector $\mathbf{m} = \mathbf{T}(\mathbf{x})$, or more generally the map $\mathbf{T}$ is a *sufficient statistic*.

10   Conclusions

We have tried to argue that there is a clear distinction between the symmetry of belief models, and models of beliefs of symmetry, and that both notions can be distinguished between when indecision is taken seriously, as is the case in Walley's [1991] behavioral theory of imprecise probabilities. Our present attempt to distinguish between these notions, and capture the distinction in a formal way, is inspired by Walley's [1991, Chapter 9] discussion of the difference between permutable and exchangeable lower previsions, and Pericchi and Walley's [1991] discussion of 'classes of reasonable priors' versus 'reasonable classes of priors'.

Indeed, there seems to be a difference of type between the two notions. The former (symmetry of models) is a property that belief models may have, and we may require, as a principle of rationality, or as a principle of 'faithful modeling', that if the available evidence is symmetrical, then our corresponding belief models should be symmetrical too. A case in point is that of complete ignorance, where the 'evidence' is completely symmetrical, and we may therefore require that corresponding belief model should be completely symmetrical too. This leads to the various principles discussed in Section 5, all of which seem to single out the vacuous belief model for representing complete ignorance, and which extend Walley's [1991, Section 5.5]

---

[62]See also footnote 56.

treatment of this matter.

The latter notion (models of symmetry) is more properly related to a type of structural assessment: if a subject believes there is symmetry, how should she model that, and how should assessments of symmetry be combined with other assessments? We have tried to answer such questions in Sections 7, where we discuss the strongly invariant natural extension.

It is well-known that if we only use Bayesian, or precise, probability models, requiring invariance of the probability measures with respect to all types of symmetry in the evidence may be impossible; examples were given by Boole, Bertrand and Fisher (see Zabell [1989b] for discussion and references). This has led certain researchers to abandon requiring the above-mentioned 'faithfulness' of belief models, or to single out certain types of symmetry which are deemed to be better than others. We have tried to argue that this is unnecessary: the vacuous belief model has no such problems, and is symmetrical with respect to any transformation you care to name. And of course, our criticism of the Principle of Insufficient Reason is not new. Our ideas were heavily influenced by Walley's [1991] book on imprecise probabilities, whose Chapter 5 contains a wonderful overview of arguments against restricting ourselves to precise probability models. Zabell [1989a] also gives an excellent discussion of much older criticism, dating back to the middle of the 19th century. In particular, Ellis's *ex nihilo nihil* — you cannot make decisions or inferences based on complete ignorance, see Zabell [1989a] — finds a nice confirmation in the fact that the vacuous belief model captures complete indecision, and that updating a vacuous belief model leads to a vacuous belief model [Walley, 1991, Section 6.6.1]. But what we have tried to do here is provide a framework and mathematical apparatus that allows us to better understand and discuss the problems underlying the Principle of Insufficient Reason, and more general problems of dealing with any type of symmetry in belief models.

This study of symmetry in relation to belief models is far from being complete however, and our notions of weak and strong invariance may have to be refined, and perhaps even modified, as well as complemented by other notions of symmetry. It might for instance be of interest to study the notion of symmetry that captures the *insufficient reason to strictly prefer* that is briefly touched upon near the end of Section 4.1. Also, we may seem more certain than we actually are about the appropriateness (in terms of having a sound behavioral justification and interpretation) of our notions of weak and (especially) strong invariance for random variables that may assume an infinite number of values. This is the point where our intuition deserts us, and where a number of interesting questions and problems leave us speechless. To name but one such problem, brought to the fore by the discussion in Section 7: for certain types of monoids, it is completely irrational to impose strong invariance (because doing so makes us subject to a sure loss). We can understand why this is the case for the monoid of all transformations, even on a finite set (Theorem 9). But why, for instance, are there no (strongly) permutation invariant coherent (lower) previsions on the set of natural (and *a fortiori* real) numbers? Why are we (consequently) reduced to using (strong) shift or translation invariance of coherent (lower) previsions when we want to try and capture the idea of a uniform distribution on the set of natural (or real) numbers? And even then, why, as is hinted at in footnote 49, are there situations where updating a (strongly) shift-invariant coherent (lower) prevision produces a sure loss? Are there appropriately weakened versions of our strong invariance condition that avoid these problems?

Acknowledgments

This paper has been partially supported by research grant G.0139.01 of the Flemish Fund for Scientific Research (FWO), and projects MTM2004-01269, TSI2004-06801-C04-01.

BIBLIOGRAPHY

[Agnew and Morse, 1938] R. P. Agnew and A. P. Morse. Extensions of linear functionals, with applications to limits, integrals, measures and densities. *The Annals of Mathematics*, 39:20–30, 1938.

[Alaoglu and Birkhoff, 1940] L. Alaoglu and G. Birkhoff. General ergodic theorems. *The Annals of Mathematics*, 41:293–309, 1940.

[Bhaskara Rao and Bhaskara Rao, 1983] K. P. S. Bhaskara Rao and M. Bhaskara Rao. *Theory of Charges*. Academic Press, London, 1983.

[Day, 1942] M. M. Day. Ergodic theorems for Abelian semi-groups. *Transactions of the American Mathematical Society*, 51:399–412, 1942.

[De Cooman and Miranda, 2006] G. de Cooman and E. Miranda. Weak and strong laws of large numbers for coherent lower previsions. *Journal of Statistical Planning and Inference*, 2006. Submitted for publication.

[De Cooman and Troffaes, 2004] G. de Cooman and M. C. M. Troffaes. Coherent lower previsions in systems modelling: products and aggregation rules. *Reliability Engineering and System Safety*, 85:113–134, 2004.

[De Cooman and Zaffalon, 2004] G. de Cooman and M. Zaffalon. Updating beliefs with incomplete observations. *Artificial Intelligence*, 159:75–125, 2004.

[De Cooman et al., 2005a] G. de Cooman, M. C. M. Troffaes, and E. Miranda. *n*-Monotone lower previsions. *Journal of Intelligent and Fuzzy Systems*, 16:253–263, 2005.

[De Cooman et al., 2005b] G. de Cooman, M. C. M. Troffaes, and E. Miranda. *n*-Monotone lower previsions and lower integrals. In F. G. Cozman, R. Nau, and T. Seidenfeld, editors, *ISIPTA 2005 – Proceedings of the Fourth International Symposium on Imprecise Probabilities and Their Applications*, pages 145–154. SIPTA, 2005.

[De Cooman et al., 2006] G. de Cooman, M. C. M. Troffaes, and E. Miranda. *n*-Monotone exact functionals and their relation to lower and upper integrals. 2006. Submitted for publication.

[De Cooman, 2000] G. de Cooman. Belief models: an order-theoretic analysis. In G. de Cooman, T. L. Fine, and T. Seidenfeld, editors, *ISIPTA '01 – Proceedings of the Second International Symposium on Imprecise Probabilities and Their Applications*, pages 93–103. Shaker Publishing, Maastricht, 2000.

[De Cooman, 2001] G. de Cooman. Integration and conditioning in numerical possibility theory. *Annals of Mathematics and Artificial Intelligence*, 32:87–123, 2001.

[De Cooman, 2005] G. de Cooman. Belief models: an order-theoretic investigation. *Annals of Mathematics and Artificial Intelligence*, 45:5–34, 2005.

[de Finetti, 1937] B. de Finetti. La prévision: ses lois logiques, ses sources subjectives. *Annales de l'Institut Henri Poincaré*, 7:1–68, 1937. English translation in [Kyburg Jr. and Smokler, 1964].

[de Finetti, 1970] B. de Finetti. *Teoria delle Probabilità*. Einaudi, Turin, 1970.

[de Finetti, 1974–1975] B. de Finetti. *Theory of Probability*. John Wiley & Sons, Chichester, 1974–1975. English translation of [de Finetti, 1970], two volumes.

[Heath and Sudderth, 1976] D. C. Heath and W. D. Sudderth. De Finetti's theorem on exchangeable variables. *The American Statistician*, 30:188–189, 1976.

[Howie, 2002] D. Howie. *Interpreting Probability: Controversies and Developments in the Early Twentieth Century*. Cambridge Studies in Probability, Induction, and Decision Theory. Cambridge University Press, Cambridge, UK, 2002.

[Kadane and O'Hagan, 1995] J. B. Kadane and A. O'Hagan. Using finitely additive probability: uniform distributions on the natural numbers. *Journal of the American Statistical Association*, 90:636–631, 1995.

[Keynes, 1921] J. M. Keynes. *A Treatise on Probability*. Macmillan, London, 1921.

[Kyburg Jr. and Smokler, 1964] H. E. Kyburg Jr. and H. E. Smokler, editors. *Studies in Subjective Probability*. Wiley, New York, 1964. Second edition (with new material) 1980.

[Maaß, 2003] S. Maaß. *Exact functionals, functionals preserving linear inequalities, Lévy's metric*. PhD thesis, University of Bremen, 2003.

[Moore and Smith, 1922] E. H. Moore and H. L. Smith. A general theory of limits. *American Journal of Mathematics*, 44:102–121, 1922.

[Moral and Wilson, 1995] S. Moral and N. Wilson. Revision rules for convex sets of probabilities. In G. Coletti, D. Dubois, and R. Scozzafava, editors, *Mathematical Models for Handling Partial Knowledge in Artificial Intelligence*, pages 113–128. Plenum Press, New York, 1995.

[Pericchi and Walley, 1991] L. R. Pericchi and P. Walley. Robust Bayesian credible intervals and prior ignorance. *International Statistical Review*, 59:1–23, 1991.

[Schechter, 1997] E. Schechter. *Handbook of Analysis and Its Foundations*. Academic Press, San Diego, CA, 1997.

[Shafer, 1976] G. Shafer. *A Mathematical Theory of Evidence*. Princeton University Press, Princeton, NJ, 1976.

[Walley and Bernard, 1999] P. Walley and J.-M. Bernard. Imprecise probabilistic prediction for categorical data. Technical Report CAF-9901, Laboratoire Cognition et Activitées Finalisés, Université de Paris 8, January 1999.

[Walley, 1981] P. Walley. Coherent lower (and upper) probabilities. Statistics Research Report 22, University of Warwick, Coventry, 1981.

[Walley, 1991] P. Walley. *Statistical Reasoning with Imprecise Probabilities*. Chapman and Hall, London, 1991.

[Walley, 1996a] P. Walley. Inferences from multinomial data: learning about a bag of marbles. *Journal of the Royal Statistical Society, Series B*, 58:3–57, 1996. With discussion.

[Walley, 1996b] P. Walley. Measures of uncertainty in expert systems. *Artificial Intelligence*, 83:1–58, 1996.

[Walley, 2000] P. Walley. Towards a unified theory of imprecise probability. *International Journal of Approximate Reasoning*, 24:125–148, 2000.

[Zabell, 1989a] S. L. Zabell. R. A. Fisher on the history of inverse probability. *Statistical Science*, 4:247–256, 1989. Reprinted in [Zabell, 2005].

[Zabell, 1989b] S. L. Zabell. The Rule of Succession. *Erkenntnis*, 31:283–321, 1989. Reprinted in [Zabell, 2005].

[Zabell, 1992] S. L. Zabell. Predicting the unpredictable. *Synthese*, 90:205–232, 1992. Reprinted in [Zabell, 2005].

[Zabell, 2005] S. L. Zabell. *Symmetry and Its Discontents: Essays on the History of Inductive Probability*. Cambridge Studies in Probability, Induction, and Decision Theory. Cambridge University Press, Cambridge, UK, 2005.

[Zadeh, 1978] L. A. Zadeh. Fuzzy sets as a basis for a theory of possibility. *Fuzzy Sets and Systems*, 1:3–28, 1978.

Motivating Objective Bayesianism: From Empirical Constraints to Objective Probabilities

JON WILLIAMSON

ABSTRACT. Kyburg goes half-way towards objective Bayesianism. He accepts that frequencies constrain rational belief to an interval but stops short of isolating an optimal degree of belief within this interval. I examine the case for going the whole hog.

1 Partial Beliefs

Bayesians argue that an agent's degrees of belief ought to satisfy the axioms of the probability calculus. Thus for example if A is the outcome that it will snow in Stroud today, and $p(A)$ is the agent's degree of belief in A, then $p(A) + p(\bar{A}) = 1$, where $\bar{A}$ is the complement of A, i.e. the outcome that it will not snow in Stroud today.

But Bayesians differ as to whether degrees of belief should satisfy any further constraints.[1] Suppose our agents know only that the physical probability (frequency or propensity) of it snowing in Stroud on a day like today is between 0.2 and 0.3. Then the three main views can be formulated thus:

Subjective Bayesianism maintains that an agent can set her degree of belief to any value between 0 and 1. Thus one agent can choose degree of belief $p(A) = 0$, another can choose $q(A) = 0.25$ and a third can choose $r(A) = 1$ — all are equally rational.[2]

Empirically-Based Subjective Bayesianism maintains that an agent's degrees of belief ought to be constrained by empirical knowledge such as knowledge of frequencies. In our example, agents should set their

---

[1] I am concerned here with constraints on prior degrees of belief. Diachronic constraints, e.g. Bayesian conditionalisation, will be discussed in §4.

[2] Bruno de Finetti was an influential subjective Bayesian — see [de Finetti, 1937].

degrees of belief between 0.2 and 0.3, but degrees of belief $p(A) = 0.2, q(A) = 0.25, r(A) = 0.3$ are equally rational.[3]

Objective Bayesianism maintains not only that an agent's degrees of belief ought to be constrained by empirical knowledge, but also that degrees of belief should be as middling as possible — as far away as possible from the extremes of 0 and 1. In our example there are two outcomes A and $\bar{A}$, the middling assignment of belief gives $p(A) = p(\bar{A}) = 1/2$, and the value in the interval $[0.2, 0.3]$ that is closest to the middling value is 0.3. Thus an agent should assign $p(A) = 0.3$, and agents that assign other degrees of belief are irrational. The agent's degrees of belief are objectively determined by her background knowledge and there is no room for subjective choice.[4]

There is also an important non-Bayesian position that is related to empirically-based subjective Bayesianism. Under this view (advocated for instance by Henry Kyburg)[5] empirical knowledge should constrain an agent's partial beliefs, but these partial beliefs are not in general real numbers — they are intervals instead. Thus in our example an agent should adopt the interval $[0.2, 0.3]$ as her partial belief in A. Because of their common ground, I shall classify empirically-based subjective probability and Kyburg's probability-interval position as *empirical-constraint theories*.

Subjective Bayesianism is attractive because it is easy to justify: one only needs an argument that degrees of belief ought to be probabilities, and the Dutch book argument does this job quite well.[6] However, many applications of probability demand objectivity; the other positions, though philosophically more taxing, are more appealing in that respect.[7] Here

---

[3]This view was adopted by Frank Ramsey: 'it will in general be best for his degree of belief that a yellow toadstool is unwholesome to be equal to the proportion of yellow toadstools which are in fact unwholesome.' [Ramsey, 1926, p. 50]. Colin Howson is a recent proponent of this type of position — see [Howson and Urbach, 1989, §13.e] for instance. [Salmon, 1990] also advocates a version of this view.

[4]Edwin Jaynes is an influential objective Bayesian — see [Jaynes, 2003]. The essential feature of objective Bayesianism is not that degrees of belief be uniquely determined by background knowledge — this is too much to ask in infinite domains [Williamson, 2006, §19] — but that constraints on degrees of belief go beyond purely empirical constraints.

[5][Kyburg Jr and Teng, 2001]

[6][Ramsey, 1926; de Finetti, 1937]

[7]De Finetti argued that subjective Bayesians can account for objectivity since under certain conditions (e.g. degrees of belief must satisfy an *exchangeability* assumption and must be updated by *Bayesian conditionalisation*) different agents' degrees of belief will converge to a single objective value in the limit. However these conditions are controversial and are by no means guaranteed to hold. Moreover, this line of argument does nothing to allay worries about a lack of objectivity in the short run: those whose degrees of belief fail to reflect their empirical knowledge may simply do worse in the short run.

I shall take it for granted that empirical constraints on partial belief are desirable.

The aim of this chapter is to examine the motivation for moving beyond empirical constraints to the stronger constraints advocated by objective Bayesianism. Is it enough to restrict partial beliefs to probability intervals, as recommended by Kyburg and empirically-based subjective Bayesians? Or should one strive for the extra objectivity afforded by the most middling degrees of belief within such intervals?[8]

The plan is to introduce the objective Bayesian framework and its standard justifications in §2 and §3. While these standard justifications are lacking for our purposes, one can appeal to considerations of objectivity (§4), efficiency (§6) or caution (§8) to try to decide between objective Bayesianism and empirical constraint theories. As we shall see in §8, objective Bayesianism clearly surpasses empirically-based subjective Bayesianism in terms of caution. On the other hand, Kyburg's approach appears to be superior to objective Bayesianism in this respect (§11). However I shall argue that, taking several considerations into account, a case can be made for objective Bayesianism.

2   The Maximum Entropy Principle

Objective Bayesianism appeals to the *maximum entropy principle* to determine the degrees of belief that an agent ought to adopt on the basis of background knowledge β.[9] In this section I shall introduce the principle and, in §3, its key justifications.

Given a finite outcome space $\Omega = \{A_1, \ldots, A_n\}$, i.e. a set of mutually exclusive and exhaustive outcomes, the most middling probability function, the *central* function, assigns each outcome the same probability, $c(A_i) = 1/n$ for $i = 1, \ldots, n$. Let $\mathbb{P}_\beta$ be the set of all probability functions that satisfy constraints imposed by the agent's background knowledge β. Objective Bayesians suggest that the agent should adopt as a representation of her degrees of belief a probability function in $\mathbb{P}_\beta$ that is closest to the central function. Now distance between probability functions is normally measured by *cross-entropy*,

$$d(p, q) = \sum_{i=1}^{n} p(A_i) \log \frac{p(A_i)}{q(A_i)}.$$

---

[8]The focus of this chapter is purely the motivation behind objective Bayesianism. There are number of other interesting challenges facing objective Bayesianism — e.g. does it suffer from representation dependence? does it apply to infinite as well as finite domains? Interesting as they are, these questions are beyond the scope of this chapter. See [Williamson, 2006] for an overview of these other challenges.

[9][Jaynes, 1957]

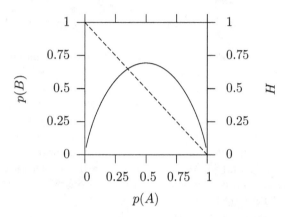

Figure 1. Probability functions (dotted line) and their entropy H (solid curve) in two dimensions.

Thus the function in $\mathbb{P}_\beta$ that is closest to c is the function that minimises

$$d(p,c) = \sum_{i=1}^{n} p(A_i) \log p(A_i) + \sum_{i=1}^{n} p(A_i) \log n = \sum_{i=1}^{n} p(A_i) \log p(A_i) + \log n.$$

This distance is minimised by the function p that has maximum *entropy*

$$H(p) = -\sum_{i=1}^{n} p(A_i) \log p(A_i).$$

Thus

Maximum Entropy Principle Suppose $\mathbb{P}_\beta$ is the set of probability functions that satisfies constraints imposed by the agent's background knowledge β. The agent should select the member of $\mathbb{P}_\beta$ that maximises entropy as her belief function.

If there are two outcomes $\Omega = \{A, B\}$ then $p(B) = 1 - p(A)$, as depicted by the dotted line in Fig. 1. Entropy H is depicted by the solid line. Clearly the closer to the centre of the dotted line, the higher the entropy.

Fig. 2 is the corresponding visualisation of the three outcome case, $\Omega = \{A, B, C\}$. The probability functions are depicted by the plane and their entropy by the curved surface.

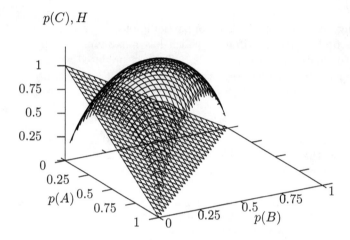

Figure 2. Probability functions (plane) and their entropy H (curved surface) in three dimensions.

3   Standard Justifications

There are two major arguments in favour of the maximum entropy principle,
but neither of these conclusively decide between objective Bayesianism and
the empirically-based approach, as we shall now see.

The original justification of the maximum entropy principle in [Jaynes,
1957] is perhaps best known. This justification appeals to Claude Shannon's
use of entropy as a measure of the uncertainty embodied in a probability
function.[10] Jaynes maintains that an agent's belief function should be informed by background knowledge but should otherwise be maximally uncertain or non-committal — thus it should have maximum entropy according
to Shannon's measure:

> in making inferences on the basis of partial information we must
> use that probability distribution which has maximum entropy
> subject to whatever is known. This is the only unbiased assign-
> ment we can make; to use any other would amount to arbitrary
> assumption of information which by hypothesis we do not have.
>
> . . .
>
> The maximum entropy distribution may be asserted for the pos-
> itive reason that it is uniquely determined as the one that is
> maximally noncommittal with regard to missing information.[11]

The gap in the argument is clear. Jaynes *assumes* that maximally non-
committal, unbiased degrees of belief are most desirable and argues that
they should then be found by maximising entropy. Even if we grant Jaynes
that entropy measures lack of commitment, we still need some reason to
accept his premiss. Why should maximally non-committal degrees of belief
be any better than, say, maximally committal degrees of belief? Why is
bias bad?

The second key line of argument in favour of the maximum entropy principle takes the form of an axiomatic derivation. There are various versions,
but the derivation of Jeff Paris and Alena Vencovská is perhaps most compelling.[12] Their argument takes the following form:

- An *inference process* is a function which maps a domain and background knowledge involving that domain to a probability function on
  the domain that satisfies the background knowledge.

---

[10][Shannon, 1948, §6]

[11][Jaynes, 1957, p. 623]

[12][Paris and Vencovská, 1990; Paris and Vencovská, 2001; Paris, 1994]

- If the selected probability function is to be construed as a representation of the degrees of belief that one ought to adopt on the basis of the background knowledge, then the inference process ought to satisfy some common-sense desiderata. For example, given two logically equivalent knowledge bases the inference process should select the same probability function.[13]

- The only inference process that satisfies these desiderata is the maximum entropy principle.

▸ The only commonsensical inference process is the maximum entropy principle.

Again, even if we grant that the only inference process satisfying the desiderata is the maximum entropy principle, this argument does not do enough for our purposes. It assumes from the outset that we need an inference process, i.e. that we need to select a single probability function that satisfies the background knowledge (Kyburg would disagree with this) and that this probability function must be uniquely determined by domain and background knowledge (empirically-based subjectivists would disagree with this, arguing that different individuals are free to choose different belief functions on the basis of the same background knowledge).

[Paris and Vencovská, 2001, §3] do relax the uniqueness requirement when they consider the case in which background knowledge imposes non-linear constraints on degrees of belief. Their modified argument is unlikely to satisfy the empirical constraint theorist, however, because some of their desiderata go significantly beyond the empirical constraints imposed by background knowledge. The Renaming Principle, for instance, dictates that (background knowledge permitting) degrees of belief should be invariant under permutations of the domain; the Independence Principle holds that in certain cases degrees of belief should be independent if there is no evidence of dependence. While these desiderata may be commonsensical, they are not merely empirical, and more justification is required to convince the proponent of an empirical constraint theory.[14]

So we see that current justifications of the maximum entropy principle do not fully motivate the move from empirical constraints to objective

---

[13]See e.g. [Paris, 1994] for a full list of the desiderata.

[14][Hosni and Paris, 2005] explore a line of justification of the desiderata, claiming that they are commonsensical because they force us to assign similar probabilities, and that conformity is some kind of rational norm. Again, many would take issue with these claims. Assigning similar probabilities may be of pragmatic advantage, but hardly seems to be requirement of rationality except in special cases (see §4 and [Gillies, 1991]). Nor need this consequence of the desiderata render the desiderata commonsensical in themselves — the end doesn't justify the means.

Bayesianism — they presume that an agent's degrees of belief should be maximally non-committal or that they should be fully determined by background knowledge and domain. Further argument is needed.

4   The Argument from Objectivity

One might try to construct an argument around the suggestion that empirically-based subjective Bayesianism is *not objective enough* for many applications of probability.[15] If applications of probability require full objectivity — i.e. a single probability function that fits available evidence — then the axiomatic justification kicks in and one can argue that the maximum entropy function is the only rational choice.

The typical Bayesian methodology involves the following process, called *Bayesian conditionalisation*. First a *prior* probability function p must be produced. Then empirical evidence E is collected. Finally predictions are drawn using the *posterior* probability function $p'(A) = p(A|E)$. Now the prior function is determined before empirical evidence is available; this is entirely a matter of subjective choice for empirically-based subjectivists. However, the ensuing conclusions and predictions may be sensitive to this initial choice of prior, rendering them subjective too. If, for example, $p(A) = 0$ then $p'(A) = p(A|E) = p(E \wedge A)/p(E) = 0$ (assuming $p(E) > 0$). On the other hand, if $q(A) = 1$ then $q'(A) = q(E|A)q(A)/(q(E|A)q(A)+q(E\wedge\bar{A})) = q(E|A)/q(E|A) = 1$.

It is plain to see that two empirically-based subjectivists can radically disagree as to the conclusions they draw from evidence. If they adopt the opposite prior beliefs they will draw the opposite conclusions. Under the empirically-based subjectivist approach, if conditionalisation is adopted then all conclusions are relative to initial opinion.

If the empirically-based subjectivist is to avoid such strong relativity, she must reject Bayesian conditionalisation as a universal rule of updating degrees of belief. The standard alternative is *cross entropy updating*.[16] Here the agent adopts prior belief function p and her posterior p' is taken to be the probability function, out of all those that are compatible with the new evidence, that is closest to p in terms of cross-entropy distance. (Note that, unlike the Bayesian conditionalisation case, the evidence does not have to be representable as a domain event E.)

Suppose $\Omega = \{A, B\}$ for instance. Two empirically-based subjectivists may set $p(A) = 0$ and $q(A) = 1$ respectively, while an objective Bayesian

---

[15]Note that [Salmon, 1990, §4] argues for a version of empirically-based subjective Bayesianism on the grounds that subjective Bayesianism is not objective enough for science.

[16][Williams, 1980]; [Paris, 1994, pp. 118–126]

is forced to set $r(A) = 1/2$ in the absence of any empirical evidence. Now suppose evidence is collected that constrains the probability of A to lie in the interval $[0.6, 0.9]$. The objective Bayesian must adopt $r'(A) = 0.6$ while the empirically-based subjectivists now adopt posteriors $p'(A) = 0.6$ and $q'(A) = 0.9$ respectively. Thus with cross entropy updating evidence can shift degrees of belief away from 0 and 1.

While cross entropy updating may be an improvement over Bayesian conditionalisation, one might argue that the remaining relativity in empirically-based subjective Bayesianism is too still much for applications of probability if applications demand full objectivity.

But such an argument would be hard to execute. First, one would expect that the amount of tolerable subjectivity would vary from application to application — it is unlikely to be the case that *all* applications of probability demand full objectivity. While objectivity of conclusions seems desirable in a computer system for controlling nuclear retaliation, it is clearly less desirable in an automated oenophile.

Second, it is difficult to judge the need for objectivity. Scientists often emphasise the objectivity of their disciplines, but it can be difficult to say whether their claims reflect their sciences' needs for objectivity or their own. (Of course, a perceived objectivity of science is rhetorically very useful to scientists — their conclusions appear more forceful.) Moreover, even if scientific methodology does assume an inherent objectivity, such an assumption may simply be erroneous. There may be less objectivity to be found than commonly supposed.[17] These difficulties have led to widespread disagreement between sociologists of science on the one hand, many of whom view scientific conclusions as highly relative, and philosophers of science and scientists on the other hand, many of whom view science as an objective activity by-and-large.[18] Until some (objective) common ground can be found in the study of science, we are a long way from determining whether the extreme objectivity of objective Bayesianism is required in science, or whether the more limited objectivity of empirically-based subjective Bayesianism is adequate.

Finally, even in cases where objectivity is required, that objectivity need not necessarily incline one towards objective Bayesianism. In order to run the axiomatic derivation of the maximum entropy principle (§3), one must first accept that the common-sense desiderata are indeed desirable. If not, one may be led to alternative implementations of the objectivity requirement. Consider the *minimum entropy principle*: if $\Omega = \{A_1, \ldots, A_n\}$ choose as belief function the probability function, out of all those that satisfy con-

---

[17][Press and Tanur, 2001] present a case for subjectivity in science.
[18][Gottfried and Wilson, 1997]

straints imposed by background knowledge β, that is as close as possible
to the function p that sets $p(A_1) = 1, p(A_2) = \cdots = p(A_n) = 0$. This is
objective in the sense that once the domain and its ordering has been fixed,
then so too is the prior probability function. Moreover this principle is not
as daft as it may first look — there may be semantic reasons for adopting
such an unbalanced prior. Consider a criminal trial setting: $\Omega = \{A_1, A_2\}$
where A_1 represents innocence and A_2 represents guilt, and where there is
no background knowledge — then the minimum entropy principle represents
a prior assumption of innocence; this is in fact the recommended prior.[19]

Note too that while the issue of objectivity might help decide between
empirically-based subjective Bayesianism and objective Bayesianism, it can
not decide between Kyburg's probability interval approach and objective
Bayesianism, since neither of these approaches permit subjective choice of
partial beliefs: background knowledge fully determines the partial beliefs
that an agent ought to adopt.

In sum, while it may be tempting to argue for objective Bayesianism
on the grounds that applications of probability demand objective conclusions, there are several hurdles to be overcome before a credible case can be
developed.

5   Rationality and Evidence

One thing should be clear by now: it can not be empirical warrant that
motivates the selection of a particular belief function from all those compatible with background knowledge, since all such belief functions are equally
warranted by available empirical evidence. If objective Bayesianism is to be
preferred over empirical-constraint approaches, it must be for non-evidential
reasons. (Equally, one can't cite evidence as a reason for abandoning objective Bayesianism in favour of an empirical-constraint approach.)

Thus the problem of deciding between objective Bayesianism and empirical-
constraint approaches hinges on the question of whether evidence exhausts
rationality. Objective Bayesianism supposes that there is more to rationality than evidence: a rational agent's degrees of belief should not only
respect empirical evidence, they should also be as middling as possible. For
empirical-constraint approaches, on the other hand, empirical warrant is
sufficient for rationality. (This puts the empirical constraint theorist at a
disadvantage, because from the empirical perspective there simply are no
considerations that can be put forward to support an empirical constraint
theory over objective Bayesianism since both are empirically optimal; in
contrast, the objective Bayesian can proffer non-empirical reasons to prefer

---

[19]I am very grateful to Amit Pundik for suggesting this example. See §8 for more on
the criminal trial setting.

objective Bayesianism over empirical constraint theories.)

If rationality goes beyond evidence, what else does it involve? We have already discussed one form of non-evidential reason that might decide between the two types of approach — a demand for *objectivity*. But there are other more overtly pragmatic reasons that can be invoked to the same end. In §8 we shall see whether *caution* can be used to motivate objective Bayesianism. But first we turn to *efficiency*.

6   The Argument from Efficiency

One might be tempted to appeal to efficiency considerations to distinguish between objective Bayesianism and empirically-based subjective Bayesianism. If objective Bayesian methods are more efficient than empirically-based subjective Bayesian methods then that would provide a reason to prefer the former over the latter.

One possible line of argument proceeds as follows. If probabilities are in any way subjective then their measurement requires finding out which particular degrees of belief have been chosen by some agent. This can only be done by elicitation: asking the agent what her degrees of belief are, or perhaps inducing them from her behaviour (e.g. her betting behaviour). But, as developers of expert systems in AI have found, elicitation and the associated consistency-checking are prohibitively time-consuming tasks (the inability of elicitation to keep pace with the demand for expert systems is known as *Feigenbaum's bottleneck*). If subjective Bayesianism or empirically-based subjective Bayesianism is to be routinely applied throughout the sciences it is likely that a similar bottleneck will be reached. For example, determining the most probable statistical model given evidence would first require eliciting model assumptions (are the agent's degrees of belief normally distributed, for instance? are certain degrees of belief probabilistically independent?) and also the agent's prior degree of belief in each model — a daunting task. On the other hand, if objective Bayesianism is adopted, degrees of belief are objectively determined by background knowledge and elicitation is not required — degrees of belief are calculated by maximising entropy. Therefore objective Bayesianism is to be preferred for reasons of efficiency.

Of course this argument fails if the objective Bayesian method is itself more computationally intractable than elicitation. Indeed Judea Pearl rejected the maximum entropy principle on the grounds that computational techniques for maximising entropy were usually intractable [Pearl, 1988, p. 463]. However, while that was indeed the case in 1988, it is not the case now. Pearl's own favoured computational tool, *Bayesian nets*, can be employed to vastly reduce the complexity of entropy maximisation, rendering the process tractable in a wide variety of natural settings — see [Williamson, 2005a, §§5.5–5.8] and [Williamson, 2005b]. Thus despite Pearl's doubts, efficiency considerations do lend support to objective Bayesianism after all.

But efficiency considerations on their own fail to distinguish between objective Bayesianism and other procedures for selecting a unique probability function. The maximum entropy principle is no more efficient than the minimum entropy principle. Worse, consider the *blind minimum entropy principle*, where one ignores background knowledge and minimises entropy straight off: if $\Omega = \{A_1, \ldots, A_n\}$ choose as belief function the probability function p such that $p(A_1) = 1$ and $p(A_2) = \cdots = p(A_n) = 0$. This modified principle avoids elicitation and is far easier to implement than the maximum entropy principle. Should one really blindly minimise entropy?

If not, efficiency can't be the whole story. At best one can say something like this: efficiency considerations tell against elicitation and motivate some procedure for mechanically determining an agent's degrees of belief; other desiderata need to be invoked to determine exactly which procedure; the axiomatic derivation can then be used to show that the maximum entropy principle is the only viable procedure.

While this is an improvement over the argument from objectivity, it is still rather inconclusive as it stands. Further work needs to be done to explain why efficiency isn't the whole story, i.e. to explain why the other desiderata override efficiency considerations. The other desiderata thus play an important role in this new argument, and stand in need of some form of justification themselves. Given that empirically-based subjectivists would find several of these desiderata dubious, their justification may turn out to be more of an ordeal than elicitation.

7   Derivation versus Interpretation

We have seen that the arguments from objectivity and efficiency require an appeal to the axiomatic derivation of the maximum entropy principle, and hence to the desiderata presupposed by that derivation. Some of these desiderata are hard to justify — empirically-based subjectivists would simply deny their desirability. Hence the arguments from objectivity and efficiency flounder.

Such difficulties are bound to beset any derivation of the maximum entropy principle. If the principle *MEP* is a logical consequence of assumptions A, i.e. $\models A \to MEP$, then the assumptions A must be at least as strong as the maximum entropy principle and are unlikely to be trivially true. Empirically based subjectivists, who reject the maximum entropy principle, can just use the contrapositive of the derivation, $\models \overline{MEP} \to \bar{A}$ as an argument for the falsity of the assumptions. Thus what is an argument in favour of the maximum entropy principle for the objective Bayesian is nothing of the sort for the empirically-based subjectivist. Consequently a derivation of the maximum entropy principle is unlikely to be of help to us in our quest to motivate objective Bayesianism.

But Jaynes' original justification, which proceeds by *interpreting* lack of commitment as entropy rather than *deriving* the maximum entropy principle, does not suffer from these difficulties. It requires an assumption, namely that one ought to be maximally non-committal, but that assumption is relatively weak — much of the work is being done by the act of interpretation.

Consider an analogy: the principle, adopted by all Bayesians, that degrees of belief should satisfy the axioms of probability. If we try to *derive* the axioms of probability from some assumptions then those assumptions will have to be at least as logically strong as the axioms, and hence at least as controversial.[20] Instead it is more usual to *interpret* degrees of belief as betting quotients[21] and then to show that the axioms of probability must hold on pain of incoherence.[22] This argument is not simply a drawing-out of logical consequences; the act of interpretation is doing significant work. Yet this interpretation is rather natural and can itself be justified. As Ramsey notes,

> this will not seem unreasonable when it is seen that all our lives we are in sense betting. Whenever we go to the station we are betting that a train will really run, and if we had not a sufficient degree of belief in this we should decline the bet and stay at home.[23]

Similarly, the interpretation of the uncertainty of a probability function as its entropy is, if not obvious, fully justifiable. Indeed Shannon provided two justifications, a derivation from desiderata, and — more importantly for Shannon — a pragmatic justification:

> This theorem [the derivation from desiderata], and the assump-
> tions required for its proof, are in no way necessary for the
> present theory. It is given chiefly to lend a certain plausibil-
> ity to some of our later definitions [including the interpretation

---

[20]Cox's derivation of the axioms of probability has been the topic of substantial controversy. [Paris, 1994, pp. 24–32] shows that this type of derivation requires strong assumptions; see also [Halpern, 1999a; Halpern, 1999b].

[21]A *betting quotient* for event E is a number q that the agent selects on the presumption that she will make a bet, paying qS with return S if E occurs, where the stake S is to be chosen after she selects q and may be positive or negative.

[22][Ramsey, 1926; de Finetti, 1937]

[23][Ramsey, 1926, p. 183]

of uncertainty as entropy]. The real justification of these definitions, however, will reside in their implications.[24]

Indeed Shannon's definitions have been very fruitful in communication and
information theory and are now well entrenched in several branches of science.

The interpretation of uncertainty as entropy plays a substantial role in
Jaynes' argument for the maximum entropy principle. The key assumption
(that a lack of commitment is desirable) is relatively meagre, in the sense
that it does not on its own presuppose entropy maximisation. We shall see
now that this assumption can be justified by an appeal to caution.

8   The Argument from Caution

As Ramsey notes in the above quote, our degrees of belief guide our actions
and our actions are tantamount to bets. To embark on a course of action
(such as going to the station) a degree of belief (in the train running, in this
case) must be sufficiently high. Now, every course of action has its associated risks. Going to the station only to find that the train is not running
wastes time and effort, and one may miss an important appointment. Such
a course of action is not to be embarked upon lightly, and prudence is required. The trigger for action will vary according to risk — if a lot hangs on
it, one may only go to the station if one has degree of belief at least 0.95 in
the train running, but if the consequences are less dire, a lower trigger-level,
say 0.85, may be appropriate. Suppose one knows only that the local train
operating company has passed the minimum threshold of eighty percent
of trains running. According to empirically-based subjective Bayesianism,
one's degree of belief in the train running can be chosen from anywhere in
the interval $[0.8, 1]$. According to objective Bayesianism the least extreme
value in this interval must be chosen: i.e. 0.8. So the empirically-based subjectivist may decide to go to the station while the objectivist decides not to.
Thus the objective Bayesian decision is more cautious and is to be preferred
since there is no empirical evidence to support a less cautious decision.

In sum, extreme degrees of belief trigger actions and open one up to
their associated risks. In the train example the objective Bayesian strategy
of adopting the least extreme degree of belief seems to be the most prudent.
Can one abstract from this case to argue that objective Bayesianism is to
be preferred in general? There are some potential difficulties with such a
move to generality, as we shall now see.

The first potential problem stems from the fact that it is not only extreme
degrees of belief that trigger actions — middling degrees of belief can also

---

[24][Shannon, 1948, §6]

trigger actions. Consider a case where a patient has one of two possible diseases A and B. A high degree of belief in A will trigger a course of medication to treat A. Similarly, a high degree of belief in B will trigger treatment of B. However, a more middling degree of belief — a degree of belief that triggers neither treatment of A nor treatment of B — will trigger another action, namely the gathering of more empirical evidence in order reach a conclusive diagnosis. Collecting further symptoms from the patient also has its associated costs and risks: it requires time, effort and money to perform more tests, and some tests might harm the patient. Now objective Bayesianism advocates setting middling degrees of belief, thereby exposing the diagnoser to the risks associated with collecting more symptoms. It seems that objective Bayesianism is not such a risk-averse position after all.

But this apparent problem does not in fact scupper the prospect of a general argument in favour of objective Bayesianism. While there are indeed cases in which middling degrees of belief trigger actions, these actions are always *less risky* than those triggered by extreme degrees of belief. Suppose in the above example that A and B are two types of meningitis, requiring very different treatment. The risks associated with either outcome are so high that the risks associated with collecting more symptoms pale into insignificance in comparison. Suppose on the other hand that A and B are just two strains of cold; A responds best to a nightly dose of rum toddy while B requires a whisky toddy taken three times a day; in either case a full recovery will be made even if no treatment is taken. In this case the risks associated with either diagnosis are so low that in the absence of a diagnosis it simply is not worth doing the blood test that would provide conclusive evidence: a more middling degree of belief does not trigger any action. The point is that collecting further evidence will only be an option if the resulting knowledge is worth it, i.e. if the costs and risks associated with the primary outcomes A and B outweigh those associated with collecting new evidence. Hence it will still be most cautious to have more middling degrees of belief.

It might be thought that there is a more serious variant of this problem. For a politician the risks associated with appearing non-committal outweigh those of committing to an unjustified course of action or making a promise that can't be kept. People like their politicians bold and it would seem that a non-committal objective Bayesian politician would not get elected. But this objection hinges on a mistaken conflation of appearance and reality. The fact that a politician should not *appear* non-committal does not mean that her beliefs should not actually *be* non-committal — politicians simply need to mask their beliefs. Their beliefs need to be as cautious as anyone else's though: they need to be shrewd judges of which promises the electorate will swallow, and should not commit to one lie over another unless they have a justified belief that they can get away with it.

A second type of problem confronts any attempt to generalise the argument from caution. This concerns cases in which risks are known to be imbalanced. In the diagnosis example, take A to be meningococcal meningitis and B to be 'flu. In this case, the risks associated with failing to diagnose A when it is present are so great that it may be prudent to *assume* A and prescribe antibiotics, unless there is conclusive evidence that decides in favour of B. We have already come across another example of an imbalance of risks: the risk of judging the innocent guilty is considered to outweigh that of judging the guilty innocent, and this motivates a presumption of innocence in criminal cases. Such presumptions seem far from non-committal, yet rational.

Perhaps the best way of resolving this difficulty is to distinguish appearance and reality again. It is important in these cases to *act as if* one believed one of the alternatives — to prescribe antibiotics or to release the suspect — not to actually believe that alternative. In these cases, the imbalanced risks motivate imbalanced trigger levels rather than imbalanced degrees of belief. If degree of belief in 'flu is higher than say 0.95 prescribe paracetamol, otherwise, if degree of belief in meningococcal meningitis is at least 0.05, prescribe antibiotics. If guilt is not proved beyond reasonable doubt (degree of belief of guilt is not higher than 0.99 say, in which case degree of belief in innocence is at least 0.01) then trigger action that corresponds to innocence, i.e. release the suspect. In both these cases the trigger level for one alternative is very high while the trigger level for the other alternative is very low.

One might respond to this move by accepting this proposed resolution for the diagnosis example, but rejecting it for the legal example. One might claim that in the legal case there should not only be a high standard of reasonable doubt for guilt and a corresponding low standard of doubt for innocence, but there should also be prior degree of belief 1 in innocence, in order to make it as hard as possible for the prosecution to sway degree of belief to beyond reasonable doubt. This response seems natural enough — it seems right to make it as hard as possible to trigger guilt. However, this response should not be acceptable to any Bayesian — whether subjective, empirically-based or objective — because it does not sit well with Bayesian methods of updating. Recall that in §4 we saw that there are two standard options for updating an agent's degree of belief in new evidence, Bayesian conditionalisation and cross-entropy updating. If Bayesian conditionalisation is adopted then a prior degree of belief 0 of guilt can never be raised above 0 by evidence, and it will be impossible to convict anybody. On the other hand, if cross-entropy updating is adopted then a presumption of innocence will make no difference in the legal example. A presumption of innocence corresponds to prior degree of belief 0 in guilt, while a maximally non-committal probability function will yield a prior degree of belief of $\frac{1}{2}$. In either case degree of belief can only be raised above 0.99 if empirical evidence constrains degree of belief to lie in some subset of the interval $[0.99, 1]$ (because the cross-entropy update is the degree of belief, from all those that are compatible with evidence, that is closest to $\frac{1}{2}$). As long as the prior degree of belief is lower than the trigger level for guilt, triggering is dependent only on the evidence, not on prior degree of belief. In sum, whichever method of updating one adopts there is no good Bayesian reason for setting prior degree of belief in guilt to be 0.

There is a third, more substantial, problem that besets an attempt to generalise the argument from caution: the maximum entropy principle is not *always* the most cautious policy for setting degrees of belief. Consider a case in which there are three elementary outcomes, $\Omega = \{A, B, C\}$, a risky action is triggered if $p(\{A, B\}) \geq 7/8$ and another risky action is triggered if $p(\{B, C\}) \geq 7/8$. Suppose there is background knowledge $\beta = \{p(B) \geq 3/4\}$. Then the set $\mathbb{P}_\beta$ of probability functions that are consistent with this knowledge is represented by the shaded area in Fig. 3. (The triangular region represents the set of all probability functions — these must satisfy $p(A) + p(B) + p(C) = 1$.) Now the maximum entropy principle advocates adopting the probability function p from $\mathbb{P}_\beta$ that is closest to the central probability function c which sets $c(A) = c(B) = c(C) = 1/3$. Thus the maximum entropy function sets $p(A) = 1/8, p(B) = 3/4$ and $p(C) = 1/8$. This triggers both $\{A, B\}$ and $\{B, C\}$ since $p(\{A, B\}) = 7/8 = p(\{B, C\})$. On the other hand, the minimum entropy principle advocates adopting the probability function q from $\mathbb{P}_\beta$ that is closest to the probability function a which sets $a(A) = 1$ and $a(B) = a(C) = 0$. Thus the minimum entropy principle sets $q(A) = 1/4, q(B) = 3/4, q(C) = 0$.[25] This triggers $\{A, B\}$, since $q(\{A, B\}) = 1 \geq 7/8$, but does not trigger $\{B, C\}$, since $q(\{B, C\}) = 3/4 < 7/8$. Hence in this case the minimum entropy principle licences the most cautious course of action while the maximum entropy principle seems to throw caution to the wind.

While this example shows that the most non-committal probability function is not the most cautious *in every situation*, it may yet be the most cautious *on average*. If it is most cautious when averaging over background

---

[25]N.b. the 'minimum entropy principle' is a bit of a misnomer — while the probability function q commits most to A, it does not actually have minimum entropy in this example. Here entropy is in fact minimised by the probability function r that commits most to B, defined by $r(A) = 0, r(B) = 1, r(C) = 0$.

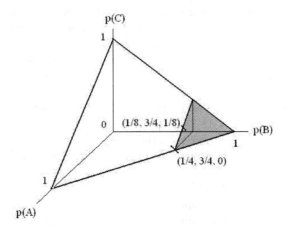

Figure 3. Maximum and minimum entropy principle probability functions.

knowledge β and decisions (i.e. $D \subseteq \Omega$ and trigger levels $\tau_D \in [0,1]$ such that a course of action is triggered if $p(D) \geq \tau_D$) then adopting the maximum entropy principle will be the best policy in the absence of any knowledge about β, D and τ_D.[26]

The average caution for a policy for setting belief function p_β can be measured by the proportion of β, D, τ_D that result in a course of action being triggered. Define the *trigger function* $T(\beta, D, \tau_D) = 1 \Leftrightarrow p_\beta(D) > \tau_D$, and 0 otherwise. Then the average caution of a policy is the measure of β, D, τ_D for which $T(\beta, D, \tau_D) = 1$, i.e.

$$C = \frac{1}{z} \sum_{D \subseteq \Omega} \int_\beta \int_{\tau_D} T(\beta, D, \tau_D) d\beta d\tau_D$$

where z is a normalising constant, $z = 2^{|\Omega|} \int_\beta \int_{\tau_D} 1 d\beta d\tau_D$. The smaller the value of C, the more cautious the policy is on average.

However it turns out that the maximum entropy principle is no more cautious than the minimum entropy principle, even if one considers average caution. By way of example consider the two-dimensional case, $\Omega = \{A, B\}$. Suppose β constrains $p(A)$ to lie in a closed interval $[x, y]$ where $0 \leq x \leq y \leq 1$, and that the trigger levels are all the same, $\tau_D = \tau \in [0,1]$. We are interested

---

[26]I leave it open here as to the mechanism that is used to select the trigger levels. These may be set by experts perhaps, by maximising expected utility, or by some other decision-theoretic procedure.

in the proportion of values of x, y, τ that trigger a decision. i.e. the volume of the part of the cube $[0,1]^3$ defined by these parameters that triggers some decision. We are only concerned with the half of the cube such that $y \geq x$, so $z = 4 \times 1/2 = 2$. Trivially, $\{A, B\}$ always gets triggered and $\emptyset$ only gets triggered if $\tau = 0$. Consider first the entropy maximisation policy. In this case $T(\beta, A, \tau) = 1 \Leftrightarrow x \geq \tau$ or $\tau \leq 1/2$ and $y \geq \tau$, while $T(\beta, B, \tau) = 1 \Leftrightarrow x \leq 1 - \tau$ or $\tau \geq 1/2$ and $y \leq 1 - \tau$. Then

$$C = z^{-1} \int_\beta \int_\tau \left[T(\beta, \{A, B\}, \tau) + T(\beta, A, \tau) + T(\beta, B, \tau) + T(\beta, \emptyset, \tau) \right] d\beta d\tau$$

$$= 2^{-1} \left[1/2 + 1/4 + 1/4 + 0 \right] = 1/2.$$

Next consider the entropy minimisation policy. In this case $T(\beta, A, \tau) = 1 \Leftrightarrow y \geq \tau$, while $T(\beta, B, \tau) = 1 \Leftrightarrow y \leq 1 - \tau$. Now

$$C = 2^{-1} \left[1/2 + 1/3 + 1/6 + 0 \right] = 1/2.$$

Thus we see that what entropy minimisation loses in caution by committing to A, it offsets by a lack of commitment to B. On average, then, entropy minimisation is just as cautious as entropy maximisation.

While it might appear that caution can not after all be used as an argument in favour of the maximum entropy principle, such a conclusion would be too hasty. In fact the maximum entropy principle *is* most cautious where it matters, namely in the face of *risky* decisions, as we shall now see.

It is important to be cautious when a course of action should not be taken lightly, and as we have seen, the importance attached to a decision tends to be reflected in its trigger level. Thus when deciding between meningococcal meningitis and 'flu, a high trigger level for 'flu indicates that the ensuing course of action (treatment of 'flu rather than meningitis) is risky. Similarly when deciding between innocence and guilt in a criminal case, there is a high trigger level for guilt because the consequences of a mistaken judgement of guilt are considered dire. Thus when we consider the average caution of a policy for setting degrees of belief, it makes sense to focus on the decisions where caution is important, namely those decisions D with a *high* trigger level τ_D.

So let us assume in our two-dimensional example that $\tau \geq 1 - \varepsilon$ where $0 \leq \varepsilon \leq 1/2$ is small. This extra constraint means that now $z = 4 \times \varepsilon/2 = 2\varepsilon$. As before $T(\beta, \emptyset, \tau) = 1$ has measure 0, but now $T(\beta, \{A, B\}, \tau) = 1$ has measure $\varepsilon/2$. In the entropy maximisation case both $T(\beta, A, \tau) = 1$ and $T(\beta, B, \tau) = 1$ have measure $\varepsilon^3/6$, and

$$C = (2\varepsilon)^{-1} \left[\varepsilon/2 + \varepsilon^3/6 + \varepsilon^3/6 + 0 \right] = (1 + 2\varepsilon^2/3)/4.$$

On the other hand in the entropy minimisation case $T(\beta, A, \tau) = 1$ has measure $\varepsilon/2 - 1/6[1 - (1 - \varepsilon)^3]$, while $T(\beta, B, \tau) = 1$ has measure $\varepsilon^3/6$, and

$$C = (2\varepsilon)^{-1} \left[\varepsilon/2 + \varepsilon/2 - 1/6[1 - (1 - \varepsilon)^3] + \varepsilon^3/6 + 0 \right] = (1 + \varepsilon)/4.$$

Thus entropy maximisation offers the smaller average caution: if $\varepsilon = 1/2$ then entropy minimisation is about 30% more cautious, while if $\varepsilon = 1/10$ it is about 10% more cautious.

In sum, one can, after all, appeal to caution to make a case for objective Bayesianism. The maximum entropy principle is *on average the more cautious policy when it comes to risky decisions*. This caution is explained by the fact that the more middling one's degrees of belief, the smaller the number of triggered decisions on average, when trigger levels are high.

9   Sensitivity Analysis

We have seen that an appeal to caution can be used to motivate the move from empirically-based subjective Bayesianism to objective Bayesianism: the latter is more cautious on average with respect to risky decisions. In this section we shall consider a possible response. Arguably there is an extension of empirically-based subjective Bayesianism that is more cautious still than objective Bayesianism.

Suppose background knowledge constrains a degree of belief to an interval. Suppose too that empirically-based subjective Bayesianism is adopted, so that an agent may choose any degree of belief within this interval. One may want to be extra cautious and avoid taking a course of action if the decision to do so depends on the degree of belief chosen. This leads to the following modification of the decision rule: instead of embarking on a course of action iff one's own degree of belief triggers the action, embark on it iff every possible agent would too, i.e. iff the whole interval of possible degrees of belief is greater than the trigger level. This decision rule might be called the *sensitivity analysis* rule — a decision is only taken if it is not sensitive to subjective choice of prior probability.[27]

Under this view, degrees of belief are partly a matter of subjective choice, but, once trigger levels are chosen,[28] it is an objective matter as to whether a decision will be triggered. In two dimensions this decision procedure exhibits the same average caution as objective Bayesianism over risky decisions.[29] However, in higher dimensions it will be more cautious in general. Recall the example of Fig. 3: here the maximum entropy principle triggers both

---

[27]This type of rule is sometimes also called a *robust Bayesian* rule.

[28]The trigger levels themselves may depend on an agent's utilities and thus be subjective.

[29]Assuming, as before, that β constrains $p(A)$ to lie in a closed interval.

decisions, the minimum entropy principle triggers one decision, while the sensitivity analysis decision procedure triggers neither.

Thus it seems that by appealing to caution the objective Bayesian is shooting herself in the foot. While such an appeal favours objective Bayesianism over empirically-based subjective Bayesianism as normally construed, it also favours the sensitivity analysis modification of empirically-based subjective Bayesianism over objective Bayesianism.

However, the sensitivity analysis approach is conceptually rather unattractive. This is because it divorces the connection between belief and action: under the sensitivity analysis approach one's degrees of belief have no bearing on whether one decides to take a course of action. It matters not a fig the extent to which one believes a patient has meningitis, because the decision as to what treatment to give is based on the *range* of permissible beliefs one might adopt, not on one's own actual beliefs. Arguably this is an unacceptable consequence of the sensitivity analysis approach which more than offsets its merit with respect to considerations of caution.

But perhaps there is some way of putting the sensitivity analysis approach on firmer footing. Perhaps one can retain its caution while re-establishing the link between belief and action. We shall investigate two possible strategies for salvaging this approach in the next two sections.

10 Higher-Order Beliefs

The sensitivity analysis approach bases decisions on the range of beliefs that an agent might adopt, rather than on the agent's own beliefs. If one wants to retain this decision mechanism but also to insist that an agent's decisions be made on the basis of her own beliefs then one must re-interpret her beliefs as somehow encapsulating this whole range.

Suppose for example that empirical constraints force $p(A) \in [0.7, 0.8]$. Under the sensitivity analysis approach the agent is free to choose whichever degree of belief she likes from this interval, but a decision on A will only be triggered if the whole interval triggers, i.e. if 0.7 is greater than the trigger level. Consider an alternative viewpoint — a Bayesian who is uncertain as to which degree of belief x to adopt from within this interval. In the face of this uncertainty higher-order degrees of belief, such as $p(x \in [0.7, 0.75])$, become relevant. Indeed the agent may form a prior belief distribution over x, and base her decision for A on various characteristics of this distribution. One decision rule, for instance, involves triggering A if $p(x \geq \tau_A) = 1$.

This alternative viewpoint yields a type of empirically-based subjective Bayesianism: an agent's degrees of belief are constrained just by empirical knowledge. It is also compatible with cautious decision rules, such as that exemplified above. Moreover by admitting higher order degrees of belief it reinstates the link between belief and action: decisions are triggered by features of these higher order beliefs. Thus this approach appears to combine the best of all worlds — perhaps for that reason it is very popular in Bayesian statistics.

But all is not plain sailing for higher-order degrees of belief. A decision rule, such as that given above, is only cautious under some priors over x. If x is uniformly distributed over $[0.7, 0.8]$ then $p(x \geq \tau_A) = 1$ iff $0.7 \geq \tau_A$, the same cautious decision rule as the sensitivity analysis approach. On the other hand, if the probability of x is concentrated on the maximal point, $p(x = 0.8) = 1$, then the decision on A triggers just when $0.8 \geq \tau_A$ — in this case the decision rule is considerably less cautious, and in particular less cautious than the maximum entropy principle. Now the empirically-based subjective Bayesian can not advocate setting one prior rather than another, since there is no extra empirical evidence to constrain choice of prior. Indeed the agent is free to choose any prior she wishes, and if she sets $p(x = 0.8) = 1$ she is far from cautious. Suggesting that the agent forms a prior over priors only defers the problem, leading to a vicious regress.

Thus we see that while the higher-order belief approach is compatible with cautious decision rules, it is also compatible with rash decision rules. It certainly can not be argued that this approach is any more cautious than the objective Bayesian methodology. Higher-order beliefs do not, after all, lead to the salvation of sensitivity analysis.

11   Interval-Valued Beliefs

There is another way one might try to salvage the cautiousness of sensitivity analysis. Again, this involves re-interpreting an agent's beliefs as encapsulating the whole range of empirically constrained values. But this time, rather than invoking uncertainty as to which degree of belief to adopt, one instead rejects the Bayesian idea that an agent's partial beliefs are numerical point-valued degrees of belief, i.e. probabilities. Under this approach an agent's partial belief in A is identified with the whole interval yielded by empirical constraints, $bel(A) = [0.7, 0.8]$ in our example. [Kyburg Jr, 2003] provides a recent exposition of this strategy.[30]

Interval-valued beliefs offer a very appealing reconstruction of the sensitivity analysis approach. A natural decision rule proceeds thus: trigger a course of events on A iff the agent's partial belief in A is entirely above the trigger level for A, in our example, iff $[0.7, 0.8] \geq \tau_A$, i.e. iff $0.7 \geq \tau_A$. Not only does the resulting framework capture the cautiousness of the sensitiv-

---

[30]See [Kyburg Jr and Teng, 2001] for the formal details of the this approach. [Kyburg Jr and Teng, 1999] argue that the interval approach performs better than the subjective Bayesian approach in the short run in a betting set-up.

ity analysis approach, it also ties the triggering of a decision to the agent's own partial belief, rather than the beliefs of other possible agents.

The crunch point is this. The partial belief approach appears to be superior to objective Bayesianism with respect to caution; does this gain outweigh any difficulties that accompany the rejection of point-valued degrees of belief in favour of interval-valued beliefs?

I would argue not. First of all, there are qualifications to be made about the cautiousness of interval-valued beliefs that diminish their supposed superiority over objective Bayesianism. Second, the problems that accompany interval-valued beliefs arguably outweigh any remaining benefit in terms of caution.

First to the qualifications. The typical way of generating interval-valued beliefs runs as follows.[31] Sample an attribute A from a population; say the sample frequency is 0.75; under certain assumptions about the sampling mechanism and the population, one might form a confidence interval, say $[0.7, 0.8]$ for the population frequency; set one's partial belief in A to this confidence interval. The key problem with this approach is that the confidence interval will depend upon the chosen confidence level as well as the sampling assumptions — thus the endpoints of the interval are somewhat arbitrary. But the decision procedure depends crucially on the endpoints: a 95% confidence interval $[0.7, 0.8]$ may trigger a course of action for A while a 99.9% confidence interval $[0.5, 1]$ may fail to trigger the action.

There is no non ad hoc way of determining a suitable confidence interval and so this approach must be abandoned if one wants an objective, cautious decision procedure. Perhaps the best alternative strategy is just to set one's partial belief in A to the sample frequency 0.75 — this is, after all, the most probable candidate for the population frequency. But then the belief is not interval-valued after all, it is point-valued. Thus the interval-valued approach loses its edge. (Arguably this is a qualification to the interval-valued belief approach rather than a reason to dismiss it altogether: one can still adopt an interval-valued belief in certain circumstances, for example if there are two samples with sample frequencies 0.75 and 0.77 respectively then it seems natural to view the whole interval $[0.75, 0.77]$ as a candidate for partial belief.)[32]

There is a further qualification to be made to the supposed superiority of the interval approach over the objective Bayesian approach. In a fully-blown decision-theoretic setting where potential losses are quantifiable, there is a sense in which interval-valued beliefs perform no better

---

[31][Kyburg Jr and Teng, 2001, §11.3] adopt this sort of approach.

[32]See [Williamson, 2005a, §5.3] for discussion of this type of constraint on rational belief.

than objective Bayesian degrees of belief. The maximum entropy principle and the sensitivity analysis approach *both behave optimally* in the sense
that they both succeed in minimising worst-case expected logarithmic loss.
(Again arguably this is merely a qualification: one may not care very much
about minimising worst-case expected logarithmic loss. On the other hand
[Grünwald and Dawid, 2004] show that a generalised version of the maximum entropy principle is equivalent to the sensitivity analysis approach in
that they both minimise loss for an arbitrary loss function, not just logarithmic loss.)

We have seen then that the advantages of interval-valued beliefs with
respect to caution are somewhat diminished. Next we turn to the problems
that accompany interval-valued beliefs. As noted in §5, it is not empirical
evidence that adjudicates between the two approaches; the problems with
interval-valued beliefs are pragmatic and conceptual.

From the pragmatic point of view, it is harder to obtain and work with
interval-valued beliefs than point-valued beliefs. Roughly speaking it is
twice as hard, since there are twice as many numbers to have to deal with:
to each point-valued degree of belief there are the two endpoints of the
corresponding interval-valued belief.[33] Intervals also make it hard to do
things that are simple using numbers. For instance, suppose one wants to
either trigger a course of action for A, or otherwise to trigger another course
of action for $\bar{A}$: to give antibiotics if meningococcal meningitis is suspected,
otherwise paracetamol. In the case of point-values degrees of belief one
simply needs to ensure that $\tau_A + \tau_{\bar{A}} = 1$, for then if $p(A) \geq \tau_A$ one action
is taken, otherwise $p(\bar{A}) = 1 - p(A) > \tau_{\bar{A}}$ and the other action is taken.[34]
On the other hand if partial beliefs are intervals then for any non-extreme
trigger levels there are partial beliefs ($[x, y]$ where $x < \tau_A$ and $y > 1 - \tau_{\bar{A}}$)
such that *neither* action triggers.[35]

There are thus pragmatic reasons to favour point-valued degrees of belief
over interval-valued beliefs. Accordingly one might reason something like
this: the formalism of point-valued degrees of belief offers a first approximation to how one should reason; the formalism of interval-valued beliefs
offers a second approximation; the first approximation tends to be easier
to use in practice and there is little to be gained (in terms of caution) by
using the second approximation; thus one should use the first approximation by default. But such a view assumes that there is essentially more to

---

[33][Cozman, 2000] develops a computational framework for interval-valued beliefs.
[34]One would of course also need a policy to decide which action is triggered if $p(A) = \tau_A$
(in which case $p(\bar{A}) = \tau_{\bar{A}}$).
[35]One can get round this problem by insisting that trigger levels be functions of the
partial beliefs themselves as well as the decision outcomes A or $\bar{A}$. The point is not that
intervals make things impossible, just that intervals make things more complicated.

interval-valued beliefs than point-valued beliefs — that conceptually they add something. There are reasons for doubting this perspective, as we shall now see.

From the conceptual point of view, the interval-valued belief approach is caught in a dilemma: it either lacks the intuitively compelling connection between beliefs and betting quotients that underpins the Bayesian approach, or it fails to add anything conceptually to the Bayesian approach. One of the key points in favour of the Bayesian approach is that an agent's partial belief in E is interpretable as a betting quotient, 'the rate p at which he would be ready to exchange the possession of an arbitrary sum S (positive or negative) dependent on the occurrence of a given event E, for the possession of the sum pS'.[36] One cannot simply identify an interval-valued partial belief with a betting quotient — a betting quotient is a single number but an interval is a set of numbers. One might try, as [Borel, 1943, §3.8] did, to interpret the interval as a *set* of acceptable betting quotients.[37] To do so one must adopt a different betting set-up to that of the Bayesian, one without the requirement that the agent buys and sells bets at the same rate, i.e. one in which S must be positive rather than either positive or negative. (Otherwise, if the agent has more than one betting quotient in the same outcome then she can simply be Dutch booked — a set of bets can be found that forces her to lose money whatever happens.) But [Adams, 1964, §7] showed formally that when this strategy is pursued one can identify a *single* probability function that can be thought of as representing betting quotients that the agent regards as fair.[38] Thus this new betting set-up is a dead end for the proponent of interval-valued beliefs: a *set* of betting quotients $[0.7, 0.8]$ for an outcome in the new set-up turns out to be equivalent to a *single* betting quotient 0.75 in the original set-up — it is just a more complicated way of saying the same thing. In sum, by trying to provide a betting interpretation for interval-valued partial beliefs one just ends up motivating point-valued degrees of belief; betting fails to provide a distinctive foundation for interval-valued partial beliefs after all, and interval-valued beliefs should not be thought of as a second approximation or refinement of point-valued beliefs. Thus the proponent of interval-valued beliefs must either accept that they are essentially just a complication of point-valued degrees of belief, or, if they are to differ conceptually, they lack any link with betting behaviour that accounts for that difference. In the absence of a viable betting interpretation, the question arises as to how interval-valued

---

[36][de Finetti, 1937, p. 62]. [Ramsey, 1926, p. 172] proposed a similar betting set-up: 'The old-established way of measuring a person's belief is to propose a bet, and see what are the lowest odds which he will accept'.

[37]See also [Smith, 1961] and [Walley, 1991, §1.6.3].

[38]See also [Koopman, 1940a; Koopman, 1940b].

beliefs are to be interpreted: what does it mean to believe A with value $[0.7, 0.8]$?[39] In either case, it is hard to see how intervals could be better candidates for partial beliefs than numbers.

The proponent of intervals may respond here that the Bayesian link between partial beliefs and betting quotients is less attractive than one might think. Of course, if there are grounds for abandoning Bayesian betting set-up then the absence of a viable betting interpretation is not such a disadvantage for interval-valued beliefs. One objection to the Bayesian betting set-up is that human agents can't always evaluate their betting quotients in terms of unique real numbers — human beliefs are simply not so precise.[40] Another is that it is rather impractical to elicit degrees of belief using a series of bets, since as noted in §6 this is a time-consuming operation, and in any case people are often either reluctant to bet at all or happy to lose money whatever happens. These points are well made, but by-the-by in our context because they only trouble the subjective and empirically-based versions of Bayesianism, not objective Bayesianism. Under objective Bayesianism, agents do not need to search their souls for real numbers to attach to beliefs, nor is the betting set-up required to measure those degrees of belief. Degrees of belief are determined by the maximum entropy principle and they are measured by maximising entropy. (In this age of mechanisation human agents and artificial agents alike can use computing power to work out the extent they ought to believe various outcomes.) Thus for objective Bayesianism the betting interpretation is only important for the *meaning* it gives to degrees of belief. The fact that the betting set-up is simplistic, or an idealisation, or impractical, is neither here nor there — the objective Bayesian does not go on to invoke the betting set-up, as Ramsey did, as an elicitation or measurement tool.

There is another conceptual problem that besets the interval-valued belief approach. One might argue, as de Cooman and Miranda do in this volume, that interval-valued beliefs are superior to point-valued beliefs because they can represent the amount of knowledge that grounds a belief. Under the objective Bayesian account, a degree of belief 0.5 that it will snow in Stroud today may be based on total ignorance, or it may be based on good evidence, e.g. the knowledge that the frequency of it snowing in Stroud on a day like today is $\frac{1}{2}$. In contrast, the interval-valued approach distinguishes between these two cases: while in the latter case the belief might have value 0.5, in the former case the belief would have value $[0, 1]$. In a sense, then, interval-

---

[39] As [Walley, 1991, p. 22] himself notes, 'unless the conclusions have a behavioural interpretation, it is not clear how they can be used for making decisions of for guiding future inquiry and experimentation.'

[40] See e.g. [Kyburg Jr, 1968].

valued beliefs tell us about the knowledge on which they are based; this is supposed to be an advantage of the interval-valued approach.

But this is evidence of a conceptual problem with the interval-valued approach, not an advantage. The problem is this. The question 'how much knowledge does the agent have that pertains to it snowing in Stroud today?' is a question about knowledge, not belief. Consequently, it should not be the model of belief that answers this question; it should be the knowledge component of the agent's epistemic state. But on the interval-valued approach, it is the belief itself that is used to answer the question (and typically there is no separate model of the agent's knowledge). Thus on this approach, the belief model conflates concerns to do with knowledge and belief. On the other hand, the objective Bayesian approach maintains a nice distinction between knowledge and belief: an agent has background knowledge which is then used to constrain her choice of degrees of belief; the former component contains information about the extent of the agent's knowledge, while the latter contains information about the strength of the agent's beliefs; neither can be used to answer questions about the other. The interval-valued approach, then, muddles issues to do with knowledge and belief, while the objective Bayesian approach keeps them apart. But the goal of both approaches is to model a rational agent's epistemic state, and this requires a sharp distinction between knowledge and belief. It is a fact of the matter that some of our full beliefs are of a higher grade than others and are more entrenched in the sense that we are less willing to revise them. I have full belief that a point chosen at random in a ball will not be its centre; I also have a full belief that I am alive; I'm quite willing to revise the former belief in the face of evidence, but not the latter — it is knowledge, not belief, that accounts for this distinction. Hence the objective Bayesian approach, by maintaining a proper distinction between knowledge and belief, offers a better model of an agent's epistemic state.

To conclude, the limited gains that interval-valued beliefs offer in terms of caution are arguably offset by the conceptual as well as pragmatic advantages of point-valued beliefs.

12   Summary

We see then that there is a case for preferring objective Bayesianism over an empirical-constraint theory of rational belief. Objective Bayesianism is more cautious than empirically-based subjective Bayesianism; while it may be less cautious than the interval-valued belief approach, it offers pragmatic advantages and a simple interpretation in terms of betting quotients.

BIBLIOGRAPHY

[Adams, 1964] Adams, E. W. (1964). On rational betting systems. *Archiv für mathematische Logik und Grundlagenforschung*, 6:7–29 and 112–128.

[Borel, 1943] Borel, E. (1943). *Probabilities and life*. Dover, New York, 1962 edition.

[Cozman, 2000] Cozman, F. G. (2000). Credal networks. *Artificial Intelligence*, 120:199–233.

[de Finetti, 1937] de Finetti, B. (1937). Foresight. its logical laws, its subjective sources. In Kyburg, H. E. and Smokler, H. E., editors, *Studies in subjective probability*, pages 53–118. Robert E. Krieger Publishing Company, Huntington, New York, second (1980) edition.

[Gillies, 1991] Gillies, D. (1991). Intersubjective probability and confirmation theory. *British Journal for the Philosophy of Science*, 42:513–533.

[Gottfried and Wilson, 1997] Gottfried, K. and Wilson, K. G. (1997). Science as a cultural construct. *Nature*, 386:545–547. With discussion in vol. 387 p. 543.

[Grünwald and Dawid, 2004] Grünwald, P. and Dawid, A. P. (2004). Game theory, maximum entropy, minimum discrepancy, and robust Bayesian decision theory. *Annals of Statistics*, 32(4):1367–1433.

[Halpern, 1999a] Halpern, J. Y. (1999a). A counterexample to theorems of Cox and Fine. *Journal of Artificial Intelligence Research*, 10:67–85.

[Halpern, 1999b] Halpern, J. Y. (1999b). Cox's theorem revisited. *Journal of Artificial Intelligence Research*, 11:429–435.

[Hosni and Paris, 2005] Hosni, H. and Paris, J. (2005). Rationality as conformity. *Synthese*, 144(3):249–285.

[Howson and Urbach, 1989] Howson, C. and Urbach, P. (1989). *Scientific reasoning: the Bayesian approach*. Open Court, Chicago IL, second (1993) edition.

[Jaynes, 1957] Jaynes, E. T. (1957). Information theory and statistical mechanics. *The Physical Review*, 106(4):620–630.

[Jaynes, 2003] Jaynes, E. T. (2003). *Probability theory: the logic of science*. Cambridge University Press, Cambridge.

[Koopman, 1940a] Koopman, B. O. (1940a). The axioms and algebra of intuitive probability. *Annals of Mathematics*, 41(2):269–292.

[Koopman, 1940b] Koopman, B. O. (1940b). The bases of probability. *Bulletin of the American Mathematical Society*, 46:763–774.

[Kyburg Jr, 1968] Kyburg Jr, H. E. (1968). Bets and beliefs. *American Philosophical Quarterly*, 5(1):54–63.

[Kyburg Jr, 2003] Kyburg Jr, H. E. (2003). Are there degrees of belief? *Journal of Applied Logic*, 1:139–149.

[Kyburg Jr and Teng, 1999] Kyburg Jr, H. E. and Teng, C. M. (1999). Choosing among interpretations of probability. In *Proceedings of the 15th Annual Conference on Uncertainty in Artificial Intelligence (UAI-99)*, pages 359–365. Morgan Kaufmann Publishers, San Francisco, CA.

[Kyburg Jr and Teng, 2001] Kyburg Jr, H. E. and Teng, C. M. (2001). *Uncertain inference*. Cambridge University Press, Cambridge.

[Paris, 1994] Paris, J. B. (1994). *The uncertain reasoner's companion*. Cambridge University Press, Cambridge.

[Paris and Vencovská, 1990] Paris, J. B. and Vencovská, A. (1990). A note on the inevitability of maximum entropy. *International Journal of Approximate Reasoning*, 4:181–223.

[Paris and Vencovská, 2001] Paris, J. B. and Vencovská, A. (2001). Common sense and stochastic independence. In Corfield, D. and Williamson, J., editors, *Foundations of Bayesianism*, pages 203–240. Kluwer, Dordrecht.

[Pearl, 1988] Pearl, J. (1988). *Probabilistic reasoning in intelligent systems: networks of plausible inference*. Morgan Kaufmann, San Mateo CA.

[Press and Tanur, 2001] Press, S. J. and Tanur, J. M. (2001). *The subjectivity of scientists and the Bayesian approach.* Chichester : Wiley.

[Ramsey, 1926] Ramsey, F. P. (1926). Truth and probability. In Kyburg, H. E. and Smokler, H. E., editors, *Studies in subjective probability*, pages 23–52. Robert E. Krieger Publishing Company, Huntington, New York, second (1980) edition.

[Salmon, 1990] Salmon, W. C. (1990). Rationality and objectivity in science, or Tom Kuhn meets Tom Bayes. In Wade Savage, C., editor, *Scientific theories*, pages 175–204. University of Minnesota Press, Minneapolis. Minnesota Studies in the Philosophy of Science 14.

[Shannon, 1948] Shannon, C. (1948). A mathematical theory of communication. *The Bell System Technical Journal*, 27:379–423 and 623–656.

[Smith, 1961] Smith, C. A. B. (1961). Consistency in statistical inference and decision. *Journal of the Royal Statistical Society Series B*, 23:1–37. With discussion.

[Walley, 1991] Walley, P. (1991). *Statistical reasoning with imprecise probabilities.* Chapman and Hall, London.

[Williams, 1980] Williams, P. M. (1980). Bayesian conditionalisation and the principle of minimum information. *British Journal for the Philosophy of Science*, 31:131–144.

[Williamson, 2005a] Williamson, J. (2005a). *Bayesian nets and causality: philosophical and computational foundations.* Oxford University Press, Oxford.

[Williamson, 2005b] Williamson, J. (2005b). Objective Bayesian nets. In Artemov, S., Barringer, H., d'Avila Garcez, A. S., Lamb, L. C., and Woods, J., editors, *We Will Show Them! Essays in Honour of Dov Gabbay*, volume 2, pages 713–730. College Publications, London.

[Williamson, 2006] Williamson, J. (2006). Philosophies of probability: objective Bayesianism and its challenges. In Irvine, A., editor, *Handbook of the philosophy of mathematics.* Elsevier. Handbook of the Philosophy of Science volume 4.

Bayesian Ptolemaic Psychology

CLARK GLYMOUR

Henry Kyburg has argued for many years *contra* Bayesian epistemology. It has spread nonetheless into psychology, where it is prominent both as a norm and a description for human judgement. I wish merely to add to the supply of arrows in Henry's quiver.

1 Ptolemy versus Copernicus in Psychology.

Between Claudius Ptolemy and Johannes Kepler there lay both 1500 years and a methodological chasm. That gap reopens from time to time, and I maintain it has in our own day in cognitive psychology, and in particular in and about Bayesian accounts of the results of various psychological experiments. That is my thesis.

Ptolemy's device, the epicycle on deferent, allowed him to account very accurately for motions with respect to the fixed stars of the sun, the moon, and the five observable planets. Indeed, thanks to Harold Bohr, we now know, as Ptolemy did not, that any periodic motions can be approximated arbitrarily well by iterations of epicycles. So far as the data concern only semi-periodic motions with respect to the fixed stars, Ptolemy's framework can fit anything. Copernicus' framework is not so generous; it requires strict relations among the observable motions of the sun and the planets, and Kepler saw in those relations the very explanatory virtues of Copernican theory, and the comparative explanatory defects of Ptolemaic theory. In Kepler's hand, the Copernican identification of observable features of solar/planetary motion — angles of separation, oppositions, etc. — with orbital features reduces various empirical regularities to mathematical necessities.

Philosophers and statisticians have often since tried without much success to formalize the general explanatory difference between theories of Ptolemy's ilk and those of Copernican explanatory style, but anyone who has spent some effort with the history of science will usually know each kind when she sees it. The difference can be seen, for example, in the 19th century atomic theory (theories, really) and the theory of equivalent weights. In contemporary psychology there are a lot of Ptolemaic cousins, and increasingly they are Bayesian.

We have seen Ptolemaic theories in various computational theories of cognition, for example, those of Allen Newell and John Anderson. Each of them postulates a universal programming system, with changing psychological glosses. The programming systems, SOAR or ACT-R, have an unlimited supply of parameters, adequate to account for any finite behavior, that is, for any behavior subjects can exhibit. The theories provide no tight constraints of the kind Kepler celebrated, or unlikely but true predictions, comparable to the phases of Venus. Any physiological or anatomical interpretations of data structures are tacked on, *post-hoc*. Nonetheless, they are celebrated theories, and that says something about the scientific sensibilities of the psychological community. Allen Newell's SOAR briefly had its own society, with limited admission, like the Viennese Psychoanalytic Society once upon a time, and the ACT programs win awards. I once asked the editor of a book describing psychological experiments and their simulation in production system programs, a book that claims in its introduction that the experiments given evidence that Production Systems are the programming system of the mind, which experiments in the book could not be simulated in BASIC. He was candid enough to say "none." In the same annoying spirit, I once asked Newell for any unlikely prediction of his theory. He produced an unconvincing paper in response. Neural net models of psychological processes in which nodes are ambiguously cells or abstract representations of concepts are in the same Ptolemaic mode. We know that neural networks form a scheme for universal approximation of any function; what we see in neural network accounts of psychological experiments is the ingenuity of the programmer, not a physiological theory except by metaphor.

In the psychology of learning and judgement, theories formulated as programming systems with a gloss are bit by bit coming to be replaced by theories formulated as Bayesian learners, with a hypothesis space, a set of parameters for each hypothesis, a likelihood function relating possible data and each hypothesis and set of parameter values, a prior probability distribution over the hypotheses and their parameters, and the standard Bayesian updating by conditionalization. They are Ptolemaic in an obvious sense: an infinity of parameters is available, and by appropriate parameter choices any consistent preferences can be represented as in accord with maximum expected utility. I will suggest later that neither are they plausibly procedural hypotheses or "rational" hypotheses. My complaint is not lodged against Bayesian statistics, or against attempts to demonstrate, both empirically and theoretically, that aspects of cellular physiology — spiking frequencies, for example–implement probabilistic calculations.

A growing number of psychologists have lately taken up representation of simple causal dependencies by graphical causal models and have investigated and debated a range of considerations about how causal relations are learned: from passive observation of associations, from associations produced by interventions, by Bayesian methods, by matching models to constraints shown by patterns in data, etc. (Glymour, 2003). As in philosophy, the psychological literature shows a predilection for accounts of learning as Bayesian updating, more I think from slavery to fashion than from consideration of what happens when the idea is applied. I will describe three recent examples, each about adult judgement. Scores of other examples are in the recent psychological literature.

2   Waving Hands at Bayes

A recent paper by Kushnir and Sobel (in press) will serve as my first stalking horse, in part because it turns on an experiment that poses inference tasks that are reasonably close in structure to real scientific inference problems, for example in genetics (Ideker, et al., 2000). The paper is a revision of an earlier conference proceeding in which the authors claimed that the experiment shows that their experimental subjects used Bayesian rather than "constraint" based inference methods, a claim downplayed in the later paper, but one of the authors (personal communication), Sobel, still maintains that the experiment provides support for the claim that subjects are doing "qualitative" Bayesian reasoning. He does not say what that is.

Kushnir and Sobel gave their subjects ("participants" is the politically correct term) the following instruction:

> "In Dr. Science's laboratory, he has created a number of games. Each game has four lights, colored red, white, blue, and yellow. Each light also has zero, one, or many sensors. Some sensors are sensitive to red light, others to blue light, others to white light, and others to yellow light and sensors will activate the light that it is connected to. For example, if the red light is connected to a yellow sensor, then whenever the yellow light activates, the red light will also activate. But, because this happens at the speed of light, all you will see is the red and yellow lights activating together. It is also possible that a light has no sensors attached to it, and therefore is not activated by any other light. Sometimes Dr. Science is very careful about how he wires the lights together. At other times, he is not as careful, and the sensors do not always work perfectly."

The subjects' task was to determine which lights had sensors for which other lights, which amounts to determining a possibly cyclic directed graph on four vertices. Subjects could push a button to illuminate any single light on any trial, and could also disable the sensors on any single light on any single trial; they were required to make at least 25 trials and, when they were satisfied that enough data had been acquired, then to report a yes/no decision (and level of confidence) for each of the 12 possible edges. Each light remained off unless its button was pushed or one of its sensors "detected" another light. The probability of a light being on given that its button was pushed was 1. We are told that "Each sensor caused activation 80% of the time" which is an insufficient specification of the joint probability distribution, given that a light button is pushed, of the activation (on/off) states of the remaining lights: it does not tell us the probability of activation — i.e., illumination– when two or more of a light's sensors receive a signal. Moreover, the instruction given the subjects contradicts the actual set-up: subjects are told that sometimes sensors are wired carefully and sometimes not, so they should expect that some sensors work 100% of the time and some sensors that do not. I leave that complication aside.

Subjects were trained with a deterministic system to recognize that use of the control — putting a "bucket" over one of the lights–which effectively disconnected a chosen single light from the system, could disambiguate some situations: "They were shown through a series of generative interventions that purple caused green and that grey caused purple, but were unable to tell whether grey was also a direct cause of green. They were then shown that by putting the bucket on purple and activating the grey light, they could distinguish whether grey was a direct cause of green or only an indirect cause of green (via the purple light)." Giving such instruction implies that the psychologists' thought they were investigating implicit features of explicit strategies.

Subjects were then tested, each with data generated from their interventions on 4 alternative structures, see Figure 1.

The psychologists' interest was in demonstrating that subjects more accurately determined structural relations if the participants themselves chose the interventions (*autonomous* condition) than if they were given instructions, i.e., ordered, as to which button to push and which control to impose (*following* condition). The sequence of interventions in the following condition were in each case those of some other autonomous subject. The psychologists suggest — indeed in a previous paper, claimed outright–that there is a Bayesian account of their results, but they give no details. I will return at the end of this essay to explanations of this effect, which almost certainly cannot be as the psychologists suggest. In any case, to give a

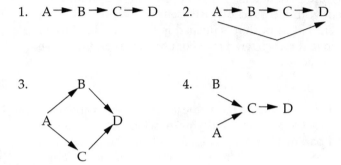

Figure 1. Lights of four colors were randomly assigned the roles of A, B, C and D.
Subjects did very well.

Bayesian account of the difference between autonomous subjects and followers, there must first be a Bayesian account of learning in the autonomous condition. What could that be?

In order to learn which graphical structure obtains, a Bayesian agent must have prior probabilities over the 4096 possible graphs on four vertices, and for each graph must specify the probability for any data sequence conditional on that graph. With these probabilities, the agent can in principle compute her probability for the data she has seen, and she can, again in principle, compute the posterior probability of each graph by Bayes Rule.

We can specify prior probabilities in many ways. For example, all graphs can be given an equal prior; or all graphs with the same number of edges can be given the same prior with the total probability for each edge number equivalence class equal, or perhaps weighted in favor of graphs with fewer edges; or we can assign a prior, k, for each edge independently, and take the prior probability of any graph to be equal to the product of the prior probabilities of its edge presences and edge absences. Since nothing is known beforehand, the edge priors should presumably be equal. So a graph with 1 edge will have a prior probability proportional to $k(1 - k)^{11}$ and a graph with two edges, any two edges, will have a prior probability of $k^2(1 - k)^{10}$, and so on. I will illustrate a Bayesian procedure (for the first trial only — for reasons to be explained) with the third way of specifying the prior probabilities.

To specify the probability of a data sequence for any particular graph requires a "parameterization" of the directed graph , with parameter values that determine the probabilities for any trial outcome, and a prior probability distribution over the parameters. I make the simplest specification I can think of consistent with the psychologists' set up: the probability of any light being on given that its button is pushed is 1. The probability of any light being on given that it is covered is 0. For any other condition, the probability that light X is on is:

(1) $Pr(X = \text{on}) = Pr(\Sigma_{Y \in \mathbf{OnPar}(X)} q_Y$

Where Y has value 1 if light Y is on and has value 0 otherwise, Σ is Boolean addition, $\mathbf{OnPar}(X)$ is the set of parents of X (lights for which X has a sensor) that are on, and q_Y is a parameter taking values in $\{0,1\}$. The parameters q_Y are independent in probability of each other and of all of the variables representing lights.

Formula (1) is the familiar noisy-or gate [Pearl, 1988]. If X has a single parent, Y, that is on, the probability that X is on is $Pr(q_y = 1)$. If X has two parents, Y and Z, that are on, then the probability that X is on is $Pr(qy = 1) + Pr(qz = 1) - Pr(qy = 1)Pr(qz = 1)$, and if three such parents, $Y, Z, W, Pr(qy = 1) + Pr(qz = 1) + Pr(q_w = 1) - Pr(qy = 1)Pr(qz = 1) - Pr(qz = 1)Pr(q_w = 1) - Pr(qy = 1)Pr(q_w = 1) + Pr(qy = 1)Pr(qz = 1)Pr(q_w = 1)$. I will assume, again for the simplest and most easily computed case, that $Pr(q_x = 1) = p$, the same for all X and for all graphs. For each graph, each prior probability for the graph and each value of p, a unique joint probability is specified for the values of all uncovered variables given that a particular button is pushed.

A Bayesian agent must combine the prior probability of each graph with a prior probability distribution *over* all possible values of p. Let $P_G(d;p)$ be a probability distribution for the outcomes on trial d, that results from a value p for the probabilities in (1), let P_G be the subjective probability distribution over values of p, and let D be the set of d — each of which is a list of which lights are on and which light is pushed — for a set of trials. Again to make matters as simple as possible, I assume P_G is a uniform distribution over p. Then:

(2) $Pr(D|G) = \Pi_{d \in D} \int_{P_G(d;p) \in P_G} P_G(d;p)dp$

(3) $Pr(D) = \Sigma_G Pr(D|G)Pr(G)$

And, by Bayes rule

(4) $Pr(G|D) = Pr(G)\Pi_{d \in D} \int_{P_G \in P_G(d)} P_G(d;p)dp / \Sigma_G Pr(D|G)Pr(G)$

And finally, for any edge E, let $G(E)$ be the set of graphs that have E as an edge. Then

(5) $Pr(E|D) = \Sigma_{G \in G(E)} Pr(G|D)$

Formula (2) is only valid if the trials are independent. Arguably, they are not. d is in effect conditioned on which buttons are pushed and which buckets are used. So the probability of D|G is in effect conditioned on the entire sequence of button pushings and bucket placements. (I don't see any way to get around this, because since people are engaged in discovery, the button pushings from trial to trial are not independent. The button pushings may depend upon the outcomes of the previous trial. Of course for each d, the only thing that matters is the buttons pushed on that individual trial.

$P(G|d1, d2) = P(G)P(d1, d2|G)/P(d1, d2)$
$P(d1, d2|G) = P(d1|G, p)P(d2|G, p, d1)$
$P(d1|G, p) = P(d1|G, p, act1)$ where act1 has probability 1
$P(d2|d1, G, p) = \Sigma P(d2|d1, G, p, act2)P(act2|d1, G, p) =$
   $= \Sigma P(d2|G, p, act2)P(act2|d1)$
$P(d1, d2|G) = P(d1|G, p, act1)\Sigma P(d2|G, p, act2)P(act2|d1)$

where act2 may be a function of $d1$.

Consider now how to compute the posterior probabilities of edges for a *first* trial in which W is covered and X is pushed with the result that Z goes on but Y does not. That enables us to compute by considering only graphs on the three variables X, Y, Z, with six possible edges.

Since Z but not Y is on when X is pushed, and W is covered, only graphs with an $X \rightarrow Z$ edge are consistent with the data. Since W is covered, we need attend only to subgraphs on the variables, X, Y, Z. There are 32 graphs on X, Y, Z containing an $X \rightarrow Z$ edge. I divide them into 3 classes as follows: Class 1 contains neither $X \rightarrow Y$ nor $Z \rightarrow Y$; Class 2 contains $X \rightarrow Y$ but not $Z \rightarrow Y$, or $Z \rightarrow Y$ but not $X \rightarrow Y$; Class 3 contains both $X \rightarrow Y$ and $Z \rightarrow Y$.

For each value of p, each Class 1 graph implies that the probability that Z, X and only Z, X are illuminated when X is pushed and W is covered is p. Class 1 contains

1 graph with 1 edge and has prior probability $\approx k(1 - k)^5$
3 graphs with 2 edges, each graph with prior probability $\approx k^2(1 - k)^4$
3 graphs with 3 edges, each graph with prior probability $\approx k^3(1 - k)^3$
1 graph with 4 edges, with prior probability proportional to $k^4(1 - k)^2$

For each value of p, each class 2 graph implies that the probability that Z, X and only Z, X is illuminated when X is pushed and W is covered is $p(1 - p)$. Class 2 contains

2 graphs with 2 edges, each graph with prior probability proportional to $k^2(1 - k)^4$

6 graphs with 3 edges, each graph with prior probability proportional to $k^3(1-k)^3$

6 graphs with 4 edges, each graph with prior probability proportional to $k^4(1-k)^2$

2 graphs with 5 edges, each graph with prior probability proportional to $k^5(1-k)^1$

For each value of p, each Class 3 graph implies that the probability that Z, X and only Z, X are illuminated when X is pushed and W is covered is $p(1-p)^2$. Class 1 contains

1 graph with 3 edges, with prior probability proportional to $k^3(1-k)^3$

3 graphs with 4 edges, each graph with prior probability proportional to $k^4(1-k)^2$

3 graphs with 5, edges, each graph with prior probability proportional to $k^5(1-k)^1$

1 graphs with 6, edges, with prior probability proportional to $k^6(1-k)^0$

Now we can compute the probabilities of various edges. For an edge such as $Z \rightarrow X$, or $Y \rightarrow Z$ or $Y \rightarrow X$ that make no new pathways from source to Y save through X, the posterior probability is the prior probability. For an edge such as $X \rightarrow Y$, since half of the probability of Class 2 graphs is for graphs with an $X \rightarrow Y$ edge, it follows that the posterior probability of the $X \rightarrow Y$ edge is equal to:

$$\frac{\int_0^1 \{1/2(p(1-p)[2k^2(1-k)^4 + 6k^3(1-k)^3 + 6k^4(1-k)^2 + 2k^5(1-k)]) + (p(1-p)^2[k^3(1-k)^3 + k^4(1-k)^2 + k^5(1-k) + k^6)\}dp}{\int_0^1 \{(p[k(1-k)^5 + 3k^2(1-k)^4 + 3k^3(1-k)^3 + k^4(1-k)^2]) + (p(1-p)[2k^2(1-k)^4 + 6k^3(1-k)^3 + 6k^4(1-k)^2 + 2k^5(1-k)]) + (p(1--p)^2[k^3(1-k)^3 + k^4(1-k)^2 + k^5(1-k) + k^6])\}dp}$$

The integrals are straightforward but tedious. The posterior does not simplify to any fraction independent of k, since it equals a ratio of the form

$$\frac{B+C}{A+2B+C}$$

where A, B, C are positive polynomials in k. The posterior of $Z \rightarrow Y$ is similar. Note that if each graph, rather than each edge, were equally likely, the computation would be slightly simpler. For the numerator and denominator in the posterior, one would need to count the number of graphs in each of the relevant classes above and multiply the cardinality of the graph class by the appropriate function of p, integrate over p, and sum.

If both Y and Z are on when X is pushed and W is covered, then the data require at least one of the following: an $X \to Y$ and $X \to Z$ edge; an $X \to Y$ and $Y \to Z$ edge; or an $X \to Z$ and $Z - Y$ edge. I will pass on the calculation.

If neither Y nor Z are illuminated, every graph on X, Y, Z is consistent with the data, and the graphs fall into several distinct classes with the likelihood as a function of p the same within each class and different between classes. Again, I skip the calculation.

The calculation of posterior probabilities from succeeding trials can become more intricate, as the probabilities of edges cease to be uniform and independent. Further, if the choice of which button to push and light to cover depends on the outcomes of previous trials, a Bayesian agent is essentially engaged in a sequential decision problem in which she is obliged to calculate the most informative next experiment. That means, for each trial, taking account of the previous trial, she must calculate the expected change in her degrees of belief from each of the 16 possible experiments she can make. Even for this simple problem, those calculations are not easy and there are a lot of them [Tong and Koller, 2001].

There are only three resources available to provide a Bayesian explanation of the difference between followers and autonomous subjects: they differ in their prior probabilities for the edges; they differ in the likelihoods — the probabilities of data given each graph and parameter probabilities — or they differ in the relevant data. The likelihood of the data given a graph, a value of p, and a choice of button to push and light to cover are a logical matter, the same in both cases; the relevant data the subjects see are the same — which lights illuminate when which buttons are pushed when which lights are covered. Two differences remain possible. One is the possibility of differences in prior probabilities in the two conditions. One source — the only one I can think of — for different prior probabilities in the two conditions is the suspicion, or degree of belief, in the following condition that the selection of button to be pushed and illumination of other lights is confounded by unknown factors. That is, something unrevealed causes the selection of the button for X and the illumination of light Z when the button for X is pushed. (A related explanation has been offered for our belief in freedom of the will [Glymour, 2004]). The effect of allowing this hypothesis in the case in which W is covered and Y does not go on, calculated above, is to decrease the prior probability of the $X \to Z$ edge compared with the value in the autonomous condition, other things equal. While such an hypothesis predicts a difference in posterior probabilities for followers and autonomous subjects, it does not suffice to explain why the followers are less accurate unless, by sheer chance (whatever that is for a Bayesian) the followers priors were sufficiently farther than the autonomous from the true frequencies of edges. Finally, followers and autonomous subjects could differ in their probability distribution for choice of actions. For example, if someone concocts a known strategy for choosing button pushings, it could be incorporated into Bayesian calculations. A set of button pushings not under their control could have a different probability distribution, perhaps leading to greater uncertainty. Each subject has her own preferred strategy in which there is a unique set of best decisions for the next trial given the outcomes of previous trials. Autonomous subjects, but usually not followers, get to use their preferred strategy. Followers instead must use a probability distribution over the strategies of whichever unknown autonomous subject they shadow. It is not clear, however, that, since the follower sees (indeed carries out) the actual sequence of decisions made by the autonomous subject she shadows, that the probabilities she has over the autonomous strategies make any difference (compared to the autonomous subject she is shadowing) in the end when she must report her judgements of the probabilities of edges.

The complexity of calculations required for a simple, coherent Bayesian account that is reliable over the general problem the psychologists pose seems to me make it implausible that their subjects instantiate a Bayesian procedure. Of itself, a Bayesian account offers no evident means to explain the superior performance of subjects in the autonomous condition. To merit the title of explanation, the suggestion that subjects use "qualitative" Bayesian procedures is obliged to show a procedure that is not "quantitative" and its connection with some real (i.e., quantitative) Bayesian procedure. The possibilities are myriad: Use only inequalities; assign incoherent priors or compute incoherent posteriors; use some alternative to conditioning for updating

What is a "constraint based" learning procedure? Some data mining algorithms for causal relations build a causal structure, or class of such structures from estimates of probability relations among the observed variables obtained from the sample data, for example from results of tests of hypotheses of conditional independence. But the psychologists' problem is special in that aspects of the structure are revealed trivially, or by simple counts, without recourse to probability estimates. For example, assuming the probability of sensor error is the same for all sensors, all of the following rules are valid or hold almost certainly in the limit as the number of trials N increases without bound. They make good decision procedures even for small N when p is $> \frac{1}{2}$.

Rule 1. If X is pushed several times and Y is illuminated more often than is Z, and W is covered, then there is a directed edge from X to Y.

Rule 2. If X is pushed, and Y is illuminated, then there is a directed path from X to Y not through any non-illuminated light.

Rule 3. If when X is pushed repeatedly, Z is never on unless Y (respectively, Y or W) is on, but not vice versa, then every directed path from X to Z is through Y (respectively Y or W).

Rule 4. If in trials which X is pushed, neither Y nor other lights are illuminated, then there is not a directed edge from X to Y.

The rules are obvious or should become obvious with a few experimental button pushes. They can be applied in a variety of strategies to solve the problems the psychologists pose, with small memory and computational requirements and with or without the use of the covering control. Merely for example, the following procedure:

> Note every possible edge as "can't tell."
>
> Choose a button at random and push it 5 times;
>
> Apply each of the 5 rules to the result and remember the required and prohibited directed edges and directed paths;
>
> Do the same with another button
>
> Until all buttons are tested
>
> For each "can't tell" edges that remains, say $X \rightarrow Z$, cover each of W, Y in succession and push X 5 times. Apply the rules; if neither covering eliminates the $X \rightarrow Z$ edge by the rules, then mark the $X \rightarrow Z$ edge required.

E.g., for structure 3, starting with pushing D, then C then B then A, we should find:

> Push D: By rule 4, eliminate all edges out of D into A, B, C
>
> Push C: By rule 4, eliminate all edges out of C into A, B; and by rule 1 $C \rightarrow D$ is required
>
> Push B: parallel to C
>
> Push A: By rule 2 and results already obtained, $A \rightarrow B$ and $A \rightarrow C$ are required.
>
> There remains a "can't tell" $A \rightarrow D$ edge.
>
> Cover B and push A: By rule 3, eliminate $A \rightarrow D$

Of course, I don't claim that the psychologists' subjects, or any one of them, used just this procedure. The point is that their problem is reliably solved by simple, non-Bayesian, low memory and low computational counting procedures. One way or another, I bet that is what their subjects are doing.[1]

3   Common Cause versus Common Effect

The first of a series of interesting experiments by Steyvers, et al. investigated how well subjects could use co-occurrences to distinguish between a system in which two of three variables were independent causes of a third and a system in which a single variable was a common cause of two others. Subjects were told that there are three aliens who communicate telepathically, that telepathic communications usually succeed but sometimes fail, and were asked to determine whether a particular alien (C) receives words from both of the others or sends words to both of the others. They were told the two structures are equally likely[2].

Subjects were given a block of eight trials of word occurrences for a given structure, and were given multiple blocks of trials with structure randomized across blocks. After each trial, subjects were required to provide a

---

[1]What then explains the difference in accuracy between the following and autonomous subjects? Lots of things, quite possibly. Commitment, perhaps. Some of the following subjects may not have embraced the goal given that they did not choose the actions. Confusion and memory limitations perhaps. Some of the following subjects may have had a fixed strategy in mind and not realized how to adapt it to the choice of interventions they were given. The set-up of Kushnir and Sobel's study does invite studies of human behavior in comparison with other norms. For example, suppose with low frequency the lights sometimes come on spontaneously, and we confine ourselves to acyclic graphs. For this setting, recent work by Frederick Eberhart suggests that the graphical causal structure can be most efficiently and accurately identified by strategies that push multiple buttons simultaneously. I should not be surprised if a slight modification of results in [Eberhardt *et al.*, 2005] applies as well to the semi-deterministic set up the psychologists use. It should be of interest whether people spontaneously use that recourse, and whether they do better if it is allowed than if they are restricted to pushing at most one button at a time.

[2]The picture is from [Steyvers, *et al.*, 2003].

conjecture as to the structure for the block in which the trial occurred. No feedback was given to the subjects. Unknown to the subjects, the transmission success probability was set at .8; if two transmissions in the same trial were received, one was chosen at random; if no successful transmission occurred to an alien, the alien displayed a word at random from among the vocabulary of 10 words.

Steyvers *et al.* found that about half of the subjects scored at chance; of the remaining, a fraction scored above chance with little variation between trials, and a highest scoring fraction scored above chance and improved across trials within a block, but not across blocks.

What explanations are possible for these results? The subjects were college students, and the half that scored at chance may simply not have cared about the problem. The half that scored above chance and did not improve on trials within a block presumably used some simple heuristic that gave them a better than $\frac{1}{2}$ chance at the right answer on any single case; the group that improved with trials presumably adapted their hypothesis to increasing evidence. A simple, reliable heuristic is available and obvious: guess that C is a common cause if $A = B$, and guess that C is a common effect otherwise. Simply guessing that C is a common cause if the number of cases in which $A = B$ is greater than the number of cases in which $A \neq B$, and guessing that C is a common effect otherwise, gives a learning rule. These simple procedures are in close agreement with the average performance of the two groups of subjects Steyvers *et al.* identify whose judgements are correct more often than chance. The simple heuristic explanation could have been tested by including blocks of trials in which C was masked. Unfortunately, that was not done.

The Steyvers *et al.* explanation is different. The subjects have prior probabilities $P(CC)$ and $P(CE)$ for the common cause and common effect models. The subjects know the two structures will be given with equal probability, so their priors are equal. They judge by the posterior ratio

(6) $\varphi = P(CC|D)/P(CE|D)$

where the trials are independent

(7) $\log(\varphi) = \sum_{t=1}^{T} \log \frac{P(D_t|CC)}{P(D_t|CE)}$

and the probabilities are determined by the true values of the probability of transmission success (α) and the number of words (n) according to

---

[4]From Steyvers *et al.*.

Table 1. The four data patterns (D) and their probabilities under a common cause (CC) and common effect model (CE) with $\alpha = 0.8$ and $n = 10$.[4]

D		$P(D|CC)$	$P(D|CE)$
1) $A = B = C$	All same	.67	.096
2) $A = C, B \neq C$ or $B = C, A \neq C$	Two adjacent same	.30	.87
3) $A = B, A \neq C$	Two outer same	.0036	.0036
4) $A \neq B, B \neq C, A \neq C$	All different	.029	.029

Further, subjects differ in their learning behavior according to two parameters. One is an exponential decay term δ for the contribution of earlier trials to the posterior:

$$(8) \quad \log(\varphi) = \sum_{t=1}^{T} \left[\log \frac{P(D_t|CC)}{P(D_t|CE)} \right] e^{-\delta(T-t)}$$

The other parameter allows decisions to be made not strictly on the odds ratio. That is, for some non-negative quantity γ:

$$(9) \quad P(CC) = \frac{1}{1 + e^{-\gamma\varphi}}$$

To carry out this computation, subjects must in each block remember which words are the same or different for each alien in the order of the trials so far, must know the true values of the transmission probability, α, the number of words, n, must use them to compute, as in table 1, the ratio of the probabilities of each trial outcome on each hypothesis, must decrement that trial term exponentially by place order, take the log, sum the results over all trials so far, take the anti-log of the sum, and substitute the result for φ in equation 8 above.

This, I submit, is not a plausible procedural hypothesis. The computational complexity is well beyond what normal people can consciously compute, and there is no evidence that participants can produce numbers corresponding to intermediate stages of the computations. Ptolemaic psychologists may argue that their models are not accounts of mental processing, but only of input/ouput relations, an issue to which I will return later.

4 Bayesian Cheng Models

Patricia Cheng has produced a series of studies exploring how large sample ("asymptotic") judgements of the strength of a putative causal factor in producing an effect are inferred from time-ordered co-occurrence data, and she has systematized the results in an interesting theory that can be reconstructed as inference to parameter values in noisy-or-gate and noisy-and-gate graphical causal models. Inevitably, her theory and some simple experimental results have been Bayesed.

Lu *et al.* consider experiments in which subjects are shown data under two conditions, before and after a treatment. The experimental display shows patients who had not (top) or had (bottom) received an allergy medicine, and who had (frown) or had not (smile) developed heachaches.

When patients were not given any medication, this is how they felt:

When patients were given medicine *A*, this is how they felt:

The point of the experiments is to demonstrate that subjects have a strong preference for attributing a causal role when the association between the putative cause and the effect is invariant, either always, or almost always, generating the effect or always, or almost always, preventing the effect. I do not contest that conclusion at all, and anyone who has spent much time with naïve subjects' judgements of cause and effect will not be surprised. Lu *et al.* claim a much stronger conclusion: subjects are doing Bayesian updating with non-uniform priors biased towards necessity or sufficiency of the absence or presence of a factor. According to their account, subjects must decide between two graphs, in the first of which *B* is a cause of *E* and in the second, *B* and *C* are causes of *E*. As usual the decision depends on the support

(10) $\log \dfrac{P(Graph1|D)}{P(Graph0|D)}$,

To obtain the ratio priors and likelihoods are required; likelihoods require integration over parameters

(11)
$$P(D|Graph1) = \int_0^1 \int_0^1 P(D|w_0, w_1, Graph1)P(w_0, w_1|Graph1)dw_0 dw_1$$

$$P(D|Graph0) = \int_0^1 P(D|w_0, Graph0)P(w_0|Graph0)dw_0$$

For Cheng noisy-or-gate models, in which b, the presence or absence of B, is 1, and the presence (1) or absence (0) of C is c, and e^+ means e occurs, and w_0 and w_1 are the causal powers of B and C, respectively, to produce e,

(12) $P(e^+|b, c; w_0, w_1) = 1 - (1 - w_0)^b (1 - w_1)^c$

Now with data in which there are $N(c^+)$ cases in which C occurs, and $N(e^+)$ cases in which e occurs, and similarly for $N(c^-)$

$P(D|w_0, w_1, Graph1)$

$$= \binom{N(c-)}{N(e+,c-)} w_0^{N(e+,c-)}(1 - w_0)^{N(e-,c-)}$$

(13) $$\binom{N(c+)}{N(e+,c+)} [1 - (1 - w_0)(1 - w_1)]^{N(e+,c+)} [(1 - w_0)(1 - w_1)]^{N(e-,c+)}$$

$P(D|w_0, Graph0)$

$$= \binom{N(c-)}{N(e+,c-)} \binom{N(c+)}{N(e+,c+)} w_0^{N(e+,c-)+N(e+,c+)}(1 - w_0)^{N(e-,c-)+N(e-,c+)}$$

One still needs $P(w_0, w_1|Graph)$, for which Hu et al. specify:

(14) $P(w_0, w_1|Graph1) = \dfrac{1}{Z} \left[e^{-\alpha w_0} e^{-\alpha(1-w_1)} + e^{-\alpha(1-w_0)} e^{-\alpha w_1} \right]$

Here α indicates the bias towards necessity and sufficiency. $\alpha = 0$ is a uniform distribution, and Hu et al. compare predictions of judgements on the uniform with $\alpha = 30$. Z is a normalizing constant.

Thus, according to Hu et al. to form a judgement as to whether C is a cause, or how likely it is to be a cause, subjects must separately multiply (14) with appropriate parameter values by each equation in (13) and integrate each result over values of w_0 and w_1. They show that the procedure nicely fits their data, but no evidence that subjects carry out any such complex calculations.

5 Modeling Versus Explaining

One sometimes hears such Bayesian models described as "rational" models of judgement, placing them as contributions to a dispute within psychology that has its modern genesis in the well-known work of Tversky, Kahneman, and others. "Rational" in this context is a misappropriated term of moral psychology, wrenched from most of its sense. "Rational" as Bayesian modelers use the term means judgements that are probabilistically coherent and updated by conditionalization. One can be rational in this sense and mad as a hatter, quite unable to learn the truth about much of anything. Simply give probabilities 0 to a lot of true stuff, and probability 1 to a lot of false stuff, and those degrees of belief will never change by conditionalization. Arguably, we cannot avoid that state. If, as I suspect and cognitive psychologists used to assume, we are bounded by Turing computability, then if we are probabilistically coherent, we must give probability 0 to an uncomputable infinity of logically contingent propositions. Turing's theorem plus the premise that our cognitions are Turing computable implies that we must be infinitely dogmatic. The newspapers confirm the hypothesis.

The connection of Bayesian coherence and updating by conditionalization with rationality as ordinarily understood is through the Dutch Book arguments, which show that if degrees of belief are betting odds on bettable propositions, then there exists a sure win combination of bets against anyone whose betting odds are not probabilistically coherent or who does not update them by conditionalization. This is a worst-case, means-ends argument. But in the worst case for the end of reliably converging to the truth, by realistically limited means, Bayesian inference methods fail. Taking computation and rationality seriously means allowing that rational means to ends must be those that can be carried out effectively, in the computational sense of "effective." The goal of inquiry is presumably interesting truth, and we know there are learning problems in which whichever of an infinite set of mutually exclusive alternative hypotheses is true can eventually be learned by Turing bounded learners, but not by any Bayesian Turing bounded learners. Whether there are interesting problems of this kind is unknown, but the very existence of any such examples shows, I think, that the equation of "rational" and Bayesian is rhetoric that could do with further reflection.

Frederick Eberhardt and David Danks [2006] have pointed out a simple and devastating objection to the claim that subjects in psychological experiments act in accord with Bayesian rationality requirements. Justified by subject training or prompting, the models typically assume participants share common priors, or closely related priors. They then argue that the *average* participant response is the maximum posterior (given the data the participants receive) probability given the prior. But, as Everhardt and Danks show, this requires either that most participants are not updating by conditionalization, or else that *most* subjects are not selecting the response that has maximum posterior probability, which, on any reasonable assumption about their utilities, Bayesian rationality would require.

Ptolemy's astronomy models the motions of the planets very well, but it doesn't correspond to what goes on. If our interest were merely in a mathematical synopsis of the data, or prediction of the distribution of values of more data of the same kind, a model like Ptolemy's is just fine. If one wants to understand what is going on, it doesn't help much. The same is true of Bayesian models of judgement that have little or no procedural plausibility or evidence. They are style — good style, nice style, style I admire for its clarity, but style not the less for that– over substance.

I can imagine objections that go like this: *Sure, Bayesian inference is an idealization, but no more an idealization than graphical causal models or causal Bayes nets as psychological models. Since you advocate the latter, how can you consistently object to the former?* My answer is that graphical causal models, and certain limited operations on them, provide qualitative descriptions of what people believe and expect when they understand causal relationships. It is not a theory of how that understanding is acquired. So the representation has only rather limited space and time requirements.

What alternatives are there? In the 1970s Rescorla and Wagner proposed a dynamics for the adjustment of association strengths in classical conditioning. Their linear dynamics requires only a simple arithmetic operation and the retention of a number representing the association strength for each factor considered, updated as trials occur. The model accounted for a great deal of the data from animal behavior in classical conditioning learning, where association strengths are somehow measured after each presentation of values of cues and target. Two odd things happened to the RW model: despite widespread use in psychology, and attempts to compare its predictions with "asymptotic" formulations of causal judgement such as Cheng's, no full, computable characterization of the equilibria (there are no real asymptotes) of the updating dynamics appeared until the dissertation work of a young philosopher, David Danks, who showed [Danks, 2003], for any joint probability distribution on cues and target, how to compute the equilibria if they exist and how to decide whether they do exist. The other odd thing is that the RW model came to be interpreted as more than a model of how association strength is learned, in fact as a model of how causal strengths are learned. In the latter role, it is essentially a kind of regression model of causal learning that does not distinguish learning from passive observations of cues and targets, learning from watching others interventions to bring about or prevent cue appearance, and learning from ones own interventions. For those reasons it became the target of many psychologists who wished to emphasize the importance of interventions in learning causes and also wished to distinguish causal beliefs from mere associations, even while allowing that associations might be evidence for causal hypotheses. The result is that the recent research on adult and children's causal judgement usually does not record trial by trial judgements — Steyvers' *et al.* paper is an exception — and does not attempt a mathematical description of the learning dynamics, except as Bayesian conditioning.

The interesting thing is that in subsequent work, Danks went on [Danks, 2003a] to provide a low memory, computationally trivial learning dynamics whose equilibria, when they exist, satisfy Cheng's hypothesis about the relation of probabilities to judgements of the power of a potential cause to produce an effect. Roughly, it looks as if given any such relation, Danks' procedures can find a dynamics that equilibrates in it. Of course, alternative dynamics may have the same equilibria, but the requirements of low computation and low memory restrict the possibilities. Rather than building computationally implausible Bayesian models of causal learning, it seems to me far more important for psychologists to investigate the learning dynamics, and how people assemble learned associations, or causal powers, into any semi-unified knowledge of causal relations.

Acknowledgements

Thanks to Choh Man Teng, David Danks, and, especially as usual, to Peter Spirtes.

BIBLIOGRAPHY

[Eberhardt and Danks, 2006] F. Eberhardt and D. Danks. Confirming Bayesian Models of Inference in Cognitive Science, preprint, Carnegie Mellon University; *Psychological Review*, submitted, 2006.

[Danks, 2003] D. Danks. Equilibria of the Rescorla-Wagner Model. *Journal of Mathematical Psychology*, 47: 109-121, 2003.

[Danks *et al.*, 2003] D. Danks, T. L. Griffiths, and J. B. Tenenbaum. Dynamical Causal Learning. In S. Becker, S. Thrun, & K. Obermayer, eds. *Advances in Neural Information Processing Systems 15*, pp. 67–74. Cambridge, Mass.: The MIT Press, 2003.

[Kushnir and Sobel, in press] T. Kushnir and D. Sobel. The importance of decision making in causal learning from interventions. (in press) *Memory and Cognition*.

[Glymour, 2003] C. Glymour. *The Mind's Arrows: Bayes Nets and Graphical Causal Models in Psychology*. Cambridge: MIT Press, 2003.

[Glymour, 2004] C. Glymour. We believe in freedom of the will so that we can learn. Comment on D. Wegner, *The Illusion of Conscious Will*, Behavioral and Brain Sciences, 2004.

[Ideker *et al.*, 2000] T. Ideker, V. Thorsson, J. A. Ranish, R. Christmas, J. Buhler, J. K. Eng, R. Bumgarner, D. R. Goodlett, R. Aebersold, and L. Hood. Integrated Genomic and Proteomic Analyses of a Systematically Perturbed Metabolic Network, *Science* 292, 929–934, 2000.

[Lu *et al.*, 2006] H. Lu, A. Yuille, M. Liljeholm, P. Cheng and K. Holyoak. Modeling Causal
    Learning Using Bayesian Generic Priors on Generative and Preventive Power. In R. Sun &
    N. Miyake (Eds.), *Proceedings of the Twenty-eighth Annual Conference of the Cognitive Science
    Society*. Mahwah, NJ: Erlbaum, 2006.
[Pearl, 1988] J. Pearl. *Probabilistic Reasoning in Intelligent Systems*. Morgan Kaufmann, 1988.
[Steyvers *et al.*, 2003] M. Steyvers, J. Tenenbaum, E. J. Wagenmakers and B. Blum. Inferring
    Causal Networks from Observations and Interventions. Cognitive Science, 27, 453–489,
    2003.
[Tong and Koller, 2001] S. Tong and D. Koller. Active Learning for Structure in Bayesian
    Networks, *Proceedings of the 2001 International Joint Conference on Artificial Intelligence*, 2001.

An Architecture for Purely Probabilistic Negotiating Agents

RONALD P. LOUI

This paper was presented at the Probability as a Guide to Life Symposium at the University of Rochester, Department of Philosophy, October 2004, in honor of my advisor, Henry E. Kyburg, Jr.

These ideas date from 1998 and result from collaboration on negotiation games with Anne Jump.

1 Introduction

Instead of viewing negotiation as a game with a solution, this paper conceives of negotiation as a process. This view of negotiation is useful both for the design of automated agents, and for the mathematical description of the social interaction. The view may be adopted as a psychological claim about the measurable forces within human agents, but that claim is not made here.

The process of negotiation is regulated by two principal kinds of linguistic act: the extension of an offer and the notification of unilateral breakdown. Other linguistic acts such as questioning, answering, focusing, and threatening, are not included here for the sake of simplicity.

At any time, an objective probability can be attached to each potential settlement, representing the probability that it will be the settlement if there is agreement. Meanwhile, an objective probability can be attached to the possibility of non-agreement, or breakdown. This objective probability is based on past observations of similarly situated agents in similar strategic circumstances, having made similar progress toward settlement. The exact ontological basis for induction and inductive method are beyond the scope of this paper, but any inductive statistical approach will suffice. It is crucial that this probability not be a subjective probability: it cannot be altered by will or desire, strategy or design, bias or dogma, or any other psychological mechanism. There is interpersonal agreement on what the probability is, once the data and the method are fixed. Furthermore, objective probability is binding upon the rational agent as a guide for action.

The process of negotiation is driven by the dynamics of objective probability assessments, conditioned on the amount of time that has elapsed since last non-trivial progress. Specifically, as time since last progress grows, the probability of breakdown rises, and the expected utility of potential settlement falls. It is an empirical hypothesis that there is a granularity of time with respect to which probability of breakdown rises monotonically as progress is not made. It is hard to imagine situations where the weight of empirical evidence does not warrant this hypothesis. The expected utility of settlement will fall as pessimism rises until expectation reaches either the best offer extended by the other agent, or else the security level of the agent.

Any offer that is slightly above security would be accepted if pessimism had no compensatory force. Of course, this behavior is not witnessed in practice, nor is it acceptable behavior for designed agents.

Whenever the other agent fails to make progress, the utility of unilateral breakdown rises. That is, the act of breaking down on the other agent becomes more valuable, the more uncooperative that agent's behavior is perceived to be. A nonstandard procedural utility is attached to the time since last progress by the other agent. The effect of this procedural utility is to create a race between the falling expectation, due to pessimism, and the rising value of breakdown, based on resentment and the prospect of punishment. It is the principal claim of this paper, as a matter of design, that these two forces suffice to govern the basic negotiating agent and produce a variety of acceptable and desirable behaviors.

The one-sided rational agent is given deadlines at which time an offer must be extended, an offer must be accepted, or unilateral breakdown must occur. There are many situations in which a one-sided bound is appropriate. For example, when a negotiating broker has the power to act on behalf of another, but does not have the full proxy to suppress independent acts by the person or persons represented by the broker. A trade representative, for example, may have the power to extend an offer, but not the exclusive power to extend offers. Similarly, the single agent may be governed by a society of minds, a large part of which will behave according to expected utility, but some of which may yet behave with independent initiative.

A two-sided rational agent is not precluded by the approach here. Such an agent would have both an objectively determined earliest time to act and an objectively determined latest time to act. However, a point-rational agent, where an exact earliest and latest time to act are prescribed, and the times coincide, violates the constructive philosophy of this approach. One-sided rational constraints provide normative guides for action, but there may be other grounds for acting "ahead of schedule."

The procedural utility here that causes rejection of an 11-cent offer in favor of a 10-cent security level at breakdown in, for example, a zero-sum split of a dollar, is not the same as a substantive utility attached to the unfairness of a split. It has become famously popular in recent years to model the outcome of negotiation in such a way that objects of value take into account perceptions of property rights or distributive justice. Fair-split effects are easily modeled prior to the process of negotiation by transforming the payoff functions. Here we are modeling the procedural effect of altering objects of value based on how they are reached through negotiation, not just what they represent to an agent in relation to other potential settlements. Here we are interested in the possibility that a unilateral 10-cent outcome has more value if the path to that outcome is torturous. Prior authors have considered the possibility that a 10-cent outcome has less value if there were an aspiration for a dollar.

In either case, whether the interest is distributional or procedural justice, the transformation of payoffs models an internal value of dignity or pride, and a willingness to invest in a social mechanism that produces external pressures on other agents to behave better.

2   Pairs of Agents with Payoffs

Two agents A and B are conjoined in action when the payoff to each agent depends on the choice pair (a, b) from the space of joint choices, Ω_A and Ω_B for A and B respectively. In strategic form, the former is the set of rows and the latter is the set of columns. In the absence of coordination, A chooses from Ω_A and B chooses from Ω_B simultaneously. In this model, the space of joint choices is also the set of potential settlements. $\Omega = \Omega_A \times \Omega_B$, but we often carry the cross-product into the expressions to emphasize the pairing of action.

The payoffs in this model are u_A and u_B.

$$u_A, u_B : \Omega_A \times \Omega_B \rightarrow \Re.$$

3   Dialogue

At any time, $\Upsilon_A(t) \subseteq (\Omega_A \times \Omega_B)$ is the offer set extended by A at t, the set of offers A has extended to B. Similarly, $\Upsilon_B(t)$ is the offer set extended by B at t. If at any time the intersection is non-null, we declare agreement and an end of dialogue, even if the intersection is non-unique.

Suppose that offer sets are monotonic in time;

$$\forall t' > t : \Upsilon(t) \subseteq \Upsilon(t')$$

i.e., offers cannot be rescinded.

Dialogue can end with the acceptance of an offer, which might as well be the reciprocal extension of an offer that has been extended by the other agent. Dialogue can also end in either agent's unilateral breakdown. This is a simple and final notification of termination, either bd_A or bd_B, depending on who declares breakdown first. Breakdown implies the absence of an agreement.

4 Security and Breakdown

In the absence of agreement, we can assume that each agent has a method for selecting an uncoordinated choice. These are a_{bd} and b_{bd},

$$bd = (a_{bd}, b_{bd}) \in (\Omega_A \times \Omega_B)$$

where the security level of A is thus $u_A(bd)$, and for B, it is $u_B(bd)$.

For monotonic payoff functions on ordered choice sets, where u_A is decreasing as choice from Ω_A or Ω_B increases, and u_B is increasing as choice from Ω_A or Ω_B increases, the security levels may as well be determined by the unique Nash equilibrium, $a_{bd} = min_{\prec}\Omega_A$, $b_{bd} = max_{\prec}\Omega_B$. In terms of strategic form, this means that the breakdown act is in the upper right corner, $bd = (1, |\Omega_B|)$, if the lowest ordered choice pair, e.g., (1,1), is in the upper left.

For all t the agent has an objective probability of breakdown, $Prob_A^{bd}(t)$, which at the moment does not distinguish between breakdown of A on B, or B on A, or simultaneous breakdown. There is a probability of settlement, distributed over all the potential settlements Ω,

$$Prob_A^s(t) = \Sigma_{\omega \in \Omega} Prob_A^s(t)(\omega) = 1 - Prob_A^{bd}(t).$$

5 Admissible and Probable Settlements

At t, the value of the best offer to A is

$$\nu_{.A}(t) = max_{\Upsilon_B(t)} u_A$$

which might also be called A's value of A's best offer from B. This is what A would get if A were to settle immediately on B's best terms at time t.

A's concession level is A's least value among A's offers to B:

$$\nu_A^{.}(t) = min_{\Upsilon_A(t)} u_A \ .$$

In this discussion, A's objective probability of settlement at t should be supported only for outcomes A values at least as much as $\nu_{.A}(t)$ and no more than $\nu_A^{.}(t)$. Furthermore, if $\nu_{.B}(t)$ and $\nu_B^{.}(t)$ are known to A, then A's probability of settlement is restricted to outcomes with u_B in that range as well:

for private payoffs (u_A only known to A, u_B only known to B),
$\forall t : \forall \omega \in (\Omega_A \times \Omega_B)$:
if ($u_A(\omega) < \nu_{.A}(t)$ or $u_A(\omega) > \nu'_A(t)$)
then $Prob^s_A(t)(\omega) = 0$.

for public payoffs (both u_A and u_B known to both A and B),
$\forall t : \forall \omega \in (\Omega_A \times \Omega_B)$:
if ($u_A(\omega) < \nu_{.A}(t)$, or $u_A(\omega) > \nu'_A(t)$,
or $u_B(\omega) < \nu_{.B}(t)$, or $u_B(\omega) > \nu'_B(t)$)
then $Prob^s_A(t)(\omega) = 0$.

The practical effect is to prune the set of admissible settlements as proposals are made. The joint pruning effect is remarkably effective in practice. For the agent guided by objective probability, this means that even a short exchange of reasonable offers can quickly identify the likely outcomes of negotiation if there is settlement, and tight bounds on resultant utilities if there is settlement.

The concession level is equal to A's value of A's best offer to B, when payoffs are monotonic:

$$\nu'_A = u_A(argmax_{\Upsilon_A(t)} u_B) .$$

For payoffs that are not monotonic, it may be better to use the latter expression as an upper bound on the support. Actual experience may include data where the values of settlement lay outside even this enlarged best-offer-to/best-offer-from range, but perhaps such data represents negotiation with agents who are not fully rational, or negotiation situations that are peculiar.

6 Expectation

As soon as there is a probability distribution over settlements,

$$Prob^s_A(t) : \Omega \rightarrow [0, 1] ,$$

there is an expected value of settlement,

$$Eu^s_A(t) = \Sigma_{\omega \in \Omega} Prob^s_A(t)(\omega) u_A(s)$$

which weighs each settlement's value with its likelihood in the usual way. Absent an actual distribution, $Prob^s_A(t)$, a convenient way to think about this is with a Hurwicz parameter, α. (We resist calling this an optimism-pessimism parameter as it is traditionally called because we are using pessimism differently below.) Presuming α_A and α_B, or just α when the distinction is not important,

$$Eu^s_A(t) = (1 - \alpha)\nu'_A(t) + \alpha\nu_{.A}(t) .$$

Expected utility takes the expected utility given settlement and the utility given breakdown, and mixes them by the probability of breakdown:

$$Eu_A(t) = Prob_A^{bd}(t)u_A(bd) + Prob_A^s(t)Eu_A^s(t) \ .$$

7 One-Sided Rational Acts

Rationality requires latest times for certain acts:

1. If at t, $\exists \omega \in \Upsilon_B(t) : Eu_A(t) \geq u_A(\omega)$, then A accepts an offer of B; else

2. If at t, $\exists \omega \in \Omega, \omega \notin \Upsilon_A(t) : Eu_A(t) \leq u_A(\omega)$, then A extends an offer to B; else

3. If at t, $Eu_A(t) \leq u_A(bd)$, then A unilaterally breaks down on B.

There is no problem ordering these so that one rule defeats another if two rules prescribe conflicting action. This is especially important with discrete time.

8 Pessimism

The probability of breakdown, $Prob_A^{bd}(t)$, naturally exhibits a dynamics we refer to as pessimism: the longer the time since non-trivial progress was made, the greater the probability of breakdown. This is an empirical hypothesis for our agents, since their assessment of probability is induced by their actual experience. Furthermore, this hypothesis seems plausible only after an initial reactionary period, post progress, during which the probability of breakdown might actually fall. Basically, if one waits long enough and nothing has happened, the probability of an agreement becomes more remote.

It is sufficient here to theorize about a granularity of time, δ, and to pay attention only to times since progess was made by the other agent. Non-trivial progress here is any offer that improves $\nu_{.A}$, though others could define non-triviality more stringently. Others might also drive agent behavior using the probabilities conditioned on time since either agent makes progress, but this choice is not important here.

Pessimism. (Empirical Hypothesis) Agents are likely to discover (for some $\delta > 0$) that

$$Prob_A^{bd}(t^{lp} + k + \delta) > Prob_A^{bd}(t^{lp} + k) \text{ for all } k > 0,$$

where t^{lp} is the time of the last non-trivial progress by the other agent, i.e., t^{lp} is the earliest time at which $\nu_{.A}$ took on its current value at time t

$$t^{lp} = argmin_z \nu_{.A}(z) = \nu_{.A}(t) \text{ and}$$

A linear pessimism is one in which $Prob^{bd}(t^{lp} + k + \delta) = (k + \delta)\pi$ for all $k > 0$, for a fixed parameter π, so long as this value is less than or equal to one. Note that linear pessimism implies a resetting of $Prob^{bd}$ to zero whenever there is non-trivial progress by the other agent.

Agents who experience unbounded pessimism are manipulable by agents who simply wait to make concessions. By withholding progress, the mendacious agent can cause the pessimism-driven agent to have a falling expectation until, eventually, $Eu_A(t)$ becomes not $Eu_A(t) = Prob_A^{bd}(t)u_A(bd) + Prob_A^s(t)Eu_A^s(t)$, but simply $u_A(bd)$.

If the manipulating agent has made an offer worth any ϵ greater than $u_A(bd)$, then the unbounded pessimistic agent will eventually accept that offer rather than break down. This is simply the "rational" behavior of taking a $u_A(bd)+\epsilon$ "sure thing" over an expecation of $u_A(bd)$. This behavior is undesirable even if there is no intention to manipulate by the agent who witholds offers while the other agent's pessimism rises.

At this point, a game theorist might try to release the agent from even the weak, one-sided rationality constraints and make room for a counter-strategy of adversarial waiting. We take a different tack. The agent who waits for offers is also the agent who experiences a rise in the procedural utility attached to unilateral breakdown. Put simply, if one agent offends the other by making the other agent wait, the second agent will find breakdown sweeter.

9 Resentment and Punishment

A procedural utility $u_A^{bd}(t)$ is attached to the act of unilateral breakdown, so that the actual value of breakdown, when initiated by A, is not $u_A(bd)$ but is instead $u_A(bd) + u_A^{bd}(t)$. For simplicity, assume the procedural effect resets, $u_A^{bd}(t) = 0$, whenever B makes non-trivial progress. That is, there is no memory of prior offense. It is just as easy to imagine a memory parameter, μ, such that $u^{bd}(t') = \mu u^{bd}(t)$ when B makes non-trivial progress at t.

u^{bd} is a source of utility which presupposes that value can be created by process.

There are many ways to understand u^{bd}. It can be a model of dignity, pride, or social investment. We were probably all taught to attach a u^{bd} value to unilateral breakdown, and would instruct our pupils to do the same; we can do no differently for our artificial agents. There may be a negative u^{bd} reflecting a penalty for breakdown, when agents have been especially cooperative. We do not theorize about negative u^{bd} here. u^{bd} is introduced solely as a force to compensate for pessimism.

There is the possibility that the agent does not actually feel the value u^{bd} in the same way that the agent feels u. There are two responses. First, artificial agents don't feel either, so they can presumably be designed with arbitrary choices for both u and u^{bd}. Second, the agent who does not actually feel u^{bd} should either not use it in calculating the value of breakdown, or else should seek more self-awareness.

The presence of this measure amends one of the rationality constraints,

3'. If at t, $Eu_A(t) \leq u_A(bd) + u_A^{bd}(t)$, then A unilaterally breaks down on B.

There are also subtle changes the equation for $Eu_A(t)$ which do not change the statement of rationality constraints 1 and 2, but which change their dynamics:

$$Eu_A(t) = Prob_A^{bd}(t)(u_A(bd) + u_A^{bd}(t)) + Prob_A^s(t)Eu_A^s(t)$$

This equation permits the value of the negotiation to rise based on speculation about the value of eventual breakdown. If one charts the Eu curve against time, it can sometimes turn upwards when the agent who is close to breaking down decides to postpone breakdown (as well as capitulation) until u^{bd} has risen.

Even more subtly, for agents who distinguish between unilaterally breaking down and being unilaterally broken down upon, the expression requires yet another parameter which weighs the probability of the former, e.g.

$$Eu_A(t) = Prob_A^{bd}(t)(u_A(bd) + (.5)u_A^{bd}(t)) + Prob_A^s(t)Eu_A^s(t)$$

when each is equally likely.

The agent who witholds an offer now creates a race between falling expectation due to pessimism and rising value of breakdown, due to resentment. Since acting on resentment is a kind of punishment for the agent who does not make progress, we refer to these agents as pessimism-punishment agents (PP agents).

A linear resentment (with no memory) is one in which $u^{bd}(t^{lp} + k) = \rho k$. This implies that $u^{bd} = 0$ when there is progress, and it implies there is no bound on how much u^{bd} can grow.

A linear PP agent is thus specified by two parameters, π and ρ.

10   Laissez-Faire Paths

For two PP agents, each agent is governed by a $Prob^{bd}$ and a u^{bd}. In the case of linear PP agents, there are four parameters to note: π_A, π_B, ρ_A, and ρ_B.

A closed offer set is one in which all settlements $s \in \Omega(t)$ that A values no more than $\nu_A^{\cdot}(t)$ are in $\Upsilon_A(t)$, and all settlements that B values no more than $\nu_B^{\cdot}(t)$ are in $\Upsilon_B(t)$.

Closed offer sets can therefore be characterized uniquely by the concession levels of each agent, $(\nu_A^{\cdot}(t), \nu_B^{\cdot}(t))$, from which one can infer $(\nu_{.A}(t), \nu_{.B}(t))$. For agents who close their offer sets, this point is like a state of the negotiation at time t. There are actually two more values to add to the state, the time since A's last progress, and the time since B's last progress. The state of negotiation for two PP agents is thus

$$\sigma(t) = (\nu_{.A}(t), \nu_{.B}(t), t_A^{lp}, t_B^{lp}) \ .$$

A successor state to $\sigma(t)$ is either a change in $\nu_{.A}$, because B makes a substantive new proposal, which resets t_B^{lp} to zero, or a change in $\nu_{.B}$, because A makes a new proposal, resetting t_A^{lp} to zero. In discrete time, A and B can both make progress in the same time step. There are two other possibilities: agreement and breakdown. Agreement can happen if A's expectation falls to $\nu_{.A}$, or if B's expectation falls to $\nu_{.B}$. Breakdown can happen if A's expectation falls below $u_A(bd) + \rho_A t_B^{lp}$, or if B's expectation falls below $u_B(bd) + \rho_B t_A^{bd}$.

A series of successor states is a laissez-faire path. All laissez faire paths end in an agreement or a breakdown. Given initial levels of concession, public valuations u_A and u_B, and public announcement of linear PP parameters, the trajectory and final state of a laissez-faire path is determinable.

The path of negotiation can be controlled by accelerating offers so that the agents move from one laissez-faire path to another.

11 Discussion

In this paper we have attempted three things. First, an expectation-driven negotiating agent is contemplated. The insistence on probabilities that are objective and the estimation of a probability of breakdown made it possible to force the agent to make rational concessions, so long as the expectation is falling due to empirically-based pessimism. The compensatory force was the willingness to punish based on resentment born of the procedural utility attached to unilateral breakdown on noncooperative negotiating partners.

Second, the resulting PP agents are decoupled from speculation about each other's mental states or solution strategies. PP agents can even announce that they are playing according to their π and ρ and are only slightly manipulable since they are essentially treating the game as a game against nature. An adversary can postpone offers until the latest time before breakdown, thus receiving the maximal offer set from the PP agent; but eventually the PP agent will break down unless concessions are matched. Memory parameters further protect the PP agent from manipulation.

Third, the picture of negotiation becomes a picture of path-selection among various laissez-faire paths, at least for a pair of PP agents negotiating with each other. The problem is not one-shot coordination within the set of possible settlements, e.g., the identification of a unique equilibrium. The problem is to identify and assess the acceptability of the path that the process of negotiation is taking, and to decide whether it is worth controlling that path with alternately motivated or alternately justified action.

Navigation: An Engineer's Perspective

MARIAM THALOS

There is a certain tangle of philosophical questions around which much philosophy, especially in our time, has circled, to the point where now there is something that deserves to be called a holding pattern around these issues: What are causes? How do they compare with reasons? What is Reason, with a capital R? How does it participate in the production of intentions that lead to action, particularly in arenas rife with uncertainty? Where do formal systems of symbols come into all of this? And how — if at all — can formal methods be harnessed to serve science and public policy, through guiding belief formation and decision-making? Henry Kyburg, Jr., has himself circled around this tangle of questions, at least once or twice. And so in his honor, and in the no-nonsense spirit of empiricism that marks his work as the work of a scientist, I'd like to sketch a way to cut through some of the Gordian knots at the center of this tangle. The theme will be that natural science has much of value to offer that has been willfully neglected by philosophers. This by itself is nowise surprising, as philosophy, particularly in the most exclusive parts of the academy, has suffered from an excessive transcendentalism — a theme to which I will return in this piece periodically. Now Kyburg has not been guilty of contributing to the causes for the decline of philosophy. He has, instead, been courageously working out the implications of his convictions regarding the virtues of vigorous formal systems, contributing to the advancement of empiricist methodologies, and generously supporting the causes of realism. In emulation of that courage, I offer this essay in the service of bold and vigorous formal systems, realism and — most emphatically — empiricism.

1   Life

Life would be so very awesome if one had only to produce a consistent set of wishes, and all would be ordered accordingly (whatever "accordingly" might come to). Well, maybe it wouldn't be so awesome, and instead chaotic, if the rules applied to every creature with wishes to put in order, however benighted; but life would sure be different, and no mistake. For, whether fortunately or not, simply pronouncing "Let there be the necessities of life, and surplus besides!" does not result in their materialization. In real time, one has to expend resources, effort and time in pursuit of the most basic needs, to say nothing at all of the dreams that make life worth living. And part of a well-ordered life is taking account, in planning, of what lies outside the ambit of what one can lay plans about — what is at least from one's own perspective, a matter of chance. But when it comes to counting costs, how exactly should risks figure in? How should we, as rational entities, arrange matters so as to take risk into account — wisely?

A preponderance of opinion has maintained, axiomatically, that separating cases involving risk off from cases where risk is not a consideration — where the agent in question knows, for every option, what will transpire if it is elected — is a wise move. And that this will help with untangling the knot of issues with which we are concerned. (However this strategy, as an axiom, is nowhere the subject of philosophical attention.) Consequently there are those who make a living by inquiring into how action is reached in cases involving none of these pesky uncertainties (a mob of philosophers advancing the study of so-called Practical Reason). And arrayed at another end are those who conceive of the case of certainty as a limiting case of that which forms the subject matter of their studies, taking for granted — axiomatically — that proper handling of this zeroeth case is a simple matter of means-ends calculus, roughly in line with what Jeremy Bentham called the Principle of Utility. This latter mob of theorists proceeds to perform what they take to be appropriate generalizations to cases with considerable risk, and in doing so pushing back the frontiers of formal decision theory.[1]

Now, thinking about risk — as such — has a long and illustrious history: the roots go back to the ancients. But the notion that chance can be quantified is a modern one, with a history that winds through the annals of gaming for fun and profit. The very idea that chance might be subject to quantification is founded in the aspiration for putting such knowledge — if it could be had — in service of one's advantage — as part of a well-ordered life in the gaming hall. And this, fundamentally, is an engineering idea.

Engineering ideas have enjoyed a long and very distinguished history all their own. It is my contention that indeed something of the engineer's concerns lie at the heart of the 20th century movement now routinely (and in some circles sneeringly) referred to as "logical positivism," with its animus for metaphysical excesses. The positivists' impulses for formal methods, lying at the foundation of contemporary decision theory, have the spirit of "let's get the thing organized such that anyone or anything — howsoever deficient of insight, but so long as he, she or it can follow explicit, step-by-step instructions — can perform it" that is so characteristic of the engineer.

---

[1]For more on these different mobs, see [Thalos, 2005].

But in their zeal for formal methods, the positivists lost sight of the fact that taking a formal approach nowise absolves one of having to do metaphysics. And indeed that utilization of formal methods creates the obligation of articulating just how formal systems have application in the world of real-time entities and their real-time affairs.

Some philosophical investigators — among them those with engineering instincts — take for granted that the question of how our concepts apply to the real world is quite unproblematic. We, as architects of our language, are in the authoritative position of specifying the contents of our concepts: we are therefore also the authorities on how to apply them. And so all we need do is simply identify or specify the content of our concepts, and then look in the world to see what, if anything, satisfies them. This, according to some pure analytic philosophers, is how concepts apply to the world. But we have known for rather a long time that this way of articulating how segments of natural language make contact with the world is entirely too simple. For example, we have known for some time that at least proper names and the names of natural kinds, marked as such in natural language, do not operate this way; and that at least in relation to these terms, application precedes content specification — if there is anything at all deserving of the name of content in these cases. Everyday users of linguistic media are indeed the least authoritative on how such terms make contact; and equally these users are least authoritative on their content, whatsoever this is. And this fact raises the very perplexing philosophical issues of what exactly is in the head of the ordinary user of symbols and symbol systems, and how the end user of such systems manage to do anything at all with them. The latter Wittgenstein was famously exercised by these perplexing issues, maintaining that meaning, whatever it is, is not in the end user's head. But if not meaning, then what? And how does it do its job?

Of course the engineer will interject that this is why we require the devising of artificial, as contrasted with natural languages. But this is a mistake. One cannot make the specification of meaning for an artificial language, without somewhere using natural language — that ill-understood beast — to do it.

The argument I've just laid out applies to more than the content of our concepts. It applies also to the very workings of our minds. Some people think that the first person is authoritative in some way, at least over the relations or successions among their thoughts — their judgments and intentions, and their reasons for them — simply because this self-same first person is the one who "makes up" her or his mind. [2] And yet this is known to be false in numerous cases. First, the evidence is that people's reasoning

---

[2]This idea is explicitly advanced by [Moran, 2001].

about their attitudes and preferences is far less a matter of attention to "what my attitudes ought to be, given my concerns and reasons" than it is a matter of what I happen to have available to reflection at the moment. But, even more importantly, my attitudes now, even those we should like to refer to as avowals, need have no special rank, status or veto-capability in relation to other factors that might affect my behavior: there is no assurance in avowals, as such, that they characterize the workings of an authoritative entity. Different folks are organized differently; in some, avowals enjoy a ranking office, whereas in others they do not. Some are so organized that avowals have only marginally more than a chance relation to the would-be actions they avow. That is what is so uncanny about the weakness of will.

And so we are overcome by the questions: Over what aspects of itself does the self hold sway? What kind of authority is that authority? In what sense, and to what extent, is the self a self-invented entity?

Of course these are very fundamental philosophical questions, at the heart of all humanistic inquiry, and as fundamental to Plato's inquiries as to those of existential and post-modern philosophers. But to the extent that empirical science can shed light on them, these are not questions for philosophers alone. And we must adopt a cooperative strategy for approaching and handling such questions. This shall be the theme of my essay: that there is quite a lot we can learn from empirical science in making our way with the knot of questions with which we began. And the real work lies in how to learn from empirical science in a way that facilitates philosophical inquiry. This is something Henry Kyburg has not forgotten.

2   Life post-positivism

There has been much lamenting in recent decades the sorry legacy of positivism, that bastard philosophy of the engineer. But I think that its appearance on the philosophical stage has had an important and salutary effect: it has made us cognizant of the benefits of formalization, and of being honest about one's intellectual objectives and intellectual aspirations. It requires us to put engineering concerns up front, and to be honest about the engineering objectives. At the same time, though, it allows for the self-deception that a formal system — if only we can have one — will answer all important foundational and philosophical questions, even in the absence of a philosophical treatment of the question of how our formal system makes contact with the world as it is in itself — in other words, it makes it easy to demand substitution of formal analytical methods for serious metaphysics. And this is just as much a legacy of what came before positivism, and made it possible — namely, rationalism — as it is a legacy of positivism itself. Let's back up a bit and make this point properly.

To be practically wise is to be disposed to guidance — to be appropriately sensitive or appreciative of features of the world that can serve as useful signposts — in such a way as serves a purpose, goal or end. The end may be a goal that the wise one aspires to inherently, or an end that the wise one comes to aspire to *through* possessing that self-same susceptibility to guidance. But what exactly does susceptibility to guidance amount to? I will refer to sensitivity to guidance as the capacity for *navigation*. An account of this capacity ought to be a central goal of a number of overlapping philosophical projects, tangled up in our Gordian knot.

But first notice that accounting for the capacity for navigation has two parts: the first consists of identifying appropriate principles of navigation; and the second consists in providing assurance that the principles of navigation so identified, are actually ones to which the entities they are designed to guide — in this instance, human beings — are capable of implementing; in other words, that the principles so identified can actually serve the purpose of navigation for members of our species. However one regards the first part of the problem, the second part is definitely empirical and ontological: it concerns how matters stand with us. And it is in a fundamental sense an engineering problem as well. Engineering problems typically have these two dimensions: the first deserves calling the *practical* problem, and the second the *theoretical* or *metaphysical* problem.

Framing the issues so as to illuminate these two questions, illuminates also what proves to be an unavoidable division of the labor performed in the cockpit (so to speak) of a human life, and effects a division of attentions to the (different) roles that navigator unit (plan-maker) and pilot unit (action-taker) enjoy. It also prompts inquiry into how performance of these tasks is coordinated to the point that it can make some kind of sense in the evolutionary scheme of things.

By breaking the project down this way into two parts, we also become sensitized to the possibility that there might be no solution to it — no way of providing an engineered method of navigation. Because there might be no principles of navigation that the "navigator" can apprehend, which would also be such that the "pilot" can put them into action. This would constitute a kind of species-wide weakness of will. It might turn out, in other words, that members of our species are insusceptible to guidance *of the form that can be produced through the kinds of artifice that members of the species are capable of devising.* Or at least, that an engineered or artificial form of navigation would be overall inefficient, perhaps because implementing it would involve bypassing another and more efficient form of navigation that evolution has also instilled in the organism. To know whether this is actually the case, we would have to have a clear grasp of the variety of forms of navigation that humans naturally enjoy, and to be able to compare them with what the engineer is proposing. And this is at least partly an empirical undertaking.

Except for David Hume, philosophers from the modern period onwards have generally not appreciated the sheer magnitude of this problem. Nor have they appreciated that the project must break down into the two parts I mentioned. For a long time now philosophers have been substituting for analysis of what it is to be susceptible to guidance, analysis of how to proceed in life in a rational manner. In other words, they've sought to produce a guide to life in general, a standard to which would be held anyone in any circumstance, without regard for that someone's specific concerns, without regard for the specific resources at their disposal. Philosophers have thereby taken for granted that susceptibility to guidance must itself be susceptibility to *their preferred* program or form of guidance. By addressing the question of guidance–of norms–as they have, they have made very substantive (and, in my view, totally indefensible) assumptions about the relationship between the nature of real-time sensitivity to guidance, and their own preferred ideals of susceptibility to guidance, tailor made to suit the resources of persons who enjoy a certain excess of leisure. Their admonitions have been contemptuous of those, in the vast majority of humanity, hard at labor and with reduced life prospects.

Kant — and indeed those who weed and prune in the garden of respect-for-autonomy that he was first to cultivate — are egregiously guilty of this error. Having entirely separated the world into two realms — the realm of freedom, on the one hand, and the realm of bondage on the other — they have no way of arranging for collaboration between the navigator and the pilot. Unfortunately, their rationalist methodology, shorn of scientific aspirations, is a disease that has plagued the Anglo-American tradition in philosophy for more than a century now — much longer than any ill effects of positivism itself. And I don't see many signs of its abatement. Here is what I mean.

Plato and Kant, each in his own way, bequeathed to us a methodology of withdrawal from all the influences of an unpredictable world of purportedly deceptive and corrupting *appearances*, in pursuit of *first principles* for every major intellectual and moral enterprise. Now familiar under the labels of *rationalism* and *transcendentalism*, this methodology commends a swift retreat to the safe haven of Reason, as the only worthy shore and soil for all that is reliable, good and worthy of affirmation. Reason in this tradition is conceived, not as an aspect of physical or embodied functioning — a facet of so-called *cognition*, whose exploits include the processing of numerous varieties of sensory stimuli. It is conceived, rather, in terms that *contrast*

with — and indeed are intended to transcend analysis of — such functioning. In this tradition, Reason is a realm unto itself, a realm of both inquiry
and value — of both knowledge and goodness — the source of all norms.
Rationalism treats the world of experience as a realm of bondage, where one
is subject only to forces — that one can neither know straightforwardly nor
control; starkly contrasting with the life of the mind as a realm of freedom,
wherein everything is transparently known and where the human being —
and being human — really comes into its own. The rationalist therefore
diagnoses a tension between what we see and experience, on the one hand,
and on the other hand what things — including ourselves — really are,
fundamentally, in a truer realm. Furthermore, the rationalist bids us, like
the prophets, retreat from the world of superficial appearances, nevermore
to return to that unworthy place out of which Reason calls us. The true
human realm is a *transcendental* realm — the realm of norms, the realm
where adherence to norms is the fundamental fact of human life.[3]

It is possible that Descartes — philosophy's cinematic archetype of the
rationalist — applied a rationalist methodology in the interests of science —
in the interests of a true description of the world.[4] Still, Plato and Kant had
entirely different aims: they were as much occupied with moral and political
objectives as they were with topics that we now think of as belonging to the
empirical sciences. And so those who followed their lead took one of two
pathways: those with a heart for science took the positivist/conventionalist
road, and the others continued along a pathway of conceptual analysis as
the single tool of the philosophical enterprise, leaving metaphysics (or so
they hoped) far behind.

The developing human sciences, meanwhile, have had a rather different
and mixed story to tell on the subject of human behavior. It is this story
that richly deserves to be brought into the limelight now, in the interests of
a better philosophical account of human navigation.

Social psychology since the 60's and 70's has demonstrated nothing if
not that human judgment and behavior is enormously sensitive to — indeed, profoundly shaped and contoured by — the presence of other people,
who those people are in our social economy, and what they are doing and
saying. For example, it should come as old news that people take cues
from the behavior of others when taking stock of their circumstances (their
"interpretations" of what they see are dependent upon how they see other
people as interpreting the situation), that the influence of women is subor-

---

[3]This is certainly the agenda we see in [Brandom, 1994], following a lead he discerns
in Wilfred Sellars.

[4][Nelson, 1995; 2005] indicate that this is not entirely correct, and — we must add —
even if it were correct, then Descartes' methodology did not enjoy as much success as it
aspired to.

dinate to the influence of men (over both men and women), that men and women exert influence in different ways, and that their group behavior is different, among many other things. (Differences in group behavior between American men and American women are especially fascinating, but not the subject of the present paper.) What's more, people's behaviors vary quite predictably with circumstance: circumstance — and sometimes rather trivial circumstance — shapes behavior in ways unrecognized in common or folk psychology.[5]

On the very cutting edge of this research is the work of social psychologist Richard Nisbett and his many collaborators [Nisbett, 2003]. They are demonstrating in a wide range of studies how culture impacts how we judge, and even how we perceive. They are showing that differences in culture between East and West result in — and indeed are sustained by — differences in cognition. Importantly, they are demonstrating that there is nothing universal about the western ways of categorizing objects, or engaging in debate over the worthiness of a hypothesis; that, indeed, there are at least two discernible styles of judgment and intellection, that might or might not admit of combination in any given context.

Therefore, to the extent that forces of various kinds affect human judgment, they leave much less room for people to apply principles of reasoning reliably. And even if we could be got to reason correctly, this may not be enough to move us all the way to action that accords with that correct reasoning. For example, economists tell us that people's valuations are subject to undesirable effects: appetite for the immediately present negatively impacts people's valuations, with the twin ill effects of weak will and discounting of the future; what's more, contrary forces tend to make us fail in the other direction: we are also, and contrarily, prone to honor sunk costs.

3   First landmark

There are developments afoot warranting optimism that soon we shall be on a road to untangling the Gordian knot I described at the outset. Things are now happening to put right some of what went wrong when philosophy invented the realm of transcendence. Here now is my Whiggish history.

One hero of the tale I shall be spinning, and founder of the Berlin circle, is Hans Reichenbach. His most important contribution, to my mind, is formulation of the Principle of the Common Cause. This principle was key to launching the enterprise known as probabilistic causality. The pinnacle of that enterprise is Judea Pearl's most recent pair of opera: *Probabilitstic Reasoning in Intelligent Systems* [1988] and *Causality: Models, Reasoning*

---

[5][Nisbett and Ross, 1991] is one of the best treatments of a range of such facts. A more recent but also more contentious treatment is [Doris, 2002].

and Inference [1990]. Pearl, who is a professor of computer science, trained as an engineer, writes that the project is to "treat causation as a summary of behavior under intervention." Underlying that enterprise is the idea that probabilities are what is changed under intervention. The perspective is this: there is at the bottom of it all, a layer of fundamental, irredeemable, uncontrollable stochasticity. And causes shift probabilities, by elevating or lowering them, in something like quantum jumps. *Causation* is the name we give to these shifts in probability.

As Kyburg is right to point out in a review of the Pearl's most recent volume [Kyburg, 2005], this idea treats probability as a most objective matter, and causation as the instrument or simply another name for the fact of its modification over time. Causality is the name by which we call those interventions that alter probabilities in time.

These ideas harmonize very well with Hume's observations that we extract probabilities from facts of correlation, and make good sense too of how we manage to master the concept of causality at an early age. These ideas elegantly advance the enterprise of probabilistic causality with a reductionist agenda — the agenda of replacing causal formulae entirely by formulae in a language with well articulated rules of inference — the language of probabilities. This agenda is justifiable, according to proponents of probabilistic causality, because there is nothing more to causal facts — at least for engineering purposes — than facts of probability. Why has this line of inquiry proven useful? Medicine and engineering have profited enormously from these ideas; and I predict that the discipline of statistics will benefit equally from it, when once statisticians recognize that this is the route to making sense of how to use the language of probability in support of causal hypotheses. And of course the analysis of causation is at one focus of the tangle of questions before us. And so progress is made on our knot of question. [6]

On the strength and vigor of probabilistic causality, a conclusion is now warranted:

- If causes are objective, then probabilities are objective.

As we will see, this is a much mightier proposition than its converse

---

[6]Once again philosophy makes progress. For progress in philosophy does not look like progress in other fields. It does not look like problem-solving, in the style of the empirical sciences. Consequently the contention that philosophy makes no progress is woefully mistaken. Philosophy is a midwife: its progress is measured by the number of new discipline births, each of which is distinguished by proprietary investigatory methods, that examine questions in workman-like fashion. Conceived this way, philosophy has made enormous progress: just look at the number of disciplines — all going under the banner of sciences — from physics to linguistics — that owe their livelihoods to philosophy.

(which is the one usually celebrated among philosophers).

4   Second landmark: Buying and selling risk

Suppose we are considering entering the insurance market. We are looking at the question of whether to sell Jones $100K of insurance for year Y, at the rate of $3000.

	Jones dies in Y $P = [0.01, 0.05]$	Jones survives Y $P = [0.95, 0.99]$
Sell	$(-97K)	$ 3K
Don't sell	0	0

Here, the numbers in brackets denote the endpoints of a probability range in which we are guessing that the true probability lies. (This is Kyburg's proposal for how best to represent our knowledge about probabilities, and I will be following it here, without comment.)

We can calculated expected monetary values (EMVs) for selling and not selling, respectively, as follows:

$$\text{EMV(selling to Jones)} = [\text{-2K}, +2K]$$
$$\text{EMV(not selling to Jones)} = 0$$

An insurance premium of $3000/yr is where we'll break even. So if we jump into the market of insuring Joneses, at a premium of say $3500/yr, and manage to sell insurance to 1000 such Joneses in the first year, then we can be almost certain that 10-50 of them will have died by the end of the year. That puts us down $1-5M, but we will have collected $3.5M in premiums, for a rough annual income of $1M, if we're allowed simply for the sake of argument to compute an average — a (roughly) million-dollar business. By the end of the decade, we will have paid out $10-50M, but taken in a certain $35M. Of course, some years we'll be down, and others we'll be up. The more Joneses we insure, the more we can be certain that our figures at year's end will be in the black.

So it can be rational for us to buy the Joneses' risk, if we buy enough of it, because by doing so we offset uncertainty to some extent. By the same token, it can be rational to sell one's risk, for exactly the same reason. Consider Jones's own table:

	Jones dies in Y $P = [0.01, 0.05]$	Jones survives Y $P = [0.95, 0.99]$
Buy	Joneses are secure	$ 3K
Don't buy	Joneses are destitute	0

The Joneses might reason as follows: We want to make sure that our family does not suffer the destitution that is certain to follow if Jones dies. And so, while the EMV of buying insurance might be roughly equal to or even less than the EMV of not doing so, it might nonetheless be worth it to us to adopt the course of action that eliminates the risk of destitution entirely. When someone reasons in such a way as to find buying insurance attractive, it's usually because they're using something besides EMVs as the final basis of their decision — something more like the maxi-min principle. Application of maxi-min is acceptance of risk-intolerance as rational. And there is a market in life insurance only if risk-intolerance is sufficiently widespread.

Redistribution of risk can be good for all concerned. But it's good in different ways for the different parties to the would-be transactions. Saying this amounts to taking a global perspective on all the decision making entities involved — a scientific perspective in the science of economics, that enables reference to what is being exchanged in the same terms. The insurance company must be offering for sale what Jones has an interest in acquiring. Alternatively, Jones must be wishing or willing to part with what the insurance company has an interest in acquiring if offset by a payoff. In the former instance, we call it insurance; in the latter, we call it risk (with compensation). (Compare: When a grocer sells you a piece of produce, money changes hands; she is offering for sale what you have an interest in acquiring, and you are in return prepared to part with what she has an interest in acquiring. In the first instance we call it a banana; in the second instance, we call it capital, or in different terms, an opportunity for further exchange.)

The point being, in an economic transaction, we have an exchange of goods between two parties; where each party is bringing to the table something the other has an interest in acquiring. The parties do not have to value the objects in question the same way: if they did, there surely would be no exchange. But the parties have to have the same understanding, at some level, about what is being exchanged. That's why the language of contract make so much sense in these contexts. The parties to an exchange in the insurance industry happen to understand that what they are exchanging is money for risk. And so just as money is objective, so must risk be as well:

- For there to be a genuine market in insurance, there must be something to buy and sell that makes the offer of transactions profitable.

- For there to be a genuine market in risk, risk has to be objective — for nothing could be more objective than what can be traded on a large scale.

- Risk is just the chance of loss.

- So, because there is a genuine market in risk, chance of loss is objective.

- If we're allowed to interpret chance of loss as probability of loss, probability of loss is objective.

- The probability of something is objective; therefore probability is objective.

5   Convergence

We now have two converging lines of inquiry, converging on the idea that, whatever the status of Reasons, probability is objective and capable of founding an analysis of causation suitable for (at least) one aspect of a theory of guidance.

The significance we place upon this convergence depends upon two fundamental moves: (1) we take seriously and at face value the language of science, for the sake of a guide to metaphysics; (2) we judge that progress has been made in each of the two lines of inquiry, at least vis--vis probability.

These are obviously not uncontroversial issues in the larger space of philosophy of science. (Indeed they might be matters subject to more controversy than the very conclusion I am saying they point to — the matter of the objectivity of probability.) But to pose and address these issues exemplifies what I call "organic philosophy." This is an approach that views all other inquiries — and particularly scientific inquiries — as having potential bearing on the present one (whatsoever it is). Organic philosophy as I am conceiving it is the progeny of the refusal to acknowledge such things as spheres of disciplinary sovereignty or autonomy, over which reign proprietary methodologies, and where only those trained or initiated in the appropriate methodologies have dominion. Organic philosophy refuses to cede, *ab initio*, any thesis to the effect that findings or methodologies in a certain field or on any one given topic, have no bearing on certain others.

Some philosophers adhere to the doctrine of reflective equilibrium, and might regard organic philosophy a sister doctrine. The method of reflective equilibrium consists in working back and forth between specific judgments (some use the term "intuitions") about particular instances or cases, the principles or rules that are conceived as governing them, and the theoretical considerations that bear upon them, revising any of these elements wherever necessary in order to achieve an acceptable harmony in the overall body of judgments, and adding to it where the occasion demands. The method succeeds and reflective equilibrium is reached when we arrive at an acceptable coherence among these things (judgments about specific cases, principles and theoretical considerations). An acceptable coherence requires not only consistency among these things, but also that some of them provide support or explanation for others. Practitioners of reflective equilibrium advocate local forms of organic philosophy, confining their conceptions of bearing and relevance to specific disciplines or to even more local areas of inquiry. (Nelson Goodman [1955], who introduced the method if not the name, practiced it on the topic of justifying inference rule in deductive and inductive logic.)

But organic philosophy is in principle global. It is born of the empiricism that has been the guiding force for enlightenment in the history of thought — Kyburg's included. To the organicist's way of thinking, a judgment on any issue should be judged by its scientific fruit — by the vigor and health of the body of scientific explanations to which it contributes.

The smaller point I wish to press here is this: that the fact of convergence just traced out illuminates the larger puzzle — the gordion knot at whose center lie puzzles concerning how practical reasoning participates in a process that leads to action: puzzles of real-time navigation in a perilous world.

The usual path in contemporary Anglo-American philosophy is quite different from this one: proponents usually just muster intuition in favor of favored metaphysical position, or produce argumentation to the effect that denying some position P on some metaphysical question results in having to deny some more cherished intuitions or ideas or linguistic practices. This is a rationalistic methodology. The path we've traveled here is quite different. It leads to convergence upon how to think about interconnected topics implicated in the puzzle of navigation.

We can pursue the same strategy in connection with the larger puzzle of navigation itself. We can further explore empirical avenues that will help us with cracking that puzzle from different angles. For example, we can inquire empirically about how people navigate their lives. This need be no rhetorical question. Here are some findings.

6 Navigation: The Science — *Or, practical reasoning for philosophers*

Not so very well known even among psychologists is the fact that people's behavior is shaped by their level — and type — of attention to themselves, a factor often discussed in the field of social psychology as self-focus. For example, it is well known that the presence of a simple mirror enhances task performance. (Subjects were given, for example, a foreign-language text copying task; when there's a mirror or camera in the room subjects produce better copies than when there's not. Interestingly, a similar effect can be produced by giving the subject a minority status in a group: subjects with minority status perform better on neutral tasks than subjects who are not in the minority.) Why does attention to self produce better task performance? The standard answer among social psychologists is that self-focus produces better self-evaluation of one's performance: it triggers on-line self-assessment.

Word of caution: the effects of self-focus are not simply that a personal standard of excellence comes into play when self-focus is brought to bear. It is rather that self-focus produces the application of *some* standard (or other) of assessment — and there are many forms of assessment that can be triggered by a mirror or a camera — and that this triggering has a profound effect on the control of behavior. So what is crucial is not that self-focus brings a proprietary standard (the "agent's avowals") into play, although it can do this. But rather that it triggers an evaluative form of executive control. (Which standard will be triggered by self-focus, of the many that might be available to an agent, is at this point in time, unknown.) In summary: self-focus is a way of triggering a standard of evaluation, and can be such that subjects are not aware of this triggering. (For more on this, see the many interesting essays in [Wegner and Vallacher, 1980]. And compare the role of self-focus in the account of executive control offered by Philip David Zelazo and his coworkers.)

That behavior has many such sources is not necessarily a good thing — though neither is it in itself a bad thing. (Indeed one should view the mirror as a very helpful — and extremely inexpensive — behavioral program for improving school performance, for example.) A theory of the regulation of executive function would (among other things) delineate cases where self-focus is good for performance from cases where it is not. And this is what an empiricist account of practical reason can be good for: articulating the principles that apply to regulation of executive function.

I therefore want to express these now well-worn results of social psychology as follows: self-focus is a means of putting or shifting executive function into an evaluative mode. And that this mode of operation in which executive function is functioning evaluatively, is quite different from other modes of its operation. For example, it is quite different from modes in which executive function is being exercised only passively or in a semi-automated or caretaker mode; this is a mode in which the executive is not intervening but simply monitoring an automated (or over-learned) task in a permissive manner. And furthermore I want to insist that the exercise of executive function — whether in an evaluative or caretaker mode — is what practical reason amounts to.

Now for a crucial philosophical point: executive function is not a principle of action. It is throughly contentless, especially vis--vis what we *should* do. It is the container, subject to being filled in any way an agent likes, or simply any way whatever (whether the agent in question likes it or not). It is the "pilot" and not the "navigator." Because we articulated the pilot/navigator distinction, we are able to cast the findings of social psychology in this help-ful way — a way that allows us to steer clear of the over-intellectualization of navigation, as will become clearer presently. Here now is more on what executive function is.[7]

For the sake of articulating the nature of executive function, I will be following a lead discernible in a recent article by psychologists Robert J. Sternberg and Louise Spear-Swerling [1998]. There they are lobbying for what they refer to as a "new construct" in psychology, one whose measure-ment will be a better predictor of success in life. They suggest calling it Personal Navigation "PN", making note that in both fiction and true life, we are familiar with the idea: it is present in the notion of life as a voy-age. I like the name as well. While I take certain cues from Sternberg and Spear-Swerling, my articulation serves a distinctively philosophical project: it serves the purpose of identifying a purely practical, counter-intellectual, counter-rationalistic conception of the practical enterprise. It is an empirical conception. Here now is a sketch of it.

It is quite well documented that some people experience serious difficul-ties in early childhood, but nonetheless enjoy enviable success in adult life. Other youngsters enjoy spectacular achievement in early childhood — often as a result of carrots and sticks provided by parents — but then proceed to snatch failure out of the jaws of success. One tempting conclusion is that there are no reliable predictors whatever of success — that it's all a matter of chance. But if there is anything at all in the idea that the successful person possesses traits that enable success, one is obliged to say something, however rough, about what the sorts of trait in question might be like. I want to do just that. The analysis I am proposing is aimed at making sense of a trait that could function as such a predictor; it is not so much a study of successful individuals, as it is a study of what such individuals *might* be like: I'm now engaging in model construction, empiricist style. Whether successful individuals are indeed as I'm modeling them, is a matter for fur-ther empirical research to decide. But we make progress one step at a time. (First there is hypothesis; testing comes afterwards.)

Consider a trait that enables an agent to exercise control over his or her voyage through life. Such a trait might be implicated in the formulation of certain kinds of goals, plans and beliefs. Not just any goals, plans and beliefs. But specifically those that are involved in finding a direction in

---

[7]The material on personal navigation here is drawn from [Thalos, 2005].

life, maintaining that direction (whether actively or passively), changing direction when appropriate, moving at such a speed as befits the circumstances and with an eye towards guarding against general fatigue vis--vis the overarching goal, using navigational aids when they are available, overcoming obstacles that prove easily surmountable, avoiding those that may not be. Now, the concept of direction is itself deserving of some analysis, but I shan't provide such a thing here. I shall simply take it for granted that a life direction is a goal in the most macroscopic sense. Sternberg and Spear-Swerling talk about self-understanding in this connection, as well as the importance of fit between abilities, interests, personality characteristics and goals in the attainment of success. And — wisely — they strenuously resist the equation of the form of self-understanding that is at the heart of their would-be PN with a range of currently measurable intellectual functions ("constructs," as psychologists have it), such as planning capacities, general intelligence, and even emotional intelligence. All these other functions, as Sternberg and Spear-Swerling argue, are much too intellectual for the purposes of navigation: they have much more to do with know-that than know-how; and PN is nothing if not pure know-how. In the terms we have been using here, PN is pilot and not navigator.

We do not yet have a clear idea of what PN is, how it functions on the ground. We have no confirmed models of how it works. We're still taking baby steps toward an account of the sources of behavior. We can perhaps expect psychologists to contribute more bountifully to filling this blank. But whither *philosophy* now?

We're no closer to an understanding of how Reason, with a capital R, figures in the economy of human navigation. But to get further along we must acknowledge that conceptual analysis is inadequate. We must find out how matters stand on the ground; or at least do only such foundational work as can be done in its absence. And we can't do that by maintaining a thoroughly reclined position in the fabled armchair.

7 Reflections on Mill's organic philosophy

John Stuart Mill (in *The Logic of the Moral Sciences*) articulated the insight that morality can be treated as a science — and that the science of morality could stand on foundations as firm as any other. One simply had to go the right way about finding these foundations. His way was to begin by asking what moral reasoning is *for*.

The impulse to ask this question is a naturalistic impulse — an impulse to think of the phenomenon in natural terms, and so in terms that lend themselves to organic philosophy. (Mill, like Aristotle before him, was a consummate organic philosopher.)

Now, Mill had numerous remarks to make in answer to his question. Recently biologists and human ecologists too have had rather a lot to say on the question — none of it, to be sure, in agreement with Mill's remarks. No matter. The point is simply that Mill's posture on the subject created the potential for dialogue with the sciences — a potential that, as it happens, went unrealized in his time. But it is worth our while here to reflect on the affinities of our approach to our topic, with Mill's to his.

Like Mill, we have found a way to conceptualize something concerning which we can ask the question: what is it for? Can we take the inquiry a step further? There are those who have asked this very question about Reason. Their approach being a transcendental one — rooted in the idea that all we need to do is examine the concepts themselves for an answer[8] — it does not help foster dialogue with the sciences.

So the question we might at this point try asking is: What are reasons for? The rationalists are sure to insist that, even if we defer to natural science, the scientists cannot help but return with a theory to the effect that, whatever the *mechanism* for launching action might involve, physiologically, the whole business will be set in motion by has a slate of Reasons — entities, realized by mental states of some kind, for which or on the basis of which a person takes action.

But the facts already go against the substance of this rationalistic contention. For action does not interlock with reasons in any tight way. There are clear and numerous cases of agents taking action where there is nothing deserving of the name of a reason for it — cases that are clear counter-instances to Donald Davidson's quasi-naturalistic proposal that we can examine actions under the cipher of reason-bearing behavior [Davidson, 1980]. Davidson's bid to found the study of action in that proposition is misguided. For the transparency the proposal aspires to accommodate — the unbreakable link it wishes to forge between agency and reason-bearing behavior — makes it ineligible for handling the routine behavior we are familiar with on an hourly basis: the highly trained, "expert" behavior, which comes in many varieties (from walking to expert dancing, to driving an automobile, to speech production, to even more self-contained or parcelized stretches of behavior like personal mannerisms and driving to one's place of work — these are habituated behaviors, trained over time, and to the point where the question of reasons for their performance, or at least for performance of their nuances, subtleties, and proper parts or segments, becomes thoroughly inapt. And neither can it accommodate, on the other end of the spectrum, the more highly charged cases involving commission of atrocity, where certain agents perform heinous acts as if willingly, but without rea-

---

[8]For example, [Velleman, 2001].

sons they can identify explicitly. (The question of ownership over these pieces of behavior is enormously important.) For even in common sense, our conception of action does not rule out, as actions, behaviors in which what moves us is dark even to ourselves, and hardly deserving the name of a reason.[9] But if action does not require reasons for its launch, this poses a substantial obstacle for the proposal that reasons are for launching action. For if action can be launched some other way, why do we need reasons at all? How do reasons interact with that other way of launching action? And how can we make theoretical room, even if we retain reasons, for that other means of launching action?

Now, PN is quite capable of operating without reasons. (For by hypothesis, that's precisely the theoretical space it occupies: it is the very thing that performs executively; it is, as we've been saying, the pilot.) So what are reasons for? They are — or perhaps it is better to say that they must have been, at least originally — for the occasions when PN does not apply, is inadequate or is in some other way bypassed. What cases are these?

There is very fascinating empirical research indicating the depth of non-personal motivations — motivations that we humans (among other species) rely upon to produce social behavior in service of the public good.[10] Elizabeth Anderson [2000] refers to it as "our stock of social capital": inclinations to trust, to comply with norms of reciprocity, cooperation and honesty, and other prosocial values, that are cultivated by nonmarket institutions such as families, neighborhoods, churches and other community groups. The very idea of social capital suggests a quite different proposal on what reasons are good for. It is this: reasons are for coordinating and cooperating with other people — they are things tailor-made for exchanged between agents, for the sake of promoting coordinated behavior. Reasons, in other words, are in the first instance for public deliberations — the deliberations of a body, a collectivity — and not in the first instance for the deliberations of any undistributed acting entity. For the deliberations of a body cannot be managed without something in the way of rationale being exchanged among parties to it. One might even go so far as to say that reasons are for facilitating a process whereby individual organisms, equipped with a capacity for exchanging linguistic acts, coalesce into a single action-taking unit.[11] (And of course the process can go in reverse: an aggregate can dissolve into constituents — or, if we prefer, constituencies — in the presence of suitable reasons.)

---

[9] I discuss this matter at some length in "Sources of Behavior", forthcoming.

[10] See [Hogg and Abrams, 1993] and [Batson, 1994].

[11] This idea is developed at some length in [Thalos, 1999] and [Thalos and Andreou, to appear].

And so now practical wisdom, approached from an empirical perspective, looks more like this. Reasons come to exist for the sake of social navigation, whilst — independently — PN evolves for the purposes of individual navigation. Reasons, no less than linguistic meaning, are in the first instance public entities. Reasons, therefore, are not for PN exclusively, but PN takes advantage of them when it can or is obliged to do so. Personal and social navigation can come to resemble one another when once PN can take advantage of such reasons as it encounters on the ground — astonishingly, "out there" rather than "in here". But social navigation and individual PN come to be independently, and are under different evolutionary pressures.

This proposal should give transcendentalists some pause.

8   Kyburg's emphasis on formal systems

Kyburg is well known for emphasizing formal systems. (This comes through especially clearly in his [1968; 1983; 1984].) To him, formal systems are the methodology of choice. How, according to Kyburg, do formal systems make contact with the world that they purport to describe? I don't believe he ever offered an account. And so I'd like to offer one that I think he might welcome.

I say that formal systems must make such contact to things in the world as they in fact manage to make, in the context of the coordinative aims (conscious or otherwise) of agents who use them. I've just purported that reasons are for facilitating the collective deliberations of agents. And I shall now add that formal systems — systems utilizing symbols — is the means by which reasons are traded. This might seem a trivial result, but it is not. The positivists' answer, for one, differed: formal systems were simply for describing how the world is, with no regard whatever for the uses to which they might be put. And in this contention the positivists were — and still are — in substantial agreement with the transcendental philosophers. We shall part ways with them.

Let's take a baby step or two with our new hypothesis.

9   The experience of *aprioricity*

Consider a formal system — for the sake of a simple example, consider the system of Arabic numerals, together with language for expressing simple functions like sums, and some basic quantifiers. This simple system can be used to express simple arithmetic truths — for example, some or all of Peano's postulates. We are accustomed to conceiving of its symbols as having been attached to their meanings by fiat, and of knowledge of its axioms (knowledge of how to combine the symbols so as to formulate basic truths) as being prior to experience. I will refer to these sentiments or judgments on our part, as the *experience of aprioricity*. Why are certain formal systems so experienced? I want to advance a very simple hypothesis in answer to this question.

I speculate that the experience results from an internalization of strategic considerations that are important in that process of coordination required to select (first of all) a common set of symbols, and ultimately also to transmit these selections to the next generation of symbol users. The internalization I am proposing is something akin to the internalization of spoken language, through subvocalization, for silent uses.

When someone wants to coordinate with another, say for the sake of a business transaction, strategic issues arise. Suppose, for example, that you and I agree to an exchange, two-for-one, coconuts for melons, over a period of time. Then the need will arise — on a repeated basis — for a means whereby to verify the computations (additions, subtractions, what-have-you) to substantiate, to the satisfaction of each party, that the terms of agreement have been met. Suppose the two parties do not already share a system for displaying and verifying computations, whatever each may enjoy in private by way of arithmetic prowess. They therefore require a language that refers to, among other things, numbers. It is, I am contending, a matter purely of strategy that one should be devised. And it is also very much a matter of pure strategy how its development unfolds. The parties will take advantage of what each knows the other can attend to, when the time comes for assigning names and signs to objects and entities referred to in computation. Holding up two fingers, for example, is one way of naming the number 2; but not without a great deal of stage-setting first. (And how shall you refer to the function of multiplication, pray tell?) It is a strategic matter, not entirely a matter of fiat, that the parties involved should be able to attend to numbers, one at a time, and in such a way as facilitates progressions to more and more intangible entities. Whether the parties can make the necessary negotiations depends upon a host of strategic factors, one of the most important being whether each is able to discern the aims and cognitive capacities of the other. These things narrow down very significantly the range of available namables.

All of these considerations apply equally, and with even more force, to the adoption of axioms or rules governing the use and combination of the names so designated. We shall all have to accept them, friends and foes alike, if we hope for a durable system to facilitate exchange. So nothing can be postulated that will be the least bit suspect as to whether it gives advantage to one party over another. This is so that no party to the exchange can object to the postulate on grounds that it might disadvantage them. And so the axioms will have to have the appearance — not so much of obviousness and simplicity — but of acceptability to all parties — something more akin to fairness than to truth. The law governing sums cannot advantage me unduly, since it must govern sums of what I am transferring to you as well as what you in return are transferring to me. This fact that postulates have to apply to all the parties, guarantees not so much that they will be simple, but that they will be such as to apply, without restrictions and qualifications, to all the categories of object in which we are apt to trade. This tends to align more reliably with features of arithmetic that involve its universality than with those that involve its simplicity or any other more abstract feature.

And the story we're now telling about the experience of aprioricity holds equally well if we conceive instead of the two imagined agents, not as competitors, but rather as teacher and pupil. The strategy for targeting namables is just as demanding in that "game" as it is in the coordination-for-verification game. The successful parties to a teaching episode need to make the same negotiations as to whether each is able to discern the aims and cognitive capacities of the other. So ordinary transmission of a formal language will call up certain features of the same experience. There is some difference with teacher and pupil, however. Acceptability of axioms between teacher and student is not predicated upon whether there is the possibility of disadvantage to either party. And so teacher and pupil will enjoy a greater freedom in exploring axioms. And this fact actually explains why alternative axiom systems are more likely to make appearance in these proto-academic settings than they are to make appearance in the marketplace. This is at least one feature of the academic context that distinguishes it from the marketplace.

Are these considerations enough to guarantee the truth of axioms whose provenance is the marketplace? Perhaps. It depends on what we have in mind by truth. To my mind, there is no more to the truth of arithmetic than its universality across a range of domains. (Indeed, as we all know, arithmetic truths do not apply to the domain of chemistry, in which chemical reactions will not always uphold the truth of proposition: 2 and 2 add up to 4; sometimes 2 and 2 add up to a large explosion.) But at least we've come so far as universalizability.

But universalizability does not amount to aprioricity in the standard sense. It is not a matter of our assent to something's truth being *independent of or prior to experience*. For the process — actual or virtual — through which parties to a business deal, or pupil and student, have to find their way, relies very much upon a mutual knowledge of (or at least presumption about) what, in fact, each has had experience of. And so, while this hypothesis purports to explain the experience of aprioricity *as an ex-*

perience, it does not purport to explain that experience as an experience of aprioricity *as aprioricity* in the proper sense of the term.

What does this say about formal systems in general? Simply that our experiences of them as apriori is an artifact of the learning process. And it suggests a deflationary attitude to the aprioristic experience. To be sure we care about their truth, but this is a matter entirely of whether they depict facts concerning their application — or, at any rate, *desiderata* regarding their application — correctly. Axioms and postulates are simply either true or not, within a range of application. And it is neither here nor there whether we experience our assent to them as apriori.

I think this is a conclusion that Henry Kyburg will like. It will warm the cockles of his realistic heart, as well as stoke his empiricism.

BIBLIOGRAPHY

[Anderson, 2000] E. Anderson. Beyond *Homo Economicus*: New Developments in Theories of Social Norms, *Philosophy and Public Affairs*, 29: 170–200, 2000.

[Batson, 1994] C. D. Batson. Why act for the public good? Four answers, *Personality and Social Psychology Bulletin*, 20: 603–610, 1994.

[Brandom, 1994] R. Brandom. *Making it explicit*, Cambridge: Harvard University Press, 1994.

[Davidson, 1980] D. Davidson. Actions, Reasons and Causes, *Essays on Actions and Events*, Oxford: Clarendon Press, 2001/1980.

[Doris, 2002] F. Doris. *Lack of Character: Personality and Moral Behavior*, New York: Cambridge University Press, 2002.

[Hoog and Abrams, 1993] M. Hogg and D. Abrams. *Group Motivation*, London: Harvester Wheatsheaf, 1993.

[Goodman, 1955] N. Goodman. *Fact, Fiction and Forecasting*, Cambridge, MA: Harvard University Press, 1955.

[Kyburg, 1961] H. E. Kyburg, Jr. textitProbability and the Logic of Rational Belief, Middletown: Wesleyan University Press, 1961.

[Kyburg, 1968] H. E. Kyburg, Jr. *Philosophy of Science: A formal approach*, New York: MacMillan, 1968.

[Kyburg, 1983] H. E. Kyburg, Jr. *Epistemology and Inference*, Minneapolis: Minnesota Press, 1983.

[Kyburg, 1984] H. E. Kyburg, Jr. *Theory and Measurement*, New York: Cambridge University Press, 1984.

[Kyburg, 2005] H. E. Kyburg, Jr. Review: "Judea Pearl's *Causality*, Cambridge University Press (2000)", *Artificial Intelligence*, 169(2):174–179, 2005.

[Moran, 2001] R. Moran. *Authority and Estrangement: An Essay on Self-Knowleege*, Princeton: Princeton University Press, 2001.

[Nelson, 1995] A. Nelson. Microchaos and Idealization in Cartesian Physics, *Philosophical Studies*, 77:377–391, 1995.

[Nelson, 2005] A. Nelson. The Rationalist Impulse. In A. Nelson (ed.), *A Companion to Rationalism*, Oxford: Blackwell Press, 2005.

[Nisbett and Ross, 1991] R. Nisbett and R. Ross. *The Person and the Situation* Philadelphia: Temple University Press, 1991.

[Nisbett, 2003] R. Nisbett. *Geography of Thought*. The Free Press, 2003.

[Pearl, 1988] J. Pearl. *Probabilitstic Reasoning in Intelligent Systems*, San Mateo, Calif. : Morgan Kaufmann Publishers, 1988.

[Pearl, 1990] J. Pearl. *Causality: Models, Reasoning and Inference*. New York: Cambridge University Press, 1990.

[Sternberg and Spear-Swerling, 1998] R. Sternberg and L. Spear-Swerling. Personal Navigation. In M. Ferrari and R. Sternberg (eds), *Self-awareness : its nature and development*, New York: Guilford Press, 219-245, 1998.

[Thalos, 1999] M. Thalos. Degrees of Freedom in the Social World: Towards a Systems Analysis of Decision, *Journal of Political Philosophy*, 7, 453–77, 1999.

[Thalos, 2005] M. Thalos. What is a feminist to do with Rational Choice? In A. Nelson (ed.) *Companion to Rationalism*, Blackwell, 450-468, 2005.

[Thalso, forthcoming] M. Thalos. Sources of Behavior. In D. Ross and D. Spurrett (eds.) *Distributed Cognition and the Will*. Cambridge: MIT Press, forthcoming.

[Thalso and Andreou, in progress] M. Thalos and C. Andreou. Towards a natural history of agency, in progress.

[Velleman, 2001] J. D. Velleman. *The Possibility of Practical Reason*. New York: Oxford University Press, 2001.

[Wegner and Vallacher, 1980] D. M. Wegner and R. R. Vallacher, eds. *The Self in Social Psychology*, New York: Oxford University Press, 1980.

[Zelazo and Muller, 2002] P. Zelazo and U. Muller. Executive Function in Typical and Atypical Development. In U. Goswami, ed. *Blackwell Handbook of Childhood Cognitive Development*, Oxford: Blackwell, 2002.

[Zelazo et al., 1997] P. Zelazo, A. Carter, J. S. Reznick, and D. Frye. Early Development of Executive Function: A Problem-Solving Framework, *Review of General Psychology*, 1: 198-226, 1997.

[Zelazo, 1996] P. Zelazo. Towards a Characterization of Minimal Consciousness, *New Ideas in Psychology*, 14: 63-80, 1996.

The Y-Function

JOHN L. POLLOCK

1 Fundamentals of Direct Inference

This paper announces a new discovery which, I think, will be of fundamental importance in the application of probabilities to the real world. It concerns direct inference, so I will begin by sketching the theory of direct inference that I developed in my [1990].

There are two general approaches to probability theory. The most familiar takes "definite" or "single-case" probabilities to be basic. Definite probabilities attach to closed formulas or propositions. I write them using all caps: PROB(P) and PROB(P/Q). There are familiar difficulties for this approach, the simplest being that of making sense of the definite probabilities themselves. A serious practical difficulty is that when taking definite probabilities as basic, one must generally assume that we come to a problem equipped with a complete probability distribution. The latter is often computationally impossible for problems of realistic complexity. Like Kyburg [1974], I am convinced that the only way to construct a useful kind of definite probability is in terms of "indefinite" or "general" probabilities. The indefinite probability of *an A* being a *B* is not about any particular *A*, but rather about the property of being an *A*. In this respect, its logical form is the same as that of relative frequencies. "prob" is a variable-binding operator, binding the "x" in "prob(Bx/Ax)". I am convinced that the latter is the only approach that can provide the ultimate foundations for probability, in the sense of (1) providing a logical analysis of useful concepts of probability, and (2) explaining the epistemological foundations of probabilistic reasoning. The latter is required to make probabilities useful in science or AI.

According to this approach, statistical induction gives us knowledge of indefinite probabilities, and then direct inference gives us knowledge of definite probabilities. Reichenbach [1949] pioneered the theory of direct inference. The basic idea is that if we want to know the definite probability PROB(Fa), we look for the narrowest reference class (or reference property) G such that we know the indefinite probability prob(Fx/Gx) and we know

Ga, and then we identify PROB(*Fa*) with prob(*Fx*/*Gx*). For example, actuarial reasoning aimed at setting insurance rates proceeds in this way. Kyburg [1974] was the first to attempt to provide firm logical foundations for direct inference, and my [1990] took that as its starting point and constructed a modified theory with a more epistemological orientation.[1] I will briefly sketch my own approach, and then discuss a general problem for theories of direct inference.

The appeal to indefinite probabilities and direct inference has seemed promising for avoiding the computational difficulties attendant on the need for a complete probability distribution. Instead of assuming that we come to a problem with an antecedently given complete distribution, one can more realistically assume that we come to the problem with some limited knowledge of indefinite probabilities, and then we infer definite probabilities from the latter as we need them. Unfortunately, as I will show in section two, it is premature to suppose that existing theories of direct inference actually solve this problem. The main point of this paper, however, is to exhibit a mathematical result that makes this problem solvable by appealing to a more powerful variety of direct inference — what I will call *computational direct inference*.

Kyburg used the term "probability" to refer only to definite probabilities, but I think it is better to distinguish between definite and indefinite probabilities. Kyburg identified indefinite probabilities with relative frequencies, but I think that is inadequate for a number of reasons (detailed in my [1990]). The simplest is that we often make probability judgments that diverge from relative frequencies. For example, we can talk about a coin being a fair coin (and so the probability of a flip landing heads is 0.5) even when it is flipped only once and then destroyed. For understanding such indefinite probabilities, we need a notion of probability that talks about *possible* instances of properties as well as actual instances. My proposal in my [1990] was that we can identify the *nomic probability* prob(*Fx*/*Gx*) with the proportion of physically possible *G*s that are *F*s. A *physically possible G* is defined to be an ordered pair $\langle w, x \rangle$ such that w is a physically possible world (one compatible with all the physical laws) and x has the property *G* at w. $\mathcal{G}$ is the set of all physically possible *G*s. We must assume the existence or a proportion function $\rho(X, Y)$ for sets X and Y. I investigated the general theory of proportion in my [1990], although I no longer regard that treatment as entirely adequate, and am working on an improved version of that theory. For the moment, let us just assume we have the proportion

---

[1]Competing theories of direct inference can be found in [Levi, 1980; Bacchus, 1990; Halpern, 1990].

function ρ. Then if it is physically possible for there to be Gs, we define:

$$\text{prob}(Fx/Gx) = \rho(\mathcal{F}, \mathcal{G}).$$

In my [1990] I proposed a generalization of this definition that handles the case of counterlegal probabilities, in which it is not physically possible for there to be Gs, but I will ignore that sophistication here.

It will often be convenient to write proportions in the same logical form as probabilities, so where φ and θ are open formulas with free variable x, $\rho(\varphi/\theta) = \rho(\{x|\varphi\&\theta\}, \{x|\theta\})$. Without going into details about the proportion function, I will make two classes of assumptions. Let $\#X$ be the cardinal of X. If Y is finite, I assume:

$$\rho(X, Y) = \frac{\#X \cap Y}{\#Y}.$$

However, for present purposes the proportion function is most useful in talking about proportions among infinite sets. The sets $\mathcal{F}$ and $\mathcal{G}$ will almost invariably be infinite, if for no other reason than that there are infinitely many physically possible worlds in which there are Fs and Gs. I assume that the standard "Boolean" principles that hold for finite relative frequencies also hold for proportions among infinite sets. In my [1990], I also defended various non-Boolean principles. In particular, I assume that principles that hold for finite sets in the limit (as their size goes to $\aleph_0$) hold for infinite sets. For instance, the following is a theorem of combinatorial mathematics:

Finite Principle of Agreement
For every $\varepsilon, \delta > 0$ there is a N such that if B is finite but $\#B > N$ then

$$\rho\left(\rho(A, X) \underset{\delta}{\approx} \rho(A, B)/X \subseteq B\right) \geq 1 - \varepsilon.$$

In other words, proportions among subsets of B tend to agree with proportions in B itself, and both the strength of the tendency and the extent of the agreement increase as the size of B increases. I take this to support a corresponding principle of agreement for infinite sets:

Strong Principle of Agreement
For every $\delta > 0$, if B is infinite then

$$\rho\left(\rho(A, X) \underset{\delta}{\approx} \rho(A, B)/X \subseteq B\right) = 1.$$

In my [1990] I showed that this principle is derivable from the general theory of proportions adumbrated there, although as I remarked above, I no longer regard that theory as entirely adequate. Nomic probabilities are proportions among physically possible objects, so the Strong Principle of Agreement implies:

Principle of Agreement for Probabilities
For every $\delta > 0$,

$$\text{prob}\left(\text{prob}(Ax/Bx) \approx_{\delta} \text{prob}(Ax/Gx)/B \preccurlyeq G\right) = 1.$$

I take the Principle of Agreement for Probabilities to underlie direct inference, as I will now explain.

The theory of direct inference is an epistemological theory. It is about how to make certain kinds of inferences. In my [1990] I showed that the entire epistemological theory of nomic probability can be derived from a single epistemological principle coupled with a mathematical theory that amounts to a calculus of nomic probabilities. The latter is richer than the standard probability calculus, because nomic probabilities have more structure than definite probabilities. The relationship between indefinite probabilities and definite probabilities is roughly analogous to the difference between the predicate calculus and the propositional calculus. I won't pursue the details of the calculus here, but see my [1990]. The single epistemological principle that underlies probabilitistic reasoning is the *statistical syllogism*. The basic form of the statistical syllogism that I will employ here is the modified version defended in my [995]:

(A3)   If G is projectible with respect to both K and U and $r > 0.5$, then $\ulcorner Kc \& \text{prob}(Gx/Kx \ \& \ Ux) \geq r \urcorner$ is a defeasible reason for $\ulcorner(Uc \rightarrow Gc)\urcorner$.[2]

I assume the theory of defeasible reasoning adumbrated in my [1995].
When U is tautologous, (A3) implies the simpler:

(A1)   If G is projectible with respect to K and $r > 0.5$, then $\ulcorner Kc \& \text{prob}(Gx/Kx) \geq r \urcorner$ is a defeasible reason for $\ulcorner Gc \urcorner$.

The simplest kind of defeater is a subproperty defeater:

(D)   If H is projectible with respect to both K and U, $\ulcorner Hc \ \& \ \text{prob}(Gx/Kx \& Ux \& Hx) \neq \text{prob}(Gx/Kx \& Ux) \ \& \ \text{prob}(Ux/Kx \& Hx) \geq \text{prob}(U/K)\urcorner$ is an undercutting defeater for (A1) and (A3).[3]

---

[2] The projectibility constraint is the familiar constraint required for inductive reasoning. This is discussed at length in my [1990]. Kyburg [1974] also noted the need for some such constraint.

[3] There are two kinds of defeaters. Rebutting defeaters attack the conclusion of an inference, and rebutting defeaters attack the inference itself without attacking the conclusion [Pollock, 1995]. Here I assume some form of the OSCAR theory of defeasible reasoning.

In (A1), U is tautologous, so in that case (D) can be simplified:

(D1)   If H is projectible with respect to K, $\ulcorner Hc\&\ \mathrm{prob}(Gx/Kx\&Hx) \neq \mathrm{prob}(Gx/Kx)\urcorner$ is an undercutting defeater for (A1).

More defeaters are required to make inferences in accordance with the statistical syllogism work properly. Several are discussed in my my [1990], but for present purposes we need not pursue this.

The statistical syllogism, coupled with the principle of agreement, gives us a defeasible reason for expecting that if $\mathrm{prob}(Fx/Gx) = r$ then for every $\delta > 0, \mathrm{prob}(Fx/Gx\&Hx) \approx_\delta r$, and the latter entails that $\mathrm{prob}(Fx/Gx\&Hx) = r$. Thus we get a general principle of direct inference, which I call *nonclassical direct inference*:

Nonclassical Direct Inference:

> If F is projectible with respect to G, $\mathrm{prob}(Fx/Gx) = r$ is a defeasible reason for $\mathrm{prob}(Fx/Gx\&Hx) = r$.

This is a kind of principle of insufficient reason. It differs from classical direct inference in that it is an inference from indefinite probabilities to indefinite probabilities rather than from indefinite probabilities to definite probabilities.

Defeaters for nonclassical direct inference follow from the defeaters for the statistical syllogism. In particular, subproperty defeaters for the statistical syllogism generate subproperty defeaters for nonclassical direct inference.

Let us define:

> $F \preccurlyeq G$ if it is physically necessary that $(\forall x)(Fx \rightarrow Gx)$.
> $F < G$ iff $F \preccurlyeq G$ but $(G \npreccurlyeq F)$.

Then subproperty defeat for (A1) generates subproperty defeat for nonclassical direct inference:

Subproperty Defeat

> $\ulcorner G < J < (G\&H)$ and $\mathrm{prob}(Fx/Jx) \neq r\urcorner$ is an undercutting defeater for nonclassical direct inference.

Direct inference is normally understood as being a form of inference from indefinite probabilities to definite probabilities rather than from indefinite probabilities to other indefinite probabilities. However, I showed in my [1990] that these inferences are derivable from nonclassical direct inference if we identify definite probabilities with a special class of indefinite probabilities. Let $\mathbb{K}$ be the conjunction of the agent's justified beliefs. Then we define:

PROB$(P/Q) = r$ iff for some n, there are n-ary properties R and S and terms $a_1, \ldots, a_n$ such that

1. it is physically necessary that $(P \leftrightarrow Ra_1\ldots a_n)$ and $(Q \leftrightarrow Sa_1\ldots a_n)$, and

2. $\text{prob}(Rx_1\ldots x_n/Sx_1\ldots x_n \& x_1 = a_1 \& \ldots \& x_n = a_n \& \mathbb{K}) = r$.

It can be shown (my [1990]) that this yields a unique value for PROB(P/Q). This is a kind of "mixed physical/epistemic probability", because it combines background knowledge in the form of $\mathbb{K}$ with indefinite probabilities.

Given this definition, we can derive the *principle of classical direct inference*:

⌜$\text{prob}(Rx_1\ldots x_n/Tx_1\ldots x_n) = r$, it is physically necessary that $(P \leftrightarrow Ra_1\ldots a_n)$ and that $(Q \leftrightarrow Sa_1\ldots a_n)$, and $Sx_1\ldots x_n < Tx_1\ldots x_n$⌝ is a defeasible reason for ⌜PROB$(P/Q) = r$⌝

Similarly, we get subproperty defeaters:

⌜$S < U < T$ and $\text{prob}(Rx/Ux) \neq r$⌝ is an undercutting defeater for classical direct inferece.

All of this is only a brief sketch of the theory of direct inference developed in my [1990], but it will suffice for present purposes. In the next section I will discuss a problem for this theory. The main part of this paper will be aimed at establishing the existence of a mathematical function that provides a solution to the problem and extends the theory of direct inference in a way that, perhaps for the first time, makes it truly useful.

2   A Problem for Direct Inference

The preceding provides a foundation for a more or less standard theory of direct inference. Perhaps its main novelty is that it is formulated in terms of a background theory of defeasible reasoning. However, this and all similar theories suffer from a fundamental difficulty that greatly diminishes their practical usefulness. If we have some complex conjunction $G_1x \& \ldots \& G_nx$ of properties and we want to know the value of $\text{prob}(Fx/G_1x \& \ldots \& G_nx)$, if we know that $\text{prob}(Fx/G_1x) = r$ *and we don't know anything else of relevance*, we can infer defeasibly that $\text{prob}(Fx/G_1x \& \ldots \& G_nx) = r$. Similarly, if we know that an object a has the properties $G_1, \ldots, G_n$ and we know that $\text{prob}(Fx/G_1x) = r$ *and we don't know anything else of relevance*, we can infer defeasibly that PROB$(Fa) = r$. The difficulty is that we usually know more. We typically know the value of $\text{prob}(Fx/G_ix)$ for some $i \neq 1$. If $\text{prob}(Fx/G_ix) = s \neq r$, we have defeasible reasons for both ⌜$\text{prob}(Fx/G_1x \& \ldots \& G_nx) = r$⌝ and ⌜$\text{prob}(Fx/G_1x \& \ldots \& G_nx) = s$⌝, and also for both $\ulcorner PROB(Fa) = r \urcorner$ and $\ulcorner PROB(Fa) = s \urcorner$, but as these conclusions are incompatible they all undergo collective defeat. Thus the standard theory of direct inference leaves us without a conclusion to draw. The upshot is that the promise of direct inference to solve the computational problem of dealing with definite probabilities without having to have a complete probability distribution was premature. Direct inference will rarely give us the probabilities we need.

Direct inference would be vastly more useful in real application if there were a function $Y(r, s)$ such that, in a case like the above, we could defeasibly expect that $prob(Fx/G_1x\& \ldots \&G_nx) = Y(r, s)$, and hence (by nonclassical direct inference) that $PROB(Fa) = Y(r, s)$. I call this *computational direct inference*, because it computes a new value for $PROB(Fa)$ rather than simply taking a value from a known indefinite probability. I call the function used in such a computation "the Y-function" because its behavior would be as diagrammed in figure 1. The general presumption has been that there is no such function, but this paper presents empirical reasons for thinking that the Y-function exists. I will present these reasons and then indicate how the existence of the Y-function can give rise to a theory of computational direct inference.

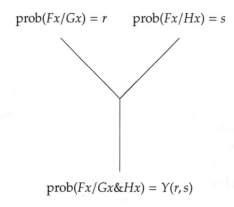

$$prob(Fx/Gx) = r \qquad prob(Fx/Hx) = s$$

$$prob(Fx/Gx\&Hx) = Y(r, s)$$

Figure 1. The Y-function

3   Discovering the Y-function

It is generally assumed that no such function as the Y-function exists. Certainly, there is no function $Y(r, s)$ such that we can conclude deductively that if $prob(Fx/Gx) = r$ and $prob(Fx/Hx) = s$ then $prob(Fx/Gx\&Hx) = Y(r, s)$. For any r and s that are neither 0 nor 1, $prob(Fx/Gx\&Hx)$ can take any value between 0 and 1. However, that is equally true for nonclassical direct inference. That is, if $\text{prob}(Fx/Gx) = r$ we cannot conclude deductively that $\text{prob}(Fx/Gx\&Hx) = r$. Nevertheless, that will tend to be the case, and we can defeasibly expect it to be the case. Might something similar be true of the Y-function? That is, could there be a function $Y(r, s)$ such that we can defeasibly expect $\text{prob}(Fx/Gx\&Hx)$ to be $Y(r, s)$?

I had always sided with the majority in supposing that there is no such function, but it occurred to me that perhaps this was premature. I did not (and do not yet) see how to prove the existence of the Y-function, but we might investigate its existence empirically. Let me explain.

On analogy with the principle of agreement, suppose there were a function $Y(r, s)$ satisfying the following principle:

Y-Principle:
For every $\varepsilon, \delta > 0$ there is a N such that if U is finite and $\#U > N$ then

$$\rho\left(f_1 \underset{\delta}{\approx} Y(f_2, f_3)/f_2 = \rho(A, B)\&f_3 = \rho(A, C)\&f_1 = \rho(A, B \cap C)\&A, B, C \subseteq U\right) \geq 1-\varepsilon.$$

Then in accordance with my general assumptions about the proportion function, it could be assumed that the following infinite generalization holds:

Strong Y-Principle:
For every $\delta > 0$, if U is infinite then

$$\rho\left(f_1 \underset{\delta}{\approx} Y(f_2, f_3)/f_2 = \rho(A, B)\&f_3 = \rho(A, C)\&f_1 = \rho(A, B \cap C)\&A, B, C \subseteq U\right) = 1.$$

Nomic probabilities are proportions among physically possible objects. There are infifnitely many physically possible objects if for no other reason than that there are infinitely many ws entering into the ordered pairs $\langle w, x \rangle$. Accordingly, we have the following:

Y-Principle for Probabilities:
For every $\delta > 0$ (where A, B and C are variables):

$$\text{prob}\left(\text{prob}(Ax/Bx\&Cx) \underset{\delta}{\approx} Y(r, s)/r = \text{prob}(Ax/Bx)\&s = \text{prob}(Ax/Cx)\right) = 1.$$

The Y-Principle combined with (A1) will yield the following principle of direct inference in the same way the Principle of Agreement combined with (A1) yields the standard principle of nonclassical direct inference:

Computational Direct Inference:
If F is projectible with respect to G and H, $\ulcorner\text{prob}(Fx/Gx) = r\&\text{prob}(Fx/Hx) = s\urcorner$ is a defeasible reason for $\ulcorner\text{prob}(Fx/Gx\&Hx) = Y(r, s)\urcorner$.

Thus to get a principle of computational direct inference, it suffices to have a function $Y(r, s)$ that satisfies the Y-principle for finite sets. Is there such a function? Not seeing how to prove that there is (or isn't) such a function, I decided to test this empirically. Given a set U, we could in principle survey all triples A, B, C of subsets of U, compute $\rho(A, C), \rho(A, C)$, and $\rho(A, B \cap C)$ and see how they are related. If the Y-Principle seems to hold for larger and larger U, that gives us an inductive reason for thinking that the Y-Principle is true. The trouble is, you can only survey all triples of U for very small U. For instance, if $\#U = 100$, the number of triples A, B, C of subsets of U is 2^{300}, which is approximately 10^{90}. This is twelve orders of magnitude greater than a recent estimate of the total number of elementary particles in the universe (10^{78}). Clearly, we cannot survey all these triples.

Although for even rather small Us, we cannot survey all of the triples of subsets, we can instead use Monte Carlo techniques. That is, we can sample the triples randomly and see what we get. Let $\text{Num}(k)$ be the set of integers $\{1, \ldots, k\}$. I wrote a program in LISP that randomly selects triples of nonempty subsets of $\text{Num}(k)$ for any k we supply, and then computes and compares the values of $\rho(A, C), \rho(A, C)$, and $\rho(A, B \cap C)$. To my surprise, this produced the plot of a very well-behaved function. The plot of 10,000 triples for $k = 10,000$ is given in figure 2. This plots the average value of $\rho(A, B \cap C)$ as a function of $\rho(A, C)$ and $\rho(A, C)$.

Although the average values are described by a well-behaved function, which I will designate *the Y-function*, this does not show that the value of $\rho(A, B \cap C)$ tends to agree with $Y(\rho(A, B), \rho(A, C))$. All it shows is that the *average value* conforms to the Y-function. However, this is also a matter we can investigate empirically. We can construct our sampling function so that it not only computes the average value of $\rho(A, B \cap C)$ for each choice of values for $\rho(A, B)$ and $\rho(A, C)$, but it also keeps track of how many of the values agree with $Y(\rho(A, B), \rho(A, C))$ to any specified degree of approximation. The results are quite striking. They are diagrammed in figures 3, 4, and 5. Inspection reveals that the envelopes containing 95% of the sampled triples get narrow rapidly.

These results give strong inductive support for the existence of a Y-function satisfying the Y-Principle. It remains a mystery what function this is. I have not been able to find any familiar function the fits the curve of figure 2. Because this function is generated by statistical processes related to normal distributions, it may have no analytic characterization.

4   Algebraic Properties of the Y-function

The most important task facing us is to identify the Y-function and prove the Y-Principle. At this point, I can do neither. However, it may be useful

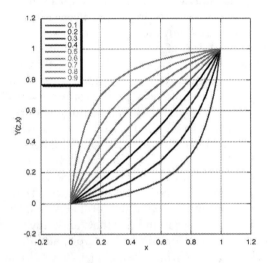

Figure 2. The Y-function, holding z constant

to investigate the algebraic properties of the Y-function that can be read off of figure 2. These may ultimately be helpful in identifying the Y-function.

Clearly, from its definition, the Y-function must be associative:

1. $Y(z, x) = Y(x, z)$.

This property is also exhibited by the plot, as illustrated in figure 6.

Principle (2) tells us that the value .5 plays a special role in the Y-function:

2. $Y(.5 + x, .5 - x) = .5$.

Equivalently:

3. $Y(z, 1 - z) = .5$

Principles (2) and (3) are illustrated by figure 7.

Principle (4) expresses another respect in which .5 plays a privileged role.

4. $Y(.5, x) = x$

Principle (5) is diagrammed in figure 8.

The curve exhibits symmetry in three directions — around both diagonals and around the center point. Symmetry around the center point is expressed by principle (6), which is diagrammed in figure 9:

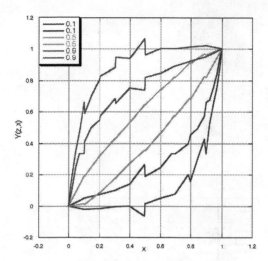

Figure 3. Upper and lower bounds for $Y(z, x)$ (for three values of z) within which 95% of the sampled triples fall for $k = 100$.

5. $Y(z, x) = 1 - Y(1 - z, 1 - x)$

Symmetry around the right-leaning diagonal is expressed by principle (6), which is diagrammed by figure 10:

6. $Y(1 - z, Y(z, x)) = x$

Principles (5) and (6) entail principle (7), which expresses symmetry around the left-leaning diagonal:

7. $Y(z, 1 - Y(z, x)) = 1 - x.$

Proof. $Y(z, 1 - Y(z, x)) = 1 - (1 - Y(1 - (1 - z), 1 - Y(z, x)) = 1 - Y(1 - z, Y(z, x)) = 1 - x.$ ∎

Principle (7), and its derivation, is diagrammed in figure 11.
A principle that will prove very important is the following:

8. $Y(z, Y(x, w)) = Y(w, Y(z, x))$

Principle (8) is illustrated by figure 12, although it does not reflect any obvious geometric properties of the curve.

The reason principle (8) is important is that, together with associativity, it entails the commutativity of the Y-function:

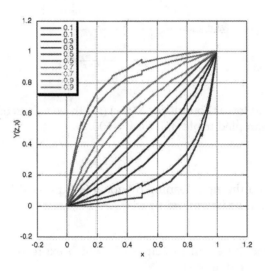

Figure 4. Upper and lower bounds for $Y(z, x)$ (for five values of z) within which 95% of the sampled triples fall for $k = 1000$.

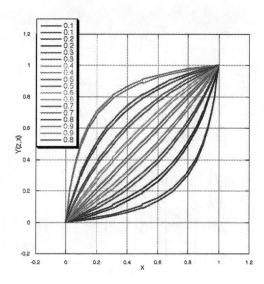

Figure 5. Upper and lower bounds for Y(z,x) within which 95% of the sampled triples fall for $k = 10,000$.

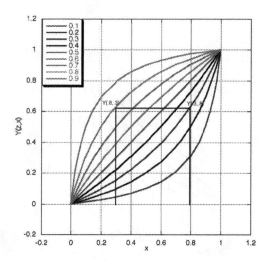

Figure 6. Associativity

John L. Pollock

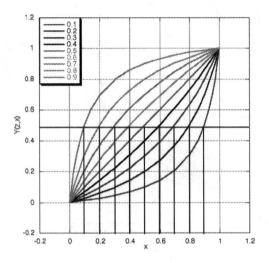

Figure 7. $Y(z, 1 - z) = .5$

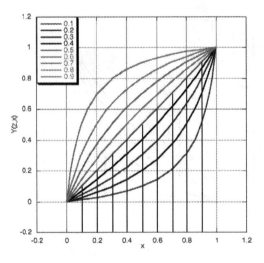

Figure 8. $Y(.5, x) = x$

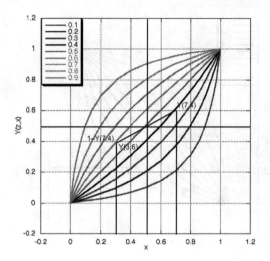

Figure 9. $Y(z, x) = 1 - Y(1 - z, 1 - x)$

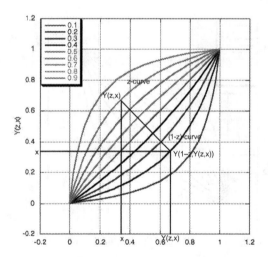

Figure 10. $Y(1 - z, Y(z, x)) = x$

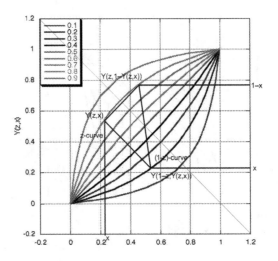

Figure 11. $Y(z, 1 - Y(z, x)) = 1 - x$

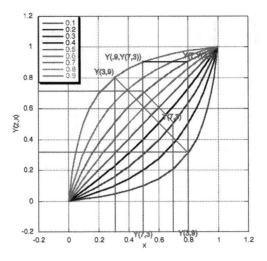

Figure 12. $Y(z, Y(x, w)) = Y(Y(z, x), w)$

9. $Y(z, Y(x, w)) = Y(Y(z, x), w)$.

Note that the Y-Principle also entails (9). So if the Y-Principle is correct, commutativity follows.

These algebraic properties describe a very well-behaved function, but it is still not clear what function it is.

5   Computational Direct Inference

The existence of the Y-function is a fundamental discovery that makes direct inference useful in ways it was never previously useful. The Y-principle tells us how to combine different probabilities in direct inference and still arrive at a univocal value. The form of the Y-function is diagrammed as in figure 13.

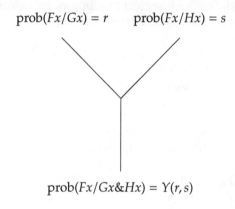

$$\text{prob}(Fx/Gx) = r \qquad \text{prob}(Fx/Hx) = s$$

$$\text{prob}(Fx/Gx\&Hx) = Y(r, s)$$

Figure 13. The Y-function

Thus, for example, if we know that the probability of Jones' dying if we shoot him is .6 and the probability of his dying if we poison him is .7, we can infer defeasibly (by A1) that the probability of his dying if we do both is $Y(.6, .7) = .77778$. This explains how built-in redundancy can increase the probability of a plan achieving its goal.

As observed in section three, the Y-Principle combined with (A1) yields the following principle of direct inference in the same way the Principle of Agreement combined with (A1) yields the standard principle of nonclassical direct inference:

Computational Direct Inference:

If F is projectible with respect to G and H, $\text{prob}(Fx/Gx) = r\ \&$

prob$(Fx/Hx) = s$ is a defeasible reason for prob$(Fx/Gx \,\&\, Hx) = Y(r,s)$.

If we know that prob$(Fx/Gx) = r$ and prob$(Fx/Hx) = s$, we can also use nonclassical direct inference to infer defeasibly that prob$(Fx/Gx\&Hx) = r$ and prob$(Fx/Gx\&Hx) = s$, and this conflicts with the conclusion that prob$(Fx/Gx\&Hx) = Y(r,s)$. However, these conflicting conclusions are obtained by applying (A1) to weaker reference properties, and so they are defeated by subproperty defeat. In general:

Computational Defeat for Classical Direct Inference:

> If F is projectible with respect to H, prob$(Fx/Hx) = s$ is an undercutting defeater for nonclassical direct inference.

Computational direct inference is subject to subproperty defeat in the same way nonclassical direct inference is, and for the same reason:

> $\ulcorner G < J < (G\&H)$ and prob$(Fx/Jx) \neq r \urcorner$ is an undercutting defeater for computational direct inference.

> $\ulcorner H < J < (G\&H)$ and prob$(x/Jx) \neq s \urcorner$ is an undercutting defeater for computational direct inference.

For its use in computing probabilities, it is very important that the Y-function is commutative. If we know that prob$(Fx/Ax) = .6$, prob$(Fx/Bx) = .7$, and prob$(Fx/Cx) = .75$, we can combine them in any order to infer defeasibly that prob$(Fx/Ax\&Bx\&Cx) = Y(.6, Y(.7, .75)) = Y(Y(.6, .7), .75) = .913043$. This makes it convenient to extend the Y-function recursively so that it can be applied to an arbitrary number of arguments (greater than or equal to 1):

$$\text{If } n > 2, Y(r_1, \ldots, r_n) = Y(r_1, Y(r_2, \ldots, r_n)).$$

Then we can strengthen computational direct inference as follows:

Computational Direct Inference:
If F is projectible with respect to $G_1, \ldots, G_n$, prob$(Fx/G_1x) = r_1$ & prob$(Fx/G_nx) = r_n$ is a defeasible reason for prob$(Fx/G_1x\&\ldots\&G_nx) = Y(r_1, \ldots, r_n)$.

Defeaters are derivable from those for "binary" computational direct inference.

6   Conclusions

The use of indefinite probabilities and direct inference seems initially to provide a computationally feasible alternative to the unrealistic requirement that we come to problems equipped with a complete distribution of definite probabilities. However, the initial promise fades with the realization that we normally have too much information, with the result that we can make conflicting inferences by nonclassical direct inference, and they collectively defeat one another. This difficulty is resolved by the discovery of the Y-function and the principle of computational direct inference, which allows us to make use of all our information to compute a single probability. As yet, we have only inductive reasons for believing that the Y-function exists, and we do not have an analytic characterization of it. Hopefully, future research will fill these lacunae.

Acknowledgements

This work was supported by NSF grant no. IIS-0412791.

BIBLIOGRAPHY

[Bacchus, 1990] F. Bacchus. *Representing and Reasoning with Probabilistic Knowledge*, MIT Press, 1990.
[Halpern, 1990] J. Y. Halpern. An analysis of first-order logics of probability, *AIJ*, **46**, 311–350, 1990.
[Kyburg, 1974] H. Kyburg. *The Logical Foundations of Statistical Inference*. Dordrecht: Reidel, 1974.
[Levi, 1980] I. Levi. *The Enterprise of Knowledge*. Cambridge, Mass.: MIT Press, 1980.
[Pollock, 1990] J. L. Pollock. *Nomic Probability and the Foundations of Induction*, New York: Oxford University Press, 1990.
[Pollock, 1995] J. L. Pollock. *Cognitive Carpentry*, Cambridge, MA: Bradford/MIT Press, 1995.

Probability Logic and Logical Probability

Isaac Levi

I have learned more about probability from reading Henry Kyburg than any other one of my contemporaries that I know. I dissent from many of the more striking and original of his thoughts. But more often than not, the disagreements have taught me more than the agreements. My teacher will no doubt declare me a dunce. But I shall, nonetheless, be eternally grateful for his instruction.

In this paper, I shall sketch an idea that, I fear, will appear to be a piece of monumental effrontery. I shall offer an account of what I once called "inductive logic" and now call "probability logic". I shall try to locate the major bone of contention between Kyburg's ideas and mine as an issue in probability logic. The effrontery comes from the fact that the version of Kyburg's ideas that shall be compared with mine is not Kyburg's but an account of probability logic that Kyburg would no doubt reject. (Kyburg's ideas are eleborated in [Kyburg, 1961; 1974] and [Kyburg and Choh Man Teng, , 2001].) There are many topics concerning which Kyburg and I disagree. My aim here is to present a version of Kyburg that brings into sharp focus the feature of his innovative approach to probability logic that expresses his deepest rejection of the ideas of Bayes, Laplace, Broad, Carnap, Jeffreys, de Finetti, and Savage. By presenting a view of Kyburg's deepest rejection that remains in agreement with Bayesian ideas as much as possible, the significance of Kyburg's anti-Bayesianism may be better appreciated.

Let me begin by saying something about logic. I am sympathetic with Arnold Kosolow's structuralist approach to what may be called "pure logic" that considers logic to be the study of structures on arbitrary domains of objects obtained by introducing an "implication" relation on these objects [Koslow, 1992]. There are many instantiations of logic so conceived. The objects can be sticks and stone, or switches in some type of computer hardware. They can be sentences in some language. And they can be beliefs. They must be organized in a manner that relates them in the manner of an implication relation.

I am interested in applications of structural logic that provide us with prescriptions for rational belief. By "belief", I mean both full belief and judgments of degree of belief and degree of belief itself in two senses: as judgments of credal probability used to evaluate expected values of options in decision problems and degrees of belief and disbelief in the sense of Shackle. I shall say no more about Shackle here; but will focus on full belief or absolute certainty and credal probability.

If I say that agent X fully beliefs h at time t, I mean that X at t is in a state of full belief $\mathbf{K}$ at t that has as a consequence another potential state of full belief – to wit, the state that h. I presuppose a Boolean algebra K of potential states of full belief that are conceptually accessible to X. X is in one of them at t – namely, $\mathbf{K}$ and $\mathbf{K}$ has consequences in accordance with the consequence relation of the Boolean algebra. The beliefs that are consequences of $\mathbf{K}$ represent commitments to judging these consequences true, ruling out as impossibilities their complements and judging potential states in the algebra that are consistent with $\mathbf{K}$ to be possibly true. Thus, if X fully believes that h, X is committed to being perfectly free of doubt that h is true at the given time. I say "committed" because X may very well fail to fulfill the commitment. This can be due to emotional frailty or lack of logical omniscience, computational capacity and memory, The Boolean algebra or, perhaps, its strengthening by allowing closure under infinitary meets and joins of various cardinalities provides the logic that characterizes what X's commitments are. Insofar as X's commitments to full belief are representable in some regimented language L, they are representable by a deductively closed theory or corpus of sentences K in L. But I wish to emphasize that from my perspective the important thing is X's state of full belief $\mathbf{K}$. How it is expressed is a secondary and relatively unimportant issue. We should not make a fetish out of language [Levi, 1991, ch2, 1997, chapters 1-3.]

On the view I am developing, change of full belief is change in doxastic commitment. Since X fails perfectly to fulfill X's commitments, X is under some obligation to improve X's performance so that X's doxastic performances undergo changes that must be kept distinct from changes in X's doxastic commitments Psychologists, social scientists and therapists are intensely interested in such changes and logicians and epistemologists have also tended to confuse such changes in commitments. Thus, traditionally beliefs have been thought to be either dispositions to behave linguistically or nonlinguistically or to be utterances, inscriptions or fits of doxastic conviction that can qualify as manifestations of such dispositions. This has led to the sad practice of trying to account for what are called the contents of beliefs by introducing propositions and, perhaps, explicating these in terms of such things as sets of possible worlds or other metaphysical artifacts. I suggest, instead, that we think of belief in the primary sense of doxastic commitment. The content of belief that h concerns the commitments undertaken by being in a state **K** that has the potential state that h as a consequence. In this sense, belief is normative to its core.

I say all this because I mean to suggest that the logic of full belief is the system of norms and prescriptions that provide a most general characterization of the commitments of full belief. This logic of full belief supports the linguistic representation of states of full belief by deductively closed sets of sentences in L or if one wishes to consider the commitments to modal judgment then by a set of sentences in the modal language ML satisfying S5 modal logic. This logic is an application to a structural logic in Koslow's sense to a particular domain – the domain of potential states of full belief. To my way of thinking, this application is far less mysterious than applications to propositions, sets of possible worlds or other such artifacts of metaphysical fantasy.

Logic as it is used here serves what Ramsey called a logic of consistency. This is a system of prescriptions or norms characterizing the doxastic commitments entailed by being in a state of full belief **K**.

According to Ramsey, the prescriptions regulate the dispositions to bet of the agent. Ramsey thought that as a matter of empirical fact, agents exhibit dispositions that to a good degree of approximation agree with the norms. I think that this is wildly at variance with the facts. On the view I favor, X's probability judgments are not dispositions but commitments to have dispositions to take risks according to norms of rationality. Probability logic specifies the features that all such commitments share in common. In this sense, they specify the invariant features of credal states. Probability logic is not logic in Koslow's sense. Probability operators are not logical or modal operators although structuralists can say that they characterize functions from elements of a Boolean algebra to real numbers satisfying appropriate conditions without compromising their structuralist view. Using such probability functions to characterize states of credal probability judgment is one such application.

Ramsey contrasted the logic of consistency with a logic of truth. In opposition to Keynes, Ramsey insisted that probability logic could not be a system of truths because credal probability judgments are neither true nor false. Much the same is true of the logic of full beliefs when this is understood to encompass judgments of possibility. Judgments of serious possibility lack truth values just as judgments of credal probability do. The logic of serious possibility can no more be a system of logical truths than the logic of probability can. Full beliefs themselves carry truth values so that there is an aspect of the logic of full belief that can be thought of as a logic of truth. But I will follow Ramsey in breaking with the Russell-Frege tradition and focus instead on the logic of consistency.

Kyburg complains and rightly so that addressing the logic of consistency for credal probability does not capture the aspect of probability logic that relates probability to evidence. That is what Kyburg's theory of epistemological probability purports to do. There are two aspects to Kyburg's thesis — one that seems right and the other mistaken.

X's credal probability judgments at t are grounded in X's evidence at t — evidence that I take to be X's state of full belief at t. That is to say, X at t is committed to a function from potential states in the algebra K to states of credal probability judgment. I call such functions confirmational commitments, $C : K \mapsto B$. That function represents X's commitment at a given time to what X's credal state should be were X in a given state of full belief or evidence $\mathbf{K}$ at that time. Applied prescriptive probability logic regulates the coherence or consistency of B through regulating the coherence or consistency of C. Any rational agent is committed to some confirmational commitment or other at a given time t. In that sense, Kyburg is right to insist that credal probability judgment is grounded in evidence.

The second thesis advanced by Kyburg in this connection is that there is a function C singled out by probability logic that all rational agents ought to embrace as their own confirmational commitment. If X and Y agree on their evidence, they ought to agree on their credal state.

I think that Kyburg is right to chastise the personalists or subjectivists including Ramsey for failing to recognize the importance of what I (not Kyburg) am calling confirmational commitments. I disagree with him, however, in his advocacy of a single standard confirmational commitment sanctioned by probability logic that qualifies as the logical probability. I concede that probability logic determines a logical probability just as the logic of full belief determines the belief state in which all and only logical truths are fully believed. But no one should ever be in either state.

There is another large respect in which I am in agreement with Kyburg. Credal states that are the values of confirmational commitments should not be required to be numerically determinate. Kyburg [1961], I.J. Good [1952; 1962], C.A.B. Smith [1961] and Daniel Ellsberg [1961; 2002] are pioneers in the period after the second world war in promoting this idea. But those who have advocated indeterminacy in probability have often differed from one another in how such indeterminacy should be understood. Although I am a fan of indeterminacy in probability judgment, Kyburg and I differ on the details.

To fix on some idea of these ideas, let us begin by specifying in more detail some of the technical notions to be deployed in the subsequent discussion [Levi, 1980].

A real valued function $Q(x/y)$ defined for all x in K and all y consistent with **K** is a *finitely additive and normalized conditional probability function defined for K* relative to **K** if and only if the following conditions are satisfied.

(1) For y consistent with **K**, $Q(x/y) \geq 0$.

(2) If $\mathbf{K} \wedge x = \mathbf{K} \wedge x'$ and $\mathbf{K} \wedge y = \mathbf{K} \wedge y'$, $Q(x/y) = Q(x'/y')$.

(3) If $x \wedge z$ is incompatible with $\mathbf{K} \wedge y$, $Q(x \vee z/y) = Q(x/y) + Q(z/y)$.

(4) If $\mathbf{K} \wedge y$ has x as a consequence, $Q(x/y) = 1$.

(5) $Q(x \wedge z/y) = Q(x/z \wedge y)Q(z/y)$.

A *credal state* B relative to state of full belief **K** is a set of conditional credal functions $Q(x/y)$ where x is any element of K and y is any element of K consistent with **K**. Members of B are *permissible credal functions* according to credal state B.

$\mathbf{C} : \mathbf{K} \mapsto \mathbf{B}$ is a **confirmational commitment** where B is a credal state relative to **K**. If X endorses C at t, it represents X's commitment to regard the credal functions in B to be permissible to use in assessing expected utilities if **K** is X's state of full belief and $C(\mathbf{K}) = B$.

Probability Logic imposes necessary conditions for the rational coherence or consistency of confirmational commitments. A *Complete Probability Logic* specifies necessary and jointly sufficient conditions for the coherence of confirmational commitments.

Here are two principles of probability logic that ought to be noncontroversial.

> **Confirmational Coherence**: If Q is a permissible credal function according to $C(\mathbf{K}) = B, Q(x/y)$ is a finitely additive and normalized probability function conditional probability function relative to **K**.

> **Confirmational Consistency**: **K** is consistent if and only if $C(\mathbf{K})$ is nonempty..

These two constraints on confirmational commitments ought to be non-controversial but Kyburg does appear to controvert them. According to Kyburg, a state of credal probability judgment is given by specifying upper and lower probabilities characterizing probability intervals and not by a set. However, an interval valued probability function might be said to determine the set of unconditional probability functions enveloped by the interval valued function and conversely a set of unconditional probability functions determines the upper and lower probability function that envelops them. To this extent, Kyburg's view could be seen to satisfy confirmational coherence and constraints that are stricter yet.

Yet, Kyburg's theory falls short of confirmational coherence according to which a credal state is representable by a set of *conditional* probability functions. Here a conditional credal probability function is required to satisfy the multiplication theorem. This makes sense when conditional probabilities determine conditional probabilities for called off bets. Here is an example:

	AE	$\neg AE$	$\neg E$
G	$S - P$	$-P$	k
R	0	0	k

Example 1.

In example 1 the decision maker X is offered a bet G on the truth of A that is called off if E should be false. I suppose that if AE is true, X receives $S - P$ utiles. X loses P utiles if $\neg AE\ E$ is true. If E is false, X receives k utiles. If X chooses R, X receives nothing in the first two states and receives the same utiles under the third eventuality.

Although the falsity of E is a serious possibility according to X's state of full belief **K**, that serious possibility may be ignored because X equiprefers G and R in case E is false so that that serious possibility is not a *relevant* possibility.

The book theorems may be extended to show that fair betting rates for called off bets cohere with betting rates for unconditional bets if and only if such fair betting rates are controlled by conditional probabilities satisfying the multiplication theorem. Of course, this is on the assumption, which is often counterfactual, that such fair betting rates exist. As Kyburg and I would both agree, they need not exist because of indeterminacy in probability judgments. Nonetheless, if credal states are intended to represent X's judgments as to how to take risks, it seems reasonable to interpret the several permissible conditional probability functions in an indeterminate credal state as satisfying the multiplication theorem. *Pace* Kyburg that is what confirmational conditionalization does.

There is another sense in which conditional probability is understood. X might consider what X's credal state should be were X's current state **K′** where **K′** is the expansion of **K** by adding E consistent with **K** to **K**– i.e., **K′** = **K** ∧ **E**. The conditional probability of A given E is understood as the probability of A on that supposition.

According to the approach taken here, the conditional probability on the supposition that E is given by the confirmational commitment $C(\mathbf{K} \wedge E)$ that specifies the set of permissible credal probabilities according to the credal state to which X is committed on the supposition that X's state of full belief is $\mathbf{K} \wedge E$.

One of the characteristic features of the Bayesian view is the insistence that the two notions of conditional probability ought to be somehow equivalent as a matter of probability logic. Kyburg dissents from this prescription. To formulate the Bayesian view more sharply, let us first introduce the notion of *conditionalization* of the credal state determined by $\mathbf{K}$ by a credal state determined by $\mathbf{K} \wedge y$.

> ***Definition of conditionalization***: $C(K \wedge y)$ is the conditionalization of $C(\mathbf{K})$ if and only if every credal probability Q^y permissible according to $C(\mathbf{K} \wedge y)$ there is a permissible Q according to $C(\mathbf{K})$ such that $Q(x/z \wedge y) = Q^y(x/z)$ and for every permissible Q there is a Q^y satisfying the same condition.

According to the Bayesian view, a confirmational commitment should satisfy the following principle as a matter of probability logic.

> **Confirmational Conditionalization**: If $\mathbf{K}$ is consistent and y is consistent with $\mathbf{K}$, $C(\mathbf{K} \wedge y)$ is the conditionalization of $C(\mathbf{K})$.

It should be clear that if confirmational conditionalization is enforced as a minimal condition of probabilistic rationality or probability logic, the two senses of conditional probability merge into one.

If confirmational conditionalization is enforced, then any confirmational commitment can be determined by specifying the credal state relative to weakest potential state of full belief $\mathbf{T}$ where X is in the state of ignorance. By confirmational conditionalization, $C(\mathbf{T} \wedge y) = C(\mathbf{K})$ is the conditionalization of $\mathbf{T}$ and permissible conditional probability functions that satisfy the multiplication theorem also qualify as suppositional conditional probabilities.

Let $CIL(\mathbf{T})$ be the set of conditional probabilities relative to the state of maximal ignorance that are permissible according to some confirmational commitment or other according to the principles of probability logic. I shall call such conditional probability functions *logically permissible*.

If CIL is recognized by probability logic to be a confirmational commitment, it qualifies as *the* logical commitment. It is far from automatic that CIL is recognized to be a confirmational commitment. Thus those who insist that confirmational commitments should be numerically determinate require confirmational commitments to recognize exactly one logically permissible conditional probability relative to **T** to be (extralogically) permissible according to **T**. This is true, for example, for Carnap's credibility functions.. Advocates of such a requirement of *confirmational uniqueness* cannot recognize CIL to be a rationally allowable confirmational commitment according to probability logic even though the function is perfectly well defined.

There is one final important point about confirmational conditionalization. It is not the same as what I call *Temporal Credal Conditionalization* and most authors call simply "conditionalization".

Temporal Credal Conditionalization (TCC) If inquirer X shifts from **K** at t to **K** $\wedge$ y at t', B_t should be the conditionalization of B.

Inverse Temporal Credal Conditionalization (ITCC) If X shifts from **K** $\wedge$ y at t to **K** at t', B_t should be the conditionalization of B_t.

Confirmational conditionalization implies TCC and ITCC provided it is supplemented with the assumption that the confirmational commitment C remain the same from t to t'. If however, C is modified in that time interval, neither TCC nor ITCC are mandated even though the confirmation commitments at t and t' both obey confirmational conditionalization.

TCC is easily recognized as "updating" by conditionalization via Bayes theorem — a principle widely used by Bayesian statisticians. Those who endore it may be interpreted within the framework adopted here as adopting confirmational commitmenets satisfying confirmational conditionalization. Harold Jeffreys [1939] and Rudolf Carnap [1937] both endorsed conceptions of "logical probability" that may be understood as functions from states of evidence to credal states obeying confirmational conditionalization that are immune to modification.

The chief difficulty with the personalist or subjectivist view as construed by many probabilists is that Temporal Credal Conditionalization is enforced as if it were a requirement of minimal rationality along with Confirmational Uniqueness even though personalists deny that there are principles of rationality that mandate one confirmational commitment rather than another. Such personalism ends up maintaining that X and Y are rationally entitled to start off with different numerically determinate confirmational commitments that both should recognize as rational. Yet both of them are required on pain of irrationality to stick with their different confirmational commitments throughout their inquiries. And aficionados of Jeffrey [1965] updating seem to embrace a similar attitude. Like Bush, Cheney and Rumsfeld we are to admire as rational the resolution to stay the course come hell or high water.

I endorse confirmational conditionalization. Kyburg does not. Yet, we both agree that TCC is not always mandatory. Neither of us endorse updating via Bayes theorem. Our differences concern the synchronic principles of probability logic and especially confirmaitonal conditionalization.

One reason for favoring confirmational conditionalization may be based on a comparison of the situation in example 1 where the choice between G and R obtains in a context where X regards the falsity of E to be serious possibility but irrelevant because of the way payoffs are evaluated and a context where X rules out the falsity of E as a serious possibility but otherwise keeps the payoffs the same. The second predicament is obtained from the first by expanding **K** by adding E and nothing else.

If you think that the comparison of the values of G and R ought not to be altered by this transformation, then confirmational conditionalization ought to be compelling to you.

Kyburg is committed to rejecting this argument just as he is committed to denying the prima facie obvious recommendation to suspend judgment when asked about the outcome of a billion ticket fair lottery. In the case of his rejection of confirmational conditionalization, his reasons seem to me far more serious than in the case of the lottery. I do not agree with him in either case; but the problems he is addressing in the case of confirmational conditionalization are to my way of thinking far more important both for statistical practice and epistemology.

There are two important problems to consider. Once concerns direct inference from information about chances or frequencies of outcomes of trials of some kind to credal judgments about the outcomes. The other concerns inference from data about outcomes of trials to conclusions about chances or population frequencies.

Consider the following example.

Example 2. Consider the following version of a familiar example: X takes for granted that 40% of the residents of the State of Satania are Protestant, 30% are Catholic, 10% are Orthodox, 15% are Muslim and 3% are Jewish and 2% are other. X also knows that person Y has been selected at random from the residents of Satania and is discovered to be a resident of the capital city of Gehinnom. X lacks precise information as the the census of religious groups in Gehinnom. However, X does know that the true frequency distribution is enveloped by [35%, 45%], [35%, 35%], [9%, 11%], [14%, 16%], [2%, 4%], [1%, 3%]. What credal state should X adopt concerning Y's religion?

According to the view I favor, X should ground his credal probability judgments on his information concerning the frequency distribution of religions according to census data about Genhinnom since that is the narrowest reference class about which X has stochastically relevant statistics. X does to be sure have more information. For example, Y was randomly selected on October 9 from the Satania population and is from Gehinnom. But X is convinced that the date of selection is stochastically irrelevant. It makes no difference to the chances of Y's being of one religious category rather than another. This conviction is not a logical truth of any sort. It is part of X's background information.

Given this recommendation, I would consider the convex hull of .45,.35, 0.09,0.16,0.02 and 0.03 and 0.35, 0.35, 0.11, 0.14, 0.04, 0.01 and maintain that any convex subset of this set of distributions would be acceptable.

Should anyone complain that demanding convexity is unreasonable, I would respond that it is absolutely crucial but would, for the sake of the argument, relax the requirement and count as acceptable *any* nonempty subset of probability distributions.

Kyburg's recommendation for this case is that the credal probability judgments should invoke the statistics for the reference class of individuals selected from Satania. The distribution of frequencies relative to the narrower reference class does not conflict with the broader one but it is less precise. So Kyburg would *mandate* the distribution 0.4, 0.3, 0.1, 0.15 and 0.02.

Let E assert that the distribution of religions in Gehinnom is given by the intervals [35,39], [31,35], [10,11], [14,15], [3,4], [1,3]. E' specifies intervals [39,45], [25,31], [9,10], [15,16], [2,3], [2,3]. If X were to find out that E is true, X would shift to using the residents of Gehinnom as reference class on Kyburg's account. The credal distributions would be the intervals just specified for E. But if X found out that E' is true, X would continue to use the reference class of residents of Satania and the precise distributions given previously.

Let us assume that confirmational conditionalization held on Kyburg's account. Given any hypothesis h concerning Y's religion from the categories specified, $Q(h) = Q(h/E \vee E') = Q(h/E)Q(E/E \vee E') + Q(h/E')Q(E'/E \vee E')$ for every permissible Q by confirmational coherence in the credal state relative to the initial state of full belief and, according to Kyburg, must be numerically determinate and equal to the uniquely permissible value relative to the expansion of **K** by adding E'. If E is added to **K**, the set of permissible values for h does not overlap that particular value. From this is follows that $Q(E'/E \vee E') = 1$ if confirmational conditionalization holds and this is unacceptable.

Thus, the two notions of conditional probability discussed earlier come apart. Part of confirmational coherence must be given up — that pertaining to the multiplication theorem for the second notion of conditional probability.

The reason that confirmational conditionalization ought to be sacrificed here is that Kyburg's account of direct inference is designed to allow for assigning relatively determinate credal probabilities to statistical hypotheses on the basis of data even though the prior credal state is indeterminate. And the motive for this is Kyburg's insistence that the only basis for assigning credal probabilities to hypotheses is through appeal to direct inference.

Kyburg's motivation for pursuing this approach is two-fold.

According to Kyburg, not only is direct inference the sole basis for assigning credal probabilities on the basis of what we take as knowledge or evidence, Kyburg insists that the probabilities assigned should be the weakest or most indeterminate credal states allowed by direct inference. If given the right reference class the set of permissible probability distributions is some subset of a set of distributions enveloped by an interval valued specification for each alternative, the set of distributions that ought to be assigned should be the largest set enveloped by the interval valued function. This too is mandated by probability logic. It recommends adopting as standard the logical confirmational commitment as the standard confirmational commitment.

The second feature is that, in constrast to Bayesian logical confirmational commitment that yields maximal indeterminacy when the prior is maximally indeterminate except for what deduction requires, the Kyburgian confirmational commitment achieves creatio ex nihilo. Under the right conditions, the introduction of new data can yield increased determinacy in probability judgment even though probability judgment prior to the data is maximally indeterminate. This is no mean trick and is the jewel in the Kyburgian crown.

As far as I can make out there is no outright contradiction in Kyburg's system although there is a certain fragility. R. A. Fisher [1956], Kyburg's predecessor in defending so-called fiducial inference thought that the fiducial probabilities obtained from a given body of data could be updated via conditionalization by additional data. His view does lead to incoherencies as Teddy Seidenfeld [1979] has observed. Kyburg has carefully avoided the temptations of Bayesianism. Whether statisticians can live with his puritanical zeal remains to be seen.

I, for my part, do not believe that there is creatio ex nihilo. There is no free lunch. Yet, Kyburg's efforts to prove these prejudices wrong are profoundly challenging. No epistemologist worth his salt can afford to ignore his contributions.

BIBLIOGRAPHY

[Carnap, 1937] R. Carnap. Testability and Meaning, 1937. Reprinted in R. Tuomela, *Dispositions*, Dordrecht: Reidel (1978), 3–16.

[Ellsberg, 1961] D. Ellsberg. Risk, Ambiguity and the Savage Axioms, *Quarterly Journal of Economics*, 75:643–69, 1961.

[Ellsberg, 2002] D. Ellsberg. *Risk, Ambiguity and Decision*, New York: Garland. (Publication of 1962 Ph.D. dissertation), 2002.

[Fisher, 1956] R. A. Fisher. *Statistical Methods and Scientific Inference*, New York: Hafter, 1956.

[Good, 1951] I. J. Good. Rational Decisions, *Journal of the Royal Statistical Society*, Ser.B. 14: 107–14, 1951.

[Good, 1961] I. J. Good. Subjective Probability as a Measure of a Nonmeasurable set. In *Logic, Methodology and Philosophy of Science, Proceedings of the 1960 International Congress*, Stanford: Stanford University Press, 319–29, 1961.

[Jeffrey, 1965] R. Jeffrey. *Logic of Decision*, New York: McGraw Hill, 1965.

[Jeffreys, 1939] H. Jeffreys. *Theory of Probability*, Oxford University Press, 1939.

[Koslow, 1992] A. Koslow. *A Structural Theory of Logic*, Cambridge: Cambridge University Press, 1992.

[Kyburg, 1961] H. E. Kyburg, Jr. *Probability and the Logic of Rational Belief*, Middletown, Conn.: Wesleyan University Press, 1961.

[Kyburg, 1974] H. E. Kyburg, Jr. *The Logical Foundations of Statistical Inference*, Dordrecht: Reidel, 1974.

[Kyburg and Teng, 2001] H. E. Kyburg, Jr and C. M. Teng. *Uncertain Inference*, Cambridge: Cambridge University Press, 2001.

[Levi, 1980] I. Levi. *The Enterprise of Knowledge*, Cambridge, Mass: MIT Press, 1980.

[Levi, 1991] I. Levi. *The Fixation of Belief and Its Undoing*, Cambridge: Cambridge University Press, 1991.

[Levi, 1997] I. Levi. *The Covenant of Reason*, Cambridge: Cambridge University Press, 1997.

[Seidenfeld, 1979] T. Seidenfeld. *Philosophical Problems of Statistical Inference: Learning from R.A. Fisher*, Dordrecht: Reidel, 1979.

[Smith, 1961] C. A. B. Smith. Consistency in statistical inference, *Journal of the Royal Statistical Society*, Ser.B. 23: 1–25, 1961.

Forbidden Fruit: When Epistemological Probability may *not* take a bite of the Bayesian apple

TEDDY SEIDENFELD

1 Elementary Probability Theory and some of its *Bayesian* fruits

1.1

Unconditional Probability $P(\bullet)$ is governed by three axioms:

Axiom 1 $PZ\bullet$) is real-valued function defined over an algebra

$$0 \le P(\bullet) \le 1$$

Axiom 2   For the sure event S, $P(S) = 1$

Axiom 3   *(additivity)* For disjoint events, where $A \cap B = \emptyset$ then

$$P(A \cup B) = P(A) + P(B)$$

Conditional Probability $P(\bullet|\bullet)$ is governed by two additional axioms:

Axiom 4 $P(A \cap B) = P(A|B) \times P(B) = P(B|A) \times P(A)$.

Axiom 5   For each $B \ne \emptyset$, $P(\bullet|B)$ is an unconditional probability.

(Aside)   Axiom 5 is of concern primarily when the conditioning event, B, is null. That is, when $P(B) = 0$ Axiom 4 fails to insure that $P(\bullet|B)$ is an unconditional probability satisfying Axioms 1-3. For discussion of some of the controversial aspects of the received theory's solution to this problem using regular conditional distributions, see my [Seidenfeld, 2001]

Let $\{H_1, H_2, \ldots, H_n\}$ be a partition into n-many pairwise disjoint and mutually exhaustive states.

The Law of Total Probability asserts

$$P(A) = \sum_i P(A \cap H_i) \tag{1}$$

which follows by additivity from the elementary identities:

$$
\begin{aligned}
A &= A \cap S \\
&= A \cap [H_1 \cup H_2 \cup \ldots \cup H_n] \\
&= [A \cap H_1] \cup [A \cap H_2] \cup \ldots \cup [A \cap H_n]
\end{aligned}
$$

By the principal axiom governing conditional probability, (1) yields

$$P(A) = \sum_i P(A|H_i) \times P(H_i) \tag{2}$$

Here, $P(A|\bullet)$ is called (a version of) the *likelihood function*.

A familiar Bayesian formulation of this law is as:

unconditonal probability	equals	*expected likelihood*
$P(A)$	$=$	$\sum_i P(A|H_i) \times P(H_i)$.

Then, unconditional probability $P(A)$ is constrained as a *convex* function of conditional probability $P(A|\bullet)$ over a partition for the argument $(\bullet)$.

It is a short step from this result to *Bayes' Theorem*. By the principal axiom of conditional probability, when $P(A) \neq 0$.

$$P(H|A) = \frac{P(A|H) \times P(H)}{P(A)}$$

And by an application of the previous law:

$$= \frac{P(A|H) \times P(H)}{\sum_i P(A|H_i) \times P(H)_i)}$$

An easy calculation then yields:

$$\frac{P(H_1|A)}{P(H_2|A)} = \frac{P(A|H_1)}{P(A|H_2)} \times \frac{P(H_1)}{P(H_2)} \tag{3}$$

1.2 Conditionalisation and three *Bayesian* fruits of these probability laws

Levi's account of why *Bayes'* Theorem creates interest in *conditional probability* is that the conditional probability, $P(\bullet|H)$, is the answer to an important hypothetical question:

"What would your probability function be were your current knowledge augmented with (consistent) H?"

Conditionalisation then fixes *Bayesian* inference, as follows. In response to the question what your uncertainty would be regarding rival hypotheses, H_1 and H_2, were you to learn that A, Bayes' theorem provides a helpful algorithm:

$$\frac{P(H_1|A)}{P(H)_2|A)} = \frac{P(A|H_1)}{P(A|H_2)} \times \frac{P(H_1)}{P(H_2)}$$

It is summarized by the familiar Bayesian mantra

posterior odds = likelihood ratio $\times$ prior odds.

This mantra is particularly useful when the rivals H_1 and H_2 are simple statistical hypotheses so that the likelihood function $P(A|\bullet)$ is fixed by non-controversial inference rules, e.g., *Direct Inference*. Here are three important fruits of *Bayesian* conditionalisation:

1st *product:* Eliminate *nuisance* parameters by averaging the likelihood function.

Suppose that, in order to make the likelihood function simple – in order to apply *Direct Inference* – additional parameters J_i are specified beyond the composite hypothesis H that is the investigator's focus of interest. These nuisances J_i can be eliminated by an application of the conditional version (2^*) of (2), given H,

$$P(A|H) = \sum_i P(A|H, J_i) \times P(J_i|H) \qquad (2^*)$$

2nd *product:* Composite data may be evaluated in any order computationally advantageous for the inference.

Suppose that the composite data are the pair (A, B) and that these are independent given the statistical hypothesis H, i.e.

$$P(A, B|H) = P(A|H) \times P(B|H)$$

or equivalently

$$P(A|H, B) = P(A|H) \text{ and } P(B|H, A) = P(B|H)$$

then

$$
\begin{aligned}
P(H|A, B) \quad &\propto \quad P(A|H) \times P(B|H) \times P(H) \\
&\propto \quad P(A|H) \times P(H|B) \\
&\propto \quad P(B|H) \times P(H|A)
\end{aligned}
$$

3rd *product:* The likelihood ratio equals the ratio of posterior odds to prior odds.

$$\frac{P(A|H_1)}{P(A|H_2)} = \frac{P(H_1|A)}{P(H_2|A)} \div \frac{P(H_1)}{P(H_2)}$$

So, the distribution of the likelihood ratio, viewed before the data are collected, is one perspective on how informative an experiment will be in changing the prior to the posterior. Unless the distribution of the likelihood ratio is the degenerate, constant = 1, the experiment has positive probability of generating evidence that, were it learned, would change the investigator's mind.

A familiar likelihood based index of *information* that measures how much a probability distribution Q differs from a distribution P , both defined on a space Ω, is *Kullback-Leiber* Information:

$$KL(Q, P) = \sum_{\omega} \log \left[\frac{Q(\omega)}{P(\omega)} \right] Q(\omega) \geq 0$$

Set P to the "prior" and Q to the "posterior" given data $X = x$, with both distributions over the common space Θ of the parameter of interest. Then:

$$KL(posterior, prior) = \sum_{\theta} \log \left[\frac{P(x|\theta)}{P(x)} \right] P(\theta|x)$$

This change is 0, i.e., there is no information gained in going from the *prior* to the *posterior* if and only if $P(x|\bullet)$, the likelihood function with respect to the parameter θ, is constant. Thus, unless an experiment is almost sure to produce irrelevant information, as indexed by a constant likelihood, the expected information gain in going from the prior to the posterior is strictly positive.

2 Epistemological Probability [*EP*] Theory and some of its original features

2.1 Epistemological Probability Theory and Direct Inference

(*Historical Aside 1*) Henry Kyburg's original theory of *Epistemological Probability* [*EP*] dates, I believe, from the final chapter in his 1956 Columbia University doctoral thesis ([Kyburg, 1956]), which was a study of the *Keynesian School* of probability, titled *Probability and Induction in the Cambridge School*. Its first full-dress, public appearance was in his [Kyburg, 1961] book, *Probability and the Logic of Rational Belief*. Even as recently as ten years ago, at a conference on Keynes at Wake Forest, Kyburg promoted *EP* as his preferred interpretation of Keynesian probability theory [Kyburg, 1995], where interval-valued probability provides a formal treatment of Keynes' important idea that not all (rational) probability judgments are comparable.

The canonical form of an *EP* statement is: $EP(\phi(s); K) = [p, q]$, where:

- *EP* is an *interval-valued* probability, $[p, q]$

- that an individual s, bear property ϕ – written $\phi(s)$;

- given background knowledge K that includes

  - the frequency information that between p and q percent of the members of the reference set R bear ϕ

  - the knowledge that individual s is a member of R

  - and for each rival reference set R' to which s is known to belong, K contains no *stronger* frequency information about ϕ

  - and except for larger sets R' ($\supset R$) to which s is known to belong, K contains no different frequency information about ϕ.

In Levi's terms, each *EP* statement is an instance of *Direct Inference*:

from the knowledge of frequencies $[p, q]$ of ϕ in a population R and that $s \in R$ *to* an interval valued probability $[p, q]$ that $\phi(s)$.

2.2 Epistemological Probability Theory and Inverse Inference

What is entirely original to *EP* is how the interplay of the *strength* and *difference* clauses for fixing the winning reference class R yields important cases of statistical *Inverse Inference* derived from *Direct Inference*.

from an interval valued probability $[p, q]$ that $\phi(s)$.

to an interval valued probability $[p, q]$ of ϕ in a population R.

EXAMPLE 1. Suppose that we have a scale on which to weigh objects. Our scale is calibrated so that, within its functioning range, if an object of μ-units mass is weighed, the readings are distributed as $X \sim N(\mu, 1)$; a Normal distribution with unit variance and mean μ.

We weigh an 1878 Indian Head penny on our scale and observe that $X = c$. This reading is a sample of one from the population of measurements taken with scales of this calibration. Our background knowledge K is otherwise uninformative about the distribution of weights of 1878 Indian Head pennies, of which about 5.8 million were minted.

What is the *EP* for statements of *Inverse Inference* about μ? For instance, what is $EP(c - 2 \le \mu \le c + 2 | X = c, K)$? The key to *EP*'s original treatment of this problem is to focus on special *pivotal* properties $\phi_r(\bullet)$ of readings from such scales.

- $\phi_r(X)$ obtains for X *if and only if* $|\mu - X| \le r (r \ge 0)$

The special feature of *pivotal* properties is that the percent of X-readings that satisfy them is known exactly, based solely on K.

$$EP(-r \le \mu - X \le r | K) = [\Phi^r_{-r}, \Phi^r_{-r}],$$

where Φ^r_{-r} is the probability that a $N(0, 1)$ variate has its value in the interval $[-r, r]$. For instance, $EP(-2 \le \mu - X \le 2 | K) \approx [.95, .95]$.

By the *Strength* rule, this yields a precise *EP Inverse* statement

$$EP(-2 \le \mu - c \le 2 | X = c, K) \approx [.95, .95]$$

or

$$EP(c - 2 \le \mu \le c + 2 | X = c, K) \approx [.95, .95]$$

EXAMPLE 2. Suppose that, as before, we have our scale with which to weigh objects. Our scale is calibrated so that, within its range, if an object of μ-units mass is weighed, the separate readings X_i of the same object are identically and independently distributed [*iid*] $X_i \sim N(\mu, \sigma^2)$, a Normal distribution with mean μ and *unspecified* variance σ^2. We take n readings $\tilde{x} = (x_1, \ldots, x_n)$ of our 1878 penny. What is the $EP(c - 2 \le \mu \le c + 2 | \tilde{x}, K)$? This problem is importantly different from the first because, though μ remains the *parameter of interest*, in this version σ^2 is a *nuisance parameter* whose value we do not know.

Again, there is a special (*Student's t*) pivotal property to deal with the inference about μ in the absence of knowledge of σ^2.

$$\phi'_r(\tilde{X}) \text{ obtains if and only if } \frac{|\mu - \tilde{X}|}{S} \le r$$

where $\bar{X} = \sum_{i=1}^n X_i$ and $S^2 = \frac{\sum_{i=1}^n (X-\bar{X})^2}{n(n-1)}$. For instance, with $n = 2$ and $r = |X_1 - X_2|/2$, we have

$$EP(X_{min} \leq \mu \leq X_{max}|\tilde{X}, K) = [.5, .5].$$

And by *EP*'s *Strength* rule for determining the reference class in a Direct Inference, we conclude the *Inverse EP* statement

$$EP(x_{min} \leq \mu \leq x_{max}|\tilde{X} = \tilde{x}, K) = [.5, .5].$$

(*Historical Aside* 2) Though Kyburg developed this mode of Inverse reasoning to show how Keynes' 1921 theory of probability – a theory that allowed (logical) probability to take non-real values – might be interpreted inside a theory of interval-valued probability, in fact, *EP* really is a wonderful and fully principled generalization of R.A.Fisher's 1930 enigmatic proposal of *fiducial* probability.

In what I think was the last of 3 rounds of correspondence exchanged with Kyburg, during Fisher's last year of life, Fisher began his letter,

> After a long while I have now succeeded in obtaining your book on Rational Belief. So far it seems to be as good as I had hoped, which would be high praise. (14 May, 1962)

But also Kyburg was mildly criticized by Fisher in a way that I suspect no other had ever been. In a 13 January 1962 letter to the Canadian Statistician D.A.Sprott, Fisher wrote,

> Do you know the name of H. Kyburg of the Rockefeller Institute 21, N.Y.? His line seems to be abstract symbolic logic, but he has recently caught fire on the fiducial argument and indeed may be exaggerating its importance"

During his long and influential career, Fisher showed no restraint criticizing many for failing to appreciate the importance to Statistics of fiducial reasoning. (See, e.g., [Fisher, 1973], section III.3.) But, Kyburg was singled out, and is unique among Fisher's targets I believe, for having committed the other error!

3 When Epistemological Probability may not go *Bayesian*!

3.1 *EP* Theory and Statistical Inference with Nuisance Parameters

Approximately 28 years ago, in an article *Direct Inference*, I. Levi demonstrated that *EP* does not satisfy *Bayesian* conditionalisation ([Levi, 1974]).

Levi's counterexample highlighted some anti-Bayesian features of the *Strength* rule: the rule to give priority to reference sets that yield precise, i.e. narrower probability intervals.

EXAMPLE 3 (Levi, 1977). Suppose we know that Petersen (denoted s) is a Swedish resident of Malmo. We are interested in the *EP* that he is a Protestant. Our rational corpus of knowledge includes the following frequency facts about the two competing reference sets: Swedes, and residents of Malmo.

- We know that 90% of Swedes are Protestants.

- But all we know about Malmo is that either

    H_1 : 85% of Malmo's residents are Protestant

    or H_2 : 91% of Malmo's residents are Protestant

    or H_3 : 95% of Malmo's residents are Protestant

    with a resulting known frequency interval [.85, .95] of % Protestant.

So,

$$EP(\phi(s); K) = [.9, .9]$$

because the *Strength* rule allows the larger reference set (Swedes) to win over the rival reference set of Malmo's residents, whose frequency interval for the property in question is less informative [.85, .95].

However, *EP* theory also entails the following statements

$$EP(\phi(s); H_1, K) = [.85, .85]$$
$$EP(\phi(s); H_2, K) = [.91, .91]$$
$$EP(\phi(s); H_3, K) = [.95, .95]$$
$$EP(\phi(s); H_1 \vee H_3, K) = [.9, .9]$$
$$EP(\phi(s); H_1 \vee H_2, K) = [.9, .9]$$
$$EP(\phi(s); H_1 \vee H_2 \vee H_3, K) = [.9, .9]$$

Each of the last three of these six *EP* statements results by an application of the *Strength* rule, which picks the larger reference class (Swedes) for determining the *Epistemological Probability* that Petersen is a Protestant.

The contradiction that results is with our first elementary law:

$$P(\phi(s)) = \sum_i P(\phi(s)|H_i) \times P(H_i)$$

There is no prior distribution $P(H_i)$ over these three simple statistical hypotheses that satisfies all six EP values. In other words, EP theory does not follow the

Bayesian law that there exists a *prior*, $P(H_i)$, against which one may average the likelihood function.

The second Bayesian version of this law is that we may eliminate *nuisance parameters* J_i by an application of the rule:

$$P(A|H) = \sum_i P(A|H, J_i) \times P(J_i|H)$$

If, to the contrary, *EP* theory followed this law, then in good Bayesian style, we could eliminate nuisance parameters by averaging them with other *EP* probabilities.

In our second example of *EP* inference, where $X \sim N(\mu, \sigma^2)$, μ is the parameter of interest, and σ^2 is the nuisance parameter. A Bayesian elimination of σ^2 can go like this:

$$p(\mu|\tilde{x}) = \int p(\mu|\tilde{x}, \sigma^2) p(\sigma^2|\tilde{x}) dp(\sigma)$$

EP theory provides precise probabilities for each of the terms on the right-hand side of this equation. But it does not take a bite of the *Bayesian* apple! This calculation is invalid. Instead, (Example 2) a direct *Student's* pivotal duplicates the conclusion of this *Bayesian* inference.

In the previous case, then, *EP* theory gets to the same place it would were it *Bayesian*. But that is not always possible, as the next example illustrates.

EXAMPLE 4 (The Behrens-Fisher problem). Let $\tilde{X}_1 = (X_{11}, \ldots, X_{1n})$ and $\tilde{X}_2 = (X_{21}, \ldots, X_{2n})$ be independent *iid* samples respectively from the two Normal distributions: $N(\mu_1, \sigma_1^2)$ and $N(\mu_2, \sigma_2^2)$. The parameter of interest is $\delta = \mu_1 - \mu_2$. The nuisance parameter is $\xi = \frac{\sigma_1}{\sigma_2}$, about which we have no frequency information.

A *Bayesian* elimination of the nuisance parameter is as follows. Let $\tilde{X} = (\tilde{X}_1, \tilde{X}_2)$.

$$p(\delta|\tilde{x}) = \int p(\delta|\tilde{x}, \xi) p(\xi|\tilde{x}) dp(\xi)$$

Again, there are pivotal variables available for *EP* to derive precise *Inverse* probabilities for each of the two terms on the right side of this equation. However, as *EP* theory is not *Bayesian*, the calculation from right to left is invalid.

Alas, there is no *direct* pivotal available, analogous to *Student's* t-pivotal, to solve the left-hand side. It appears that *EP* theory here is missing the pleasures of this *Bayesian* fruit. *EP* theory could take a bite of this Bayesian apple, but it does not.

However, the conflict between *EP* theory and these Bayesian laws is not merely a case of *EP* theory missing out some Bayesian consequences of what it already entails. We cannot graft onto *EP* theory these missing Bayesian conclusions, as the next example illustrates.

EXAMPLE 5 (The Hollow Cube).   We are interested in the volume V of a hollow cube. We have available two sources of experimental data. We may accurately fill the hollow cube with a liquid of density, 1-unit mass/unit volume, and weigh that on our scale of known precision, resulting in the random variable $X_L \sim N(V, 1)$ Alternatively, we may cut a rod of density 1-unit mass/unit length, to the edge of the cube and weigh that on our scale: $X_R \sim N(V^{1/3}, 1)$.

As in Example 1, with either observation taken alone, there is an *Inverse EP* statement about the unknown V: With $X_L = x_L$ then *EP* entails that $V \sim N(x_L, 1)$. With $X_R = x_R$ then *EP* entails that $V^{1/3} \sim N(x_R, 1)$.

Though it is invalid by *EP* standards we may try to use the 2nd set of Bayesian laws to combine the two observations. There are three approaches:

$$
\begin{aligned}
p(V | x_L, x_R) \quad &\propto \quad p(x_L | V) \times p(x_R | H) \times p(V) \\
&\propto \quad p(x_L | V) \times p(V | x_R) \\
&\propto \quad p(x_R | V) \times p(V | x_L)
\end{aligned}
$$

EP theory does not entail a precise *prior* $p(V)$ for use in the first line. Moreover, there is no direct pivotal method using (X_L, X_R). At bottom, this is because there is no common 1-dimensional sufficient statistic for V that summarizes the 2-dimensional data.

But by the preceding results, *EP* theory entails precise (point-valued) probabilities for each term in the 2nd and 3rd lines, above. But they may not be added to EP theory. *These yield contradictory results!* This is because the Bayes-model associated with the 2nd line carries a precise, different prior for V than does the Bayes-model associated with the 3rd line. Thus, *EP* theory *must not* take this bite of the Bayesian apple as a method for combining composite data.

I do not know the full *EP* solution to the problem of the Hollow Cube. I conjecture that, because there are so many competing pivotal variables available for inference about V each yielding a different interval *EP* solution, the resulting *EP* interval estimates about V are vacuous, or nearly so. For example, in addition to the two pivotal variables relating to the inference of Example 1, each of which uses only one of the two observations, also there is the pivotal variable $[(X_L + X_R) - (V + V^{1/3})]$, which is pivotal based on the fact that the random variable $(X_L + X_R)$ has a normal distribution $N(V + V^{1/3}, 2)$. These three pivotal variables generally result in competing, precise statistical statements about V that prevent each other, by *EP*'s *Difference* rule.

The open challenge I see to *EP* theory highlighted by the Hollow Cube problem is how to combine a variety of statistical data, data that do not admit a common sufficient statistic. It appears that with a variety of evidence, within *EP* theory, an increase in the variety of evidence available may decrease the informativeness of the resulting statistical conclusions. This fact provides transition to a discussion of the third and final *Bayesian* law in Section 1 of this essay concerning the informational value of new evidence. That law says, as measured by any one of a large family of indices of statistical *Information*:

> unless an experiment is almost sure to produce irrelevant evidence, it carries a positive expected *Information* gain comparing the *posterior Information* with the *prior Information*.

In short, that law promises that changes in expected *Information* that result from *conditionalization* on new evidence will not go down, and will go up unless the data are irrelevant, as judged by the likelihood. *EP* theory does not partake in this *Bayesian Tree of Knowledge*. Is that ignorance a state of statistical bliss for *EP*?

3.2   *EP* theory and *Dilation* of interval valued probabilities.

The final contrast I want to draw is with a rival position that, like *EP*, uses interval-valued probability rather than real-valued probability, but unlike *EP* it incorporates *Bayesian conditionalization*. I. Levi's *Indeterminate Probabilities* [*IP*] provides an ideal version of such a rival theory ([Levi, 1974]). In it a rational agent's degrees of belief are represented by a convex set of $\wp$ of probabilities. The agent obeys conditionalization in the sense that the corresponding set of conditional probabilities $\{P(\bullet|H) : P \in \wp\}$ answers the question,

> What would your probability be were your current knowledge augmented with (consistent) H?

In these two rivals, *EP* and *IP* Theories, by contrast with the original (Bayesian) theory, one entirely *new* aspect of the agent's *uncertainty* of an event E is captured by the range of the probability interval for E. For example, in this new sense there is maximal uncertainty about E when the probability interval is the vacuous $[0, 1]$ range, and in this same sense that uncertainty is reduced when the probability interval for E shrinks to, say, $[.4, .7]$.

The anomalous phenomenon concerning this sense of *uncertainty*, on which I close this essay is called *dilation* (See [Seidenfeld, 1993] and [Herron, 1997]). Let experiment E carry possible outcomes $\{e_1, \dots, e_n\}$. Let $\wp$ be a non-empty convex set of probabilities. And let B be some event of interest.

DEFINITION 6. E *dilates* the set of probabilities for B just in case, for $i = 1, \ldots n$,

$$inf_\wp P(B|e_i) < inf_\wp P(B) \leq sup_\wp P(B) < sup_\wp P(B|e_i)$$

In words, when *dilation* occurs, under conditionalization the hypothetical new evidence is sure to *increase* the *uncertainty* of B, in the sense just described.

EXAMPLE 7 (A *Heuristic Example of Dilation*). Let $P^*(\bullet)$ denote the upper probability and let $P_*(\bullet)$ denote the lower probability with respect to the set $\wp$. Suppose that A is a highly uncertain event. That is $P^*(A) - P_*(A) \approx 1$. Let $\{H, T\}$ indicate the flip of a fair coin with outcomes independent of A. That is, $P(A, H) = P(A)/2$ for each $P \in \wp$. Define event B by, $B = \{(A, H), (A^c, T)\}$. The situation is depicted by the familiar 2x2 table:

	H	H
A	B	B^c
A^c	B^c	B

Note that B is pivotal-like! That is, it follows, simply, that $P(B) = .5$ for each $P \in \wp$. B carries no *uncertainty* in the novel sense of uncertainty common to EP and IP.

But

$$0 \approx P_*(B|H) < P_*(B) = P^*(B) < P^*(B|H) \approx 1$$

and

$$0 \approx P_*(B|T) < P_*(B) = P^*(B) < P^*(B|T) \approx 1$$

Thus, regardless how the coin lands, the conditional probability for event B dilates to a large interval, from a precise value of .5. In the novel sense of uncertainty relevant to *IP*, the uncertainty for B increases *for certain* by conditionalizing on the outcome of the $\{H, T\}$ experiment. Thus, within *Indeterminate Probability* theory, where conditionalization obtains, new evidence may increase uncertainty for sure.

In [Seidenfeld, 1993] Theorem 4.1, we show that only the density-ratio model for statistical uncertainty among neighborhood models is immune to dilation. In that sense, dilation is not rare within *IP* theory.

Though I have not here reported the decision theory that goes together with the theory of *Indeterminate Probabilities*, it should not be surprising that a decision maker will try to avoid learning evidence that dilates probabilities. I am ready to argue that such a decision maker will pay to avoid dilation! Then, for such a decision maker, the new evidence carries negative value.

By contrast, dilation is an impossibility within *EP* theory, and for the very same reason that it resists conditionalization! The *Strength* rule, which is the culprit that prevents *EP* from being *Bayesian*, also is the reason that *EP* is immune to dilation! Within *EP* theory, the evidence that causes dilation for *Indeterminate Probability* theory, is made innocuous by *strength*. Simply put, those problematic data are treated as irrelevant! This raises the question whether ignorance of certain Bayesian methods may indeed result in a state of bliss concerning statistical inference!

BIBLIOGRAPHY

[Bennett, 1990] J.H. Bennett ed., *Statistical Inference and Analysis: selected correspondence of R. A. Fisher*. Oxford U. Press, 1990.

[Fisher, 1973] R.A. Fisher, *Statistical Methods and Scientific Inference*, (3rd ed.), Hafner Press, 1973.

[Herron, 1997] T. Herron *et. al.*, Divisive Conditioning: further results on dilation, *Philosophy of Science*. 64, pp. 411-444, 1997.

[Kyburg, 1956] H.E. Kyburg, *Probability and Induction in the Cambridge School*. Ph.D. Thesis. Columbia University: NYC, 1956.

[Kyburg, 1961] H.E. Kyburg, *Probability and the Logic of Rational Belief*. Wesleyan U Press, 1961.

[Kyburg, 1995] H.E. Kyburg, Keynes as a Philosopher, in *New Perspectives on Keynes, History of Political Economy*, Annual Supplement to Volume 27, pp. 7–32, 1995.

[Levi, 1974] I. Levi, On Indeterminate Probabilities, *J.Phil* 71, pp. 391–418, 1974.

[Levi, 1977] I. Levi, Direct Inference, *J.Phil* 74, pp. 5–29, 1977.

[Seidenfeld, 1992] T. Seidenfeld, R.A.Fisher's Fiducial Argument and Bayes' Theorem, *Statistical Science* 7: 358-368, 1992.

[Seidenfeld, 2001] T. Seidenfeld, Remarks on the Theory of Conditional Probability, in *Probability Theory: Philosophy Recent History and Relations to Science*, V.F.Hendricks, S.A.Pedersen, and K.F.Jorgensen, eds, Kluwer Academic, pp. 167–178, 2001.

[Seidenfeld, 1993] T. Seidenfeld and L. Wasserman, Dilation for Sets of Probabilities", *Annals of Statistics*, 21, pp. 1139-1154, 1993.

Bayesian Inference with Evidential Probability

HENRY E. KYBURG, JR.

ABSTRACT. Evidential probability includes principles for adjudicating conflict between potential reference classes. Included among these principles is a principle that allows the use of inferences with Bayesian structure when that is appropriate. In the history of evidential probability this principle has enjoyed several names and formulations, some of which have been abandoned as simply wrong, others of which have been superceded by formulations of allegedly greater clarity. This paper will not trace the historical backing and filling, but will seek to provide a new and clear characterization of when joint statistical knowledge should be employed, and will relate this characteriztion to earlier formulations. We will also provide examples of circumstances under which our knowledge of joint distributions may be employed in a Bayesian way.

1   Introduction

The idea was this: that our probabilistic knowledge should both be based on long run relative frequencies, *and* be based on everything we know about the instance at hand. If we want to know the probability that Alice Winters will attend the barn dance next Saturday, we want to take account of all the facts we know about her, including the fact that she is more likely to go if she is invited by Ted than if she is invited by Tom, and thus including what we know about Ted's intentions. It is natural to think of this problem in terms of possible *reference classes*. This is particularly the case if, like many writers, you would like to have your probabilities tied to objective relative frequencies.

2   *Probability and the Logic of Rational Belief*

This was the general idea underlying the system embodied in [Kyburg, 1961] as well as [Kyburg, 1974]: the basic intuition was that we could more easily specify grounds for *denying* claims about randomness than we could spell out conditions under which such claims could be accepted.

One such principle has seemed uncontroversial: if your choice is between two reference classes, one of which you know to be included in the other, it is to the more "specific" reference class that you should turn for guidance. To make this idea precise, we shall introduce some more recent notation (the notation of PLRB was dreadful!) and some terminology. We shall write "$aRANK_K(B, C)$" for the claim that a is a random member of B with respect to C, relative to our background knowledge K. We shall write $\ulcorner\%(t, r, p, q)\urcorner$ for the claim that the best information in K about the proportion of satisfiers of the matrix t among the satisfiers of the matrix r is that it lies between p and q.[1]

To say that $\ulcorner\%(t, r, p, q)\urcorner$ *differs from* $\ulcorner\%(t', r', p', q')\urcorner$ is to say that it is neither a theorem that $[p, q] \subseteq [p', q']$ nor a theorem that $[p', q'] \subseteq [p, q]$. To say that $\ulcorner\%(t, r, p, q)\urcorner$ *is stronger than* $\ulcorner\%(t', r', p', q')\urcorner$ is to say that it is a theorem that $[p, q] \subset [p', q']$.

These two cases are handled separately in PLRB. We said that B' prevented the randomness of a in B with respect to C *by difference* provided there was an a' and a C' such that the following ws known:

1. $a \in C \equiv a' \in C'$

2. $a' \in B'$

3. $\%(C, B, p, q)$ differs from $\%(C', B', p', q')$

and such that neither

1. $B' \subset B$, nor

2. for some B'', $B' \subseteq B \times B''$

was known.

The second (less interesting) case is that in which B' prevented the randomness of a in B with respect to C *by strength*. This holds when there are a', C' such that

1. $a \in C \equiv a' \in C'$

2. $a' \in B'$

3. $\%(C, B, p, q)$ is stronger than $\%(C', B', p', q')$

---

[1] We follow Quine [Quine, 1951] in using quasi-quotation (corners) to specify the *forms* of expressions in our formal language. Thus $\ulcorner S \leftrightarrow T \urcorner$ becomes a specific biconditional expression on the replacement of S and T by specific formulas of the language.

and nothing prevents the randomness of a' in B' with respect to C' by difference.

Given the observation that there is always a *smallest* potential reference class — in the example offered, the set of occasions next Saturday on which Alice either will or will not attend the barn dance (there is only one such occasion, next Saturday, and in that reference class the frequency of barn dance attendance is either 0 or 1) — and the fact that the relative frequency in that reference set is often known merely to be in the interval $[0, 1]$ (it must be 0 or 1), it is natural to consider a principle according to which vacuous information may be disregarded, or, more generally, in which less nearly vacuous information is preferred to more nearly vacuous information.

Given the definitions of prevention by difference and prevention by strength, the definition of randomness was clear: "a is a random member of B with respect to C, relative to K" just in case nothing prevents it by difference and nothing prevents it by strength. And then we have probability: the probability of S is $[p, q]$ relative to K just in case there is a triple $\langle a, B, C \rangle$ such that a is a random member of B with respect to C, relative to K and "%(C, B, p, q)" is in K.

I have mentioned these details of the system of PLRB because it was clear, even then, that conditioning was being taken account of. The second proviso of ignorance in the characterization of prevention of randomness by difference, allows us to disregard the overall proportion of black balls in an experiment of drawing first a bag and then a ball from that bag, in favor of the weighted average of proportions of black balls in the bags.

3   LFSI and "The Reference Class"

These two works [Kyburg, 1974; Kyburg, 1983] bracket Levi's 1977 demonstration that my approach to probability comes into conflict with "Bayesian conditioning." What is most noteworthy about these two works, to me, is that they both follow the lead of PLRB in seeking a single set of objects whose known statistics will bound the probability in question. I was suprised, indeed, shocked, to discover this only recently. This approach is definitely abandoned by the appearance of "Combinatorial Semantics" in *Confirmational Intelligence* [Kyburg, 1997] and [Kyburg and Teng, 2001], where probabilities are determined by frequencies in *sets* of possible reference classes. When did this big change occur? I suspect that it occurs in a conference paper in Computer Science. No matter, since it is not directly relevant to the matter at issue.

In all of these works, both the early ones and the later ones, I have made something (not much, I think) of the fact that there exists a classical real-valued probability function that conforms to the probability intervals of evidential probability. Levi's argument is: "This is as it may be, but there are cases in which we can *show* that there is no conditional probability that matches the evidential updating." The example that demonstrates this occurs in [Levi, 1977] and also in [Levi, 1980.]. Levi and I had an exchange about the example in 1983 [Kyburg, 1983; Levi, 1983], but to little effect, since the example was revived by Teddy Seidenfeld in the present volume.

Here is the simple example: We know that Peterson is a Swedish resident of Malmö. (I am following the exposition of [Levi, 1980.].) We know that 90% of Swedes are Protestant. However, of the Swedes in Malmö all we know is that the proportion of Protestants is either 0.89, 0.91 or 0.92; represent these statistical hypotheses by $h_{0.89}$, $h_{0.91}$, and $h_{0.92}$. This may seem like a strange epistemic state of affairs, but it is clear that we could easily generate a concrete real case with bags and balls. It follows easily that knowing these things, the evidential probability that Peterson is a Protestant, e, is 0.9: $\text{Prob}(e, K) = [0.9, 0.9]$. This depends on the precision of our knowledge of the frequency of Protestants among Swedes, as well as on the vagueness of our knowledge of the frequency of Protestants in Malmö. So far, so good.

Let us represent the addition of t to the evidence K in the context $\text{Prob}(S, K)$ by $\text{Prob}(S, K, t)$. Standard evidential considerations lead to $\text{Prob}(e, K, h_{0.89} \vee h_{0.91}) = \text{Prob}(e, K, h_{0.89} \vee h_{0.92}) = \text{Prob}(e, K) = [0.9, 0.9]$. Again, no argument; but no problem.

Here is the problem. Levi's ingenious idea was to apply what Seidenfeld calls "the principal axiom governing conditional probability:" $P(A) = \sum_i P(A|H_i) \times P(H_i)$. If evidential probability were real-valued, the analogue would be: $\text{Prob}(e, K) = \sum_i \text{Prob}(e, K, h_i)\text{Prob}(h_i, K)$. As a particular case, we have, $\text{Prob}(e, K) = 0.9 = \text{Prob}(e, K, h_{0.89} \vee h_{0.91}) \times \text{Prob}(h_{0.89} \vee h_{0.91}, K) + \text{Prob}(e, K, h_{0.92}) \times \text{Prob}(h_{0.92}, K)$. In view of $\text{Prob}(e, K, h_{0.89} \vee h_{0.91}) = 0.9$ and $\text{Prob}(e, K, h_{0.92}) = 0.92$, we can infer $\text{Prob}(h_{0.92}, K) = [0, 0]$. A similar calculation yields $\text{Prob}(h_{0.91}, K) = [0, 0]$. From these values the "principle axiom" will yield $\text{Prob}(e, K) = [0.89, 0.89]$, contradicting $\text{Prob}(e, K) = [0.9, 0.9]$ Where did we go wrong?

We can get a clear idea of what went wrong by reverting to the earlier-threatened balls-in-bags model. Suppose that the dictator of Sweden has ruled that the distribution of Protestants in Malmö is to satisfy either $h_{0.89}$, $h_{0.91}$ or $h_{0.92}$. That is how the odd disjunction got into our corpus of evidence. Furthermore, she has determined that which of the three hypotheses is to be true is to be determined by a chance[2] device so that $h_{0.89}$ will be true with chance 0.2, $h_{0.91}$ with chance 0.3, and $h_{0.92}$ with chance 0.5. Implementing the result is no problem for an absolute dictator.

In this case the frequency of Protestants in Malmö is $0.2 \times 0.89 + 0.3 \times$

---

[2]I do not mean to invoke any metaphysical notion of chance here!

$0.91 + 0.5 \times 0.92 = 0.911$. (We will look more closely at such inferences later.) Of course, this is just an instance of that "principal axiom" mentioned before. It applies straight-forwardly to frequencies, in particular, the frequencies of Protestants in Malmö. This conflicts with Sweden as a whole, and is the correct probability for Peterson, according to the rule of specificity.

As our knowledge of the chance device employed by the dictator to determine the composition of the population of Malmö becomes less precise, this axiom does us less good. Suppose that the chance of $h_{0.89}$ is between 0.2 and 0.6; that of $h_{0.91}$ is exactly 0.3, and that of $h_{0.92}$ is between 0.1 and 0.5. The maximum frequency of Protestants is 0.911; the minimum frequency is 0.899. Precision kicks in, and the evidential probability that Peterson is a Protestant is 0.9. (But an only slightly lower chance for the 0.89 hypothesis would lead to an interval [0.910,0.911].)

Of course we cannot infer that the probability of $h_{0.92}$ is zero from the fact that it lies between 0.1 and 0.5! In fact, in the original story we have no probability for $h_{0.92}$ other than the trivial [0,1]. There is nothing to apply the famous axiom to.

The basic thing that goes wrong is that there *are* no real values for the probabilities of the three hypotheses. Given the specified background knowledge, $\text{Prob}(h_{0.92}, K) = [0, 1]$. Period. This does not mean that someone given to fantasy, having background knowledge K, could not imagine having a degree of belief in $h_{0.92}$. On our view, such feverish imaginings have nothing to do with probability.

But of course there is more to be said. Levi wries, "Kyburg's theory entails a violation of confirmational conditionalization due to his modification of the principle of direct inference [the addition of the strength rule]." [Levi, 1980., p.385]. Seidenfeld says that evidential probability "fails conditionalization" and that it "must not follow Bayesian algorithms." Thus it appears that going the route of evidential probability, based on direct inference, is going to entail jettisoning much of the powerful machinery of modern statistical theory, based as it is on conditioning.

In what follows, I shall be concerned to show that this appearance is false. We might already suspect this, on the basis of the story of the mad dictator of Sweden, where whether the "Bayesian axiom" holds or not depends on the details of the story. In general, we will find that Bayesian arguments often work out, and, more interesting and more important, the principles of evidential probability provide clear criteria for adjudicating between classical and Bayesian statistics.

4 "Combinatorial Semantics" and *Uncertain Inference*

Barring perhaps a few conference papers, these works [Kyburg and Teng, 2001; Kyburg, 1997] represent a sharp discontinuity with what has gone before. What is curious about this is that the discontinuity was unperceived by me: I thought that the general idea was the same, and that only the details differed. But before the article, the object of the rules was to pick out a single formula that would serve to determine a reference formula (reference class) whose statistics would yield the probability we were after. After that article the point of the rules was to pick out a *set* of formulas whose statistics would *jointly* determine the probability we were after. Initially probability was determined by our statistical knowledge concerning a class of which we could say that a certain object was a *random member*. Subsesquently, there is no such class. There is a collection of relevant classes, and the probability is determined by the *cover* of the set of parameters in a set of those classes.

To put the matter more explicitly, to compute the probability of a statement S, we look at the set of all those triples of sentences $\ulcorner \tau(\alpha) \urcorner$, $\ulcorner \rho(\alpha) \urcorner$, and $\ulcorner \%x(\tau(x), \rho(x), p, q) \urcorner$ such that $\ulcorner S \leftrightarrow \tau(\alpha) \urcorner$ is in K, $\ulcorner \rho(\alpha) \urcorner$ is in K, and the statistical statement is in K where τ is a suitable target formula and ρ is a suitable reference formula. The requirement of "suitability" has been with us since [Kyburg, 1961], in one form or another. In a number of places, random quantities, functions from objects in domain to reals, are also taken account of explicitly. Both classes of formulas were intended to be given, recursively, as part of the specification of the language. Note that α may represent a tuple of terms, and x a tuple of variables.

The point of the rules is, as it has been all along, to allow us to focus on the *relevant* parts of our statistical knowledge. To this end, we want to disregard some of the triples of sentences. We say that two statistical statements $\%x(\tau(x), \rho(x), p, q)$ and $\%x(\tau'(x), \rho'(x), p', q')$ *differ* just in case neither $[p, q] \subseteq [p', q']$ nor $[p', q'] \subseteq [p, q]$ holds. We apply the following rules sequentially:

1. If two statistical statements differ and the first is based on a marginal distribution, while the second is based on the full joint distribution, ignore the first. This gives conditional probabilities pride of place *when they conflict with the equivalent unconditional probabilities*.

2. If two statistical statements differ and the second employs a reference formula that logically entails the reference formula employed by the first, ignore the first. This embodies the well-known principle of *specificity*.

Those statistical statements we are not licensed to ignore we will call *relevant*. A set of statistical statements that contains every relevant statistical statement that *differs* from a statement in it will be said to be *closed under difference*.

3. The probability of S is the shortest cover of any non-empty set of relevant statistical statements closed under difference; alternatively it is the intersection of all such covers.

An example may help. Suppose the intervals mentioned in the set of relevant statements are [0.20,0.30], [0.25,0.35], [0.22,0.37], [0.40,0.45], [0.20,0.80], [0.10,0.90],[0.10,0.70]. There are three sets that are closed under difference; {[0.20,0.30], [0.25,0.35], [0.22,0.37], [0.40,0.45]}, {[0.20,0.80], [0.10,0.70]}, and {[0.10,0.90]}. The first set contains an interval that is included in another interval, and the third set is a singleton. The probability is [0.20,0.45].

We can show that this definition and the algorithm it suggests is correct, in the sense that if there is any inference structure for S — i.e., terms τ, ρ, α, p, and q such that we know $\ulcorner S \leftrightarrow \tau(\alpha) \urcorner$, $\ulcorner \rho(\alpha) \urcorner$ and $\ulcorner \%x(\tau(x), \rho(x), p, q) \urcorner$, then the probability of S is uniquely defined. Clauses (1) and (2) cannot eliminate all the inference structures for S. (Nothing differs from $\ulcorner \%x(S, x = x, 0.0, 1.0) \urcorner$; either everything or nothing is such that S.) There must therefore be a nonempty set of relevant statistical statements relevant to S. Closure under difference yields a set of exclusive classes of relevant statistical statements: If s is in one class and t is in another, it is not the case that s and t differ; one must be more precise than the other. More than this, if c_1 and c_2 are distinct classes of statistical statements, each closed under difference, then the cover of the set of intervals in one class must include the cover of the set of intervals of the other class. To see this, let $L(c_1)$ and $U(c_1)$ be the lower and upper bounds of the cover of c_1, and similarly for $L(c_2)$ and $U(c_2)$. Suppose $L(c_2) \leq L(c_1)$ but that $U(c_2) < U(c_1)$. Then there is some interval mentioned in c_1 whose upper point is greater than any point in an interval of c_2, and whose lower point must be greater than some lower point of some interval in c_2. This interval therefore differs from an interval in c_1, and therefore belongs in c_1, contrary to our assumption.

We can also show (it is shown in [Kyburg and Teng, 2001] that if the probability of S is the interval $[p, q]$ relative to a consistent set of sentences K, then the relative frequency of models of K in which S holds lies between p and q.

5   The Hollow Cube

There has been from the beginning a close connection betweeen evidential probability and fiducial probability. The paradigm case of fiducial inference is that in which we argue from the value of an instance of a population in which we know that X is distributed normally with unknown mean μ and known variance s, to a probability about the value of μ. Suppose we measure an item o from the population and add $X(o) = 3.6$ to our body of evidence. Relative to this body of evidence, the evidential probability that "$3.6 - k \leq \mu \leq 3.6 + k$" is $\int_{-k}^{k}[X - \mu]dN(0, s)$, since if X is distributed normally $N(\mu, s)$, $X - \mu$ is distributed normally $N(0, s)$ This is also the fiducial probability. Note what it is based on. While we do not know the precise distribution of the random quantity X, we know that it is one of a set of distributions (normal distributions with standard deviation s), and we do know the exact distribution of the random quantity $X - \mu$ in this set of distributions. This serves as the basis for our probability. There is some idealization involved; what we mean is that for any reasonably large interval, the long run relative frequency with which $(X - \mu)$-values fall in that interval is approximately that given by the normal integral over that interval. There is thus a frequency basis for the evidential probability.

This holds for any *reasonable* interval, and indeed for any reasonable Borel set. Fisher [Fisher, 1956] goes further, at least as a matter of convenience, and speaks of the *fiducial distribution* of μ given $X(o)$ as a "probability distribution" on which one could condition. (He referred to this as "fiducial" probability, as opposed to ordinary probability.) There is a sense in which the fiducial distribution is an evidential probability distribution: its integral over any interval will yield the probability that μ belongs to that interval.

We are now ready to tackle Seidenfeld's hollow cube. We may measure the length of a side of this cube in two ways: by measuring the weight of a rod of unit weight per unit length matching one edge of the cube, or by measuring the weight of a mass of liquid of unit density just filling the cube. These weights are measured on a scale whose errors are distributed normally, $N(0, 1)$.

The set of weighings R on this scale is a perfectly good reference set, and the quantity $E(x)$ = the error of measurement x is a perfectly good random quantity. The last paragraph says that we know the distribution of the quantity E in the set R. Let us make a measurement, m_1 of the rod; we obtain the value $V(m_1)$. All this being in our corpus K, we can compute the probability, relative to K, that the length of a side L lies in any interval $[a, b]$ as $\int_a^b N(0, 1)$. This comes from the fact that "$a < L < b \leftrightarrow a < V(m_1) + E(m_1) < b$" and therefore "$a < L < b \leftrightarrow a - V(m_1) < E(m_1) < b - V(m_1)$", where $E(m_1)$ is the error made on measurment m_1, combined with the fact that we know

(1) "$\%x(E(x) \in \mathbf{b_1}, R(x), \int_{\mathbf{b_1}} N(0, 1), \int_{\mathbf{b_1}} N(0, 1)$",

where $\mathbf{b_1} = [a - V(m_1), b - V(m_1)]$.

Before moving on to Seidenfeld's issue, let us look at two more examples of possible evidential probability structures. Suppose we make a second measurement, m_2 of the weight of the rod, and that it yields the value $V(m_2)$. *Ignoring* the first measurement, we would get a perfectly good alternative evidential probability for the length of a side of the cube. It would be based on "$a < L < b \leftrightarrow a - V(m_2) < E(m_2) < b - V(m_2)$", where $E(m_2)$ is the error committed in the second measurement. The probability depends on our statistical knowledge

(2) "$\%x(E(x) \in \mathbf{b_2}, R(x), \int_{\mathbf{b_2}} N(0,1), \int_{\mathbf{b_2}} N(0,1)$",
    where $\mathbf{b_2} = [a - V(m_2), b - V(m_2)]$.

Of course, in real life we would not want to ignore the first measurement; the whole point of making two measurements is in the hopes of *combining* them. We can construct a third inference structure that reflects the combination of the two measurements: The object is the ordered pair of measurements $\langle m_1, m_2 \rangle$. The reference class is $R^2 - D$ — the set of all distinct pairs; this is a perfectly good reference set. The random quantity we want to look at is $E^*(m_1, m_2) = (E(m_1) + E(m_2))/2$, the average error of the average of the two measurements. The sum of two acceptable random quantities is an acceptable random quantity. Furthermore, we may take the errors of two measurements to be independent, yielding

(3) "$\%x, y(E^*(x, y) \in \mathbf{b}, R^2(x) - D, \int_{\mathbf{b}} N(0, 1/\sqrt{2}), \int_{\mathbf{b}} N(0, 1/\sqrt{2})$".

Taking account of "$a < L < b \leftrightarrow a < \frac{V(m_1) + V(m_2)}{2} + E^*(m_1, m_2) < b$", we also have "$a < L < b \leftrightarrow a - \frac{V(m_1) + V(m_2)}{2} < E^*(m_1, m_2) < b - \frac{V(m_1) + V(m_2)}{2}$". Thus for $\mathbf{b_3} = [a - \frac{V(m_1) + V(m_2)}{2}, b - \frac{V(m_1) + V(m_2)}{2}]$, the proportion given by (3) is $\int_{\mathbf{b_3}} N(0, 1/\sqrt{2})$.

Now we have a surfeit of riches: we have three reasonable numbers to associate with the sentence "$a < L < b$". Furthermore, the three (degenerate) probabilities conflict. Back to the rules: Rule (1) says that if two statistical statements conflict, and the first is based on a marginal distribution while the second is based on the full joint distribution, ignore the first (and the probability based on it). The statistical statements (1) and (3) conflict. Statement (1), however, is based on a marginal distribution (R) while statement (3) is based on the full joint distribution ($R^2 - D$). The first rule therefore directs us to ignore statement (1) in the presence of statement (3). Similarly, we should ignore (2) in the presence of statement (3). Given both measurements, the correct probability is given by $\int_{\mathbf{b_3}} N(0, 1/\sqrt{2})$ with $\mathbf{b_3} = [a - \frac{V(m_1) + V(m_2)}{2}, b - \frac{V(m_1) + V(m_2)}{2}]$. This conforms to both common sense and scientific practise.

Now consider the situation to which Seidenfeld directs our attention. As we have already seen, when we measure the weight of the rod composing the side of the cube on a scale with error distributed normally, $N(0,1)$, we obtain a perfectly good evidential probability for the length of the side. Alternatively, suppose we weigh the volume of liquid in the hollow cube. Again we get a perfectly good evidential probability structure.

The set of weighings R on this scale is a perfectly good reference set, and the quantity $E(x) =$ the error of measurment x is a perfectly good random quantity. We know the distribution of the quantity E in the set R. Let us make a measurement m_3 of the contained liquid; we obtain the value $V(m_3)$. Taking account of the biconditionals,

$$
\begin{aligned}
a < L < b \quad &\leftrightarrow \quad a < V^{1/3} < b \\
&\leftrightarrow \quad a < [V(m_3) + E(m_3)]^{1/3} < b \\
&\leftrightarrow \quad a^3 < V(m_3) + E(m_3) < b^3 \\
&\leftrightarrow \quad a^3 - V(m_3) < E(m_3) < b^3 - V(m_3).
\end{aligned}
$$

If the only data we have is the result of m_3 we can compute the probability, relative to K, that the length of a side L lies in the interval $[a, b]$ as $\int_{\mathbf{b_3}} N(0, 1)$.

(4) "$\%x, y(E(x) \in \mathbf{b_4}, R(x), \int_{\mathbf{b_4}} N(0, 1), \int_{\mathbf{b_4}} N(0, 1)$",
where $\mathbf{b_4} = [a^3 - V(m_3), b^3 - V(m_3)]$.

The problem arises when we have made both measurements; we have measured the weight of rod matching a side, and we have measured the weight of the contained liquid. As before, $R^2 - D$ is a perfectly good reference class. We even know the distribution of *pairs* of errors $\langle E(x), E(y) \rangle$ in that class. But what we need is an acceptable quantity $E^\dagger(x, y)$ such that for some Borel set $\mathbf{b}$, $a < L < b \leftrightarrow E^\dagger(m_1, m_3) \in \mathbf{b}$. When we were weighing the rod twice, we could show that the average of the two measurements minimized the error, and that the random quantity that represented that error was the average of the random quantities representing the error of each measurement.

It seems hard to come up with a function $E^\dagger(x, y)$ that is a satisfactory random quantity, and thus that no expression of the form $\%x, y(E^\dagger(x, y) \in \mathbf{b}, R^2 - D, p, q)$ can be a known statistical statement in the sense of our theory, despite the fact that $R^2 - D$ is a perfectly good reference class. Thus while weighing the rod and weighing the enclosed liquid each give perfectly good probabilities for $a < L < b$, there seems to be no way to combine the two observations to get an improved probability, as we could before. We must combine the information according to rule (3), obtaining a less precise probability: $\text{Prob}(a < L < b, K) = [\int_{\mathbf{b_1}} N(0, 1), \int_{\mathbf{b_3}} N(0, 1)]$.

6 Bayesian Inference

What is Bayesian inference? Despite the current popularity of Bayesianism, it is not easy to say. In a simple-minded form it could be taken as the recommendation to "update by Bayes' theorem" — i.e., to replace $P(H)$ by $P(H|E)$ when we have observed E. Many people, most notably Isaac Levi, have observed that this doesn't always make sense. It particularly doesn't make much sense when $P(H|E)$ is thought of as the ratio of $P(H \wedge E)$ to $P(E)$. From the present point of view, according to which probabilities are intervals, this makes even less sense.

Where conditioning via Bayes' theorem does make sense is where we have knowledge of a joint distribution, and the evidence allows us to transform that distribution. For example, if we think of the next outcome of the roll of a die, we attribute a probability of about $1/6$ to the outcome *one*. If we are given as evidence that the next outcome will be an odd number, then we are alleged to calculate $P(one|odd) = P(one \wedge odd)/P(odd) = \frac{1/6}{1/2}$. This typical example of conditioning is trivially accounted for by rule (2) of our rules for finding a reference class or set of reference classes: the rule of specificity. The guiding relative frequency of ones among odd valued tosses is a third; this is a subset of the set of all tosses, and thus, if we know that the toss in question yields an odd value, it is the relative frequency of one third that we take as measuring the probability.

A more interesting form of conditioning arises when we consider the next toss of a die chosen from a bag of dice, in which 20% have the distribution $\langle 1/12, 1/6, 1/6, 1/6, 1/6, 3/12 \rangle$ and the rest are normal. A simple question that can obviously be answered by "direct inference" is this: What is the probability of getting a *one* on the next toss of the chosen die? The answer is simple enough: 20% of the time we get a biased die, in which *one* occurs $1/12$ of the time, and 80% of the time we get a normal die, yielding *one* a sixth of the time: $0.2 \times 1/12 + 0.8 \times 1/6 = 0.15$. Surely there is no mystery here, nor any reason to deviate from direct inference.

A more characteristically "Bayesian" question is this. We choose a die from the bag, toss it once, obtain a *six*, and ask for the probability that we happen to have chosen a biased die. The Bayesian answer: The prior probability we have chosen a biased die is 0.20; the probability that we get a *six* if we have the biased die is $3/12$; otherwise the probability is $1/6$. We calculate: $P(bias|six) = P(six|bias) \times P(bias)/P(six) = \frac{3/12 \times 0.2}{3/12 \times 0.2 + 1/6 \times 0.8} = 0.28$. But again we have no need to deviate from direct inference, and in fact the first rule for probability directs us exactly to the right reference class. We ignore the fact that *in general* 20% of the draws are of a biased die, on the grounds that this statistic reflects a marginal distribution over

all outcomes, where in fact we want the frequency of biased dice in the set
of draws of dice that yield a *six* on the first toss.

7   Conditioning

In reasonable axiomatizations of classical probability, conditional probability arises (as a primitive notion; the usual "definition" is not a definition)
in the context of the *multiplication theorem*: $P(S \wedge T|K) = P(S|K) \times P(T|K, S) = P(T|K) \times P(S|K, T)$. (We write K, T as an abbreviation of
$K \cup \{T\}$.) Let us first consider the case of independence: the case in which
$P(S|K, T) = P(S, K)$. It is clear that when $P(S|K)$ and $P(T|K)$ are degenerate, point valued, probabilities, the product rule holds. What happens
when independence does not hold?

Suppose that we have probabilities $[p_S, q_S]$ for S, based on R_S, T_S, and
α_S, and probabilities $[p_T, q_T]$ for T, based on R_T, T_T, and α_T. For simplicity, suppose that there is only a single reference class for each sentence.
According to the rules of the evidential probability game, we may consider
the cross products, $R_S \times R_T$ and $T_S \times T_T$ and the pair $\langle \alpha_S, \alpha_T \rangle$. Barring difficulties of deductive closure, we have $S \wedge T \leftrightarrow \langle \alpha_S, \alpha_T \rangle \in T_S \times T_T$
and $\langle \alpha_S, \alpha_T \rangle \in R_S \times R_T$. We also have in K the statistical statements
$\%x(T_S(x), R_S(x), p_S, q_S)$ and $\%x(T_T(x), R_T(x), p_T, q_T)$.

These statistical statements give us the *marginal* distributions on $R_S \times
R_T$. Putting r, s, t, u for the *unknown* frequencies, we obtain the following:

	S	$\overline{S}$	
T	t	r	v_T
$\overline{T}$	u	s	$1 - v_T$
	v_S	$1 - v_S$	

From our assumed statistical knowledge, we can obtain constraints on
the margins $p_S \le v_S \le q_S$ and $p_T \le t_t \le q_T$. But these constraints amount
to very little. For example, the frequency within the reference class $T_T \times R_S$
of $T_T \times T_S$, corresponding to the conditional probability of S given T, can
have any value from 0 to 1, provided v_T is positive, which is only ruled out
by $q_T = 0$.

Things are not always this bad, as we have seen in the case of independence; but independence reflects a great deal of knowledge: with a
knowledge of independence we can pass from knowledge of the marginal
frequencies directly to knowledge of the frequencies in the cross product.

A far more interesting case is that which arises often in statistical inference: If T represents a statistical hypothesis and S a sampling distribution,
then $Prob(S|K, T)$ may be quite precisely known. Thus we would have,
corresponding to the table above, a table surmounted (in the finite case) by

a list of hypotheses, $H_0, \ldots, H_k$, and bordered on the left by a list of *kinds* of samples, $S_0, \ldots, S_m$. Given a hypothesis, the relative frequencies of the kinds of samples can ordinarily be calculated — that is, $\text{Prob}(S_i | K, H_j)$ is known. The S_i may be sets of subsets of the population being sampled, sets of sequences of items in that population, kinds of subsets of the population, etc. No matter. Even an approximate hypothesis H_j will determine, within limits, the relative frequency of samples of kind S_i.

What we do not have yet, are the entries in the table, since we have said nothing about the hypotheses. If we had a reference class for the hypotheses, about which we had knowledge, so that we could have useful probabilities $\text{Prob}(H_j | K) = [p_j, q_j]$, we could fill in the entries aproximately, and so compute the approximate frequency distribution for the hypotheses *given* a sample of kind S_i. There are certainly circumstances under which we do have this kind of knowledge — for example, we might know the genetic makeup of the parents of the sweet peas we are testing, and this gives us the frequencies of the kinds of genetic makeups in those sweet peas.

For example, suppose we consider a plant that is a hybrid of two pure strains: white-flowering (WW) and pink-flowering (ww). That plant may have any one of three genetic compositions: purebred white (WW), pure-bred pink (ww), or hybrid (Ww). We cross it with itself, and observe a white flower on its single offspring.

	WW	ww	Ww	
White	1/4	0	3/8	5/8
Pink	0	1/4	1/8	3/8
	1/4	1/4	1/2	

We may calculate that in 60% of the cases in which a single offspring is white, it came from a parent that was a hybrid – i.e, in which the statistical hypothesis Ww is true.

8   Extended Example

Here is an extended example in which both statistical hypotheses and individual statements obtain probabilities from the same database. Consider a laboratory in which genetics is being studied. In particular, we are examining the selective disadvantage of having short hair, which is controlled by two alleles at a single locus: H is the dominant long-haired allele, h the recessive counterpart. The parameter s represents the deviation from the Hardy-Weinberg ratios over a single generation: the chance of survival of the short-haired phenotype is $1 - s$ compared to a chance of 1 for the long-haired phenotype. The basic reference classes are the class of mice, R_1, together with R_1^n for arbitrary n, and the set of cages, C. We are concerned with several quantities: $Hairy$, which has the value 1 for hairy mice, and 0 for the others; G, which has the value 0 for HH, 1 for Hh and 2 for hh; Gen, which has the value 0 for the first generation, 1 for the second, 2 for the third, 3 for the fourth and 4 for the fifth; and F_1 which gives the proportion of hairy mice in a set of mice. $Lives(x, y)$ is a relation that holds when x is a mouse and y is a cage, and x lives in y.

The body of knowledge K that determines the probabilities we are interested in consists of the following. In addition to the biological facts already mentioned, we know that the genotypes of the initial stock of mice are distributed 0.04 HH, 0.32 Hh, 0.64 hh. It follows that the frequency of H alleles is 0.2 and the frequency of h alleles is 0.8. For any population in which the proportion of H alleles is p_H, the distribution of genotypes is multinomial, $M(p_H^2, 2p_H(1 - p_H), (1 - p_H)^2)$. The disadvantage of being short-haired (i.e., the chance of dying before reproducing) is $0.3 \leq s \leq 0.4$. With random mating, p_H is 0.8 for the first generation, between 0.248 and 0.269 for the second generation, between 0.300 and 0.342 for the third generation, between 0.352 and 0.414 for the fourth generation, and between 0.403 and 0.480 for the fifth. The initial population is 100 mice; the population is allowed to increase by 100 in each generation, so we have total of 15 cages, numbered randomly c_1 to c_{15}, each containing 100 mice. Each mouse receives a unique random number, m_1 to m_{1500}. Each cage is occupied by mice of a single generation.

In addition to this general knowledge (and its implications) we have the following additional background knowledge:

Mouse m_{455} lives in cage c_9; $Gen(m_{455}) = 2$.

Mouse m_{366} lives in some cage.

Mouse m_{412} lives in cage c_{10}.

$\{m_7, m_{1065}, m_{211}, m_{1206}, m_{87}\} \subseteq \{x : Gen(x) = 1\}$.

Mouse m_{255} lives in cage c_6; $Hairy(m_{255}) = 1$.

The label was eaten off cage c; $m_6, m_{708}, m_{422}, m_{63}, m_{1132}$ and m_{597} came from that cage, and $F_1(\{m_6, m_{708}, m_{422}, m_{63}, m_{1132}, m_{597}\}) = 0.5$.

We choose a generation at random (e.g., using a table of random numbers) and a mouse — say m^* — at random from that generation.

1. $\mathrm{Prob}(G(m_{455}) = 0, K) = [0.090, 0.117]$. This is the frequency of the HH genotype in generation 3. The frequency overall is $[0.116, 0.158]$, which conflicts with the frequency in generation 3, but our second rule directs us to ignore this frequency. Similarly, $\mathrm{Prob}(Hairy(m_{455}) = 1, K) = [0.510, 0.567]$. The frequency overall is $[0.553, 0.617]$, but specificity rules that out.

2. With regard to mouse m_{366}, all we know is that it is in the experiment,

so these overall rates *do* apply to it.

3. We do know more about mouse m_{412}, in particular that it lives in cage c_{10}, but this gives us no information about phenotype or genotype. We can say $\text{Prob}(Lives(m_{366}, c_{10}), K) = [.067, .067]$, whereas $\text{Prob}(Lives(m_{412}, c_{10}), K) = [1, 1]$.

4. Let $M_5 = \{m_6, m_{1065}, m_{211}, m_{1206}, m_{87}\}$. The probability of statement S asserting that M_5 has 2 mice of genotype HH, 2 of genotype Hh, and 1 of genotype hh, is calculated from the set of distribution functions applicable to the second generation: $\text{Prob}(S, K) = [\binom{5}{2,2,1}(0.061)^2(0.373)^2(0.566), \binom{5}{2,2,1}(0.072)^2(0.393)^2(0.534)] = [0.00879, 0.01283]$.

5. That mouse m_{255} lives in cage c_6 gives us no information about its genotype, but the fact that it is hairy does. We want the proportion of HH among hairy mice, not among mice in general. We calculate $\text{Prob}(G(m_{255}) = 0, K) = [0.2097, 0.2561]$.

6. We need to classify the mice in the cage with the missing label according to their generation; we know that of six mice from that cage, half are hairy. We know that three of fifteen cages, or 20%, contain third generation mice; but we also know that *more* than 20% of the pairs, consisting of a cage and a sample of six, of which half are hairy, are pairs in which the cages are cages of third generation mice. Specifically, we can compute that the maximum fraction of such pairs in which the mice are third generation is $\dfrac{0.20\binom{6}{3}(0.567)^3(0.433)^3}{\binom{6}{3}(0.617)^3(0.383)^3}$ (corresponding to the highest selection factor $s = 0.4$); and the minimum fraction is $\dfrac{0.20\binom{6}{3}(0.510)^3(0.490)^3}{\binom{6}{3}(0.553)^3(0.447)^3}$ corresponding to a selection factor $s = 0.3$. Thus, by the first rule, we have $\text{Prob}(\forall m(Lives(m, c) \supset Gen(m) = 2), K) = [0.2067, 0.2243]$. We also have $\text{Prob}(\forall m(Lives(m, c) \supset Gen(m) = 3), K) = [0.2312, 0.2552]$ and $\text{Prob}(\forall m(Lives(m, c) \supset Gen(m) = 4), K) = [0.1934, 0.2659]$. This reflects the fact that later generations are more frequently represented among the fifteen cages.

7. Finally, the probability that m is hairy is the average of the probabilities over each of the five generations; it is not the proportion of hairy mice in the laboratory — again, by rule one — since that proportion (which we know to lie between 0.553 and 0.617) conflicts with proportion of pairs, consisting of a generation and a mouse in that

generation in which the mouse is hairy: $\mathrm{Prob}(Hairy(m) = 1, K) = [0.5056, 0.5554]$

We might make this situation more concrete by discussing illustrative bets among the lab technicians. This would serve the additional purpose of illustrating that this is a situation in which some "opinions" are simply *wrong*.

9   Conclusion

As Levi and Seidenfeld have shown, there is an incompatibility between confirmational conditionalization and evidential probability. Since Levi, at least, does not accept the universal applicability of Baysian conditionalization (temporal credal conditionalization), it is not clear how seriously one should take this complaint. It is clear, however, that Bayesian arguments that are based on knowledge of frequencies (as opposed to fantasies about subjective degrees of belief) not only admit an interpretation in terms of evidential probability, but are *demanded* by it, in accord with rule (1) for computing probabilities. Legitimate inferences — that is, inferences based on our knowledge of approximate frequencies — have nothing to fear from evidential probability.

BIBLIOGRAPHY

[Fisher, 1956] Ronald A. Fisher. *Statistical Methods and Scientific Inference*. Hafner Publishing Co., New York, 1956.
[Kyburg and Teng, 2001] Henry E. Kyburg, Jr. and Choh Man Teng. *Uncertain Inference*. Cambridge University Press, New York, 2001.
[Kyburg, 1961] Henry E. Kyburg, Jr. *Probability and the Logic of Rational Belief*. Wesleyan University Press, Middletown, 1961.
[Kyburg, 1974] Henry E. Jr. Kyburg. *The Logical Foundations of Statistical Inference Reidel*. Reidel, Dordrecht, 1974.
[Kyburg, 1983] Henry E. Kyburg, Jr. The reference class. *Philosophy of Science*, 50:374–397., 1983.
[Kyburg, 1997] Henry E. Kyburg, Jr. Combinatorial semantics. *Computational Intelligence*, 13:215–257, 1997.
[Levi, 1977] Isaac Levi. Epistemic utility and the evaluation of experiments. *Philosophy of Science*, 44:368–386., 1977.
[Levi, 1980.] Isaac Levi. *The Enterprise of Knowledge,*. MIT Press, Cambridge, 1980.
[Levi, 1983] Isaac Levi. Kyburg on random designators. *Philosophy of Science*, 50: 635–642, 1983.
[Quine, 1951] W. V. O. Quine. *Mathematical Logic*. Harvard Press, Cambridge, 1951.

INDEX

www.ingramcontent.com/pod-product-compliance
Lightning Source LLC
Chambersburg PA
CBHW071104050326
40690CB00008B/1108